Magnetic Bearings and Bearingless Drives

Magnetic Bearings and Bearingless Drives

Akira Chiba, Tadashi Fukao,
Osamu Ichikawa, Masahide Oshima,
Masatsugu Takemoto and David G. Dorrell

AMSTERDAM • BOSTON • HEIDELBERG • LONDON
NEW YORK • OXFORD • PARIS • SAN DIEGO
SAN FRANCISCO • SINGAPORE • SYDNEY • TOKYO

ELSEVIER

Newnes is an imprint of Elsevier

Newnes

Newnes
An imprint of Elsevier
Linacre House, Jordan Hill, Oxford OX2 8DP
30 Corporate Drive, Burlington, MA 01803

First published 2005

British Library Cataloguing in Publication Data
A catalogue record for this book is available from the British Library

Library of Congress Cataloguing in Publication Data
A catalogue record for this book is available from the Library of Congress

ISBN 0 7506 5727 8

For information on all Newnes publications visit our website at
www.newnespress.com

Typeset by Integra Software Services Pvt. Ltd, Pondicherry, India
www.integra-india.com

Front cover picture reproduced by kind permission of Ebara Research Co. Ltd

Working together to grow
libraries in developing countries

www.elsevier.com | www.bookaid.org | www.sabre.org

ELSEVIER   BOOK AID
             International   Sabre Foundation

Transferred to Digital Printing in 2008

Contents

List of Contributors

Akira Chiba
Tokyo University of Science
2641 Yamazaki Noda
Chiba
Japan

David G. Dorrell
University of Glasgow
Glasgow
UK

Tadashi Fukao
Musashi Institute of Technology
1-28-1 Tamatsutsumi Setagaya
Tokyo
Japan

Osamu Ichikawa
Polytechnic University
4-1-1 Hashimotodai Sagamihara
Kanagawa
Japan

Masahide Oshima
Tokyo University of Science Suwa
5000-1 Toyohira Chino
Nagano
Japan

Masatsugu Takemoto
Musashi Institute of Technology
1-28-1 Tamatsutsumi Setagaya
Tokyo
Japan

Foreword

This book is primarily about bearingless machines, where the magnetic levitation of the rotor is provided from within the main stator bore (in addition to torque production if motoring, or electrical power production if generating). The bearingless type of motor is still, to some extent, in the development stage and therefore still at an early point of evolution. The basic theory, design and control rules are being researched and investigated and there is still much work to be done. The authors have had many years of experience in the design and operation of bearingless machines and this book represents their combined experience as pioneers in this area. Therefore this book describes the leading edge of the technology as it stands today.

However, this is not an advanced textbook suitable only for engineers with a good working knowledge of the subject. It is also aimed at both mechanical and electrical engineers with little prior knowledge and who wish to get an understanding of the subject. Therefore the first few chapters describe the basic operation of the bearingless machine, which will help strip away the initial thought that a self-levitating rotor is a very difficult problem that may be too difficult to understand and construct. The book then goes on to describe the operation of different types of bearingless machine, the control techniques required and then gives the specification and performance figures for actual machines that the authors have developed and studied. The last chapter describes examples of bearingless machines that others have developed. The only theory that is necessary for good understanding is control, which spans both electrical and mechanical engineering, though a basic knowledge of magnetic circuits and the operation of standard brushless permanent magnet machines, synchronous- and switched-reluctance machines and induction motors would be highly advantageous.

The types of bearingless machine covered in this book are brushless permanent magnet ac machines, synchronous- and switched-reluctance machines and induction motors. These represent the main motor topologies that can be used in bearingless machines. In addition, homopolar motors can also be used (which are really a derivative of stepping motors). There are also design compromises to consider. In a permanent magnet machine, thick magnets give good torque production and robust operation; however, thin magnets lead to reduced levitation winding current. Some of the solutions that are implemented to obtain stable and controllable levitation in these machines are quite ingenious; for instance, the use of a consequent-pole rotor in a permanent magnet machine to modulate the suspension winding MMF in order to generate airgap flux waves of the correct pole number, phase and frequency. A second example is the use of multiple rotor conductor paths in the cage induction motor to decouple the suspension winding from the rotor. These subtleties are all described in this book. The control strategies for each machine are also described. These can be quite complex since the position and magnitude of the airgap flux has to be carefully controlled so that torque is produced and the rotor is magnetically suspended in the centre of the stator bore. The aim is often to use a standard inverter for the main motor winding with additional power-electronic and control circuitry for the suspension winding. In all cases it is necessary to have radial position sensors to control the levitation.

The applications of bearingless motors are many and varied. As an engineer interested in drive motors and electric power applications, I can immediately see the relevance of this book to very high speed drives (in order to greatly reduce the motor volume) and also to flywheel energy storage. Currently, it is common to take a standard motor and replace the mechanical bearings with magnetic bearings to obtain very high speed operation; or in a flywheel, to have separate magnetic suspension and motor/generator components – this book will give engineers the basic knowledge to go one step further and consider the motor and bearings as an integral unit.

I was first asked to referee the initial proposal for this book as a specialist in the design of electric machines (primarily induction and permanent magnet machines) and then requested to help check the technical descriptions as a native English-speaking engineer. The authors have written the book in a second language, rather than writing it in Japanese and having it translated, in an attempt to reach a wider audience – they are to be congratulated in undertaking such a task. I have found the book enlightening and it has opened up a new sphere of interest and research.

Dr David G. Dorrell
University of Glasgow
September 2004

Preface

As we progress into the twenty-first century global warming has become an important issue. CO_2 gas emissions must be reduced to preserve the correct air content. Modern man, who accepts the fruits of recent energy-based technologies (such as mass transport using cars, trains and aeroplanes, and climate-controlled work and home environments via air-conditioners) as necessities, cannot put up with the inconveniences of even the not-too-distant past. To maintain and develop these energy-consuming technologies, alternative energy resources and efficiency improvements are necessary.

Effective efficiency improvements have been, and are being, actively invest-igated in Japan and Europe. For example, the integration of an electric motor and generator system into a gasoline-powered vehicle (hybrid vehicle) provides excellent fuel economy and harmful gas reduction. Air-conditioners are another example of an application that has seen substantial efficiency improvement. These are now driven by variable-speed motor drives with rare-earth permanent-magnet motors and power electronics inverters. The cost of these advanced motor drives is decreasing thanks to mass production. As can be seen in these examples, the integration of high-efficiency drives into power mechatronics systems is one of the keys for improving system efficiency and hence reducing CO_2 gas emission.

In recent years, magnetic levitation and suspension systems have become a realistic proposition. For example, the Transrapid Train system developed by German engineers has been installed in Shanghai, China. In this train system, the bogies are levitated by electro-magnets. Therefore the operating speed is considerably higher than that for conventional high-speed trains such as the Shinkansen, TGV and ICE. The fastest passenger train in the world runs on the Yamanashi Maglev test line which has been developed in Japan. The bogies are again magnetically levitated by the interaction between superconducting coils and ambient-temperature coils installed along the track side walls. At a speed of more than 200 km/h, the wheels are lifted in a similar manner to an aeroplane so that the train is magnetically suspended and near-silence is experienced in the cabin. A top speed of more than 580 km/h has been recorded thanks to the magnetic levitation.

In addition to train systems, magnetic levitation and suspension has also been applied in other industries. Some applications, such as compressors, refrigerators, spindles and generators, require high rotational speed to minimize the weight, dimensions and cost, and to maximize the efficiency of the whole system. Low loss and maintenance-free operation are also required by the shaft-support bearings. However, chemical pumps, turbo-molecular pumps, blood pumps, bio-reactors, semiconductor processes, blowers and refrigerators operate under harsh environmental conditions, such as vacuum and extremely low and high temperatures, and in the presence of explosive, poisonous and bio-chemical fluids. Conventional mechanical bearings cannot be installed so that one solution is to suspend the machine shaft by magnetic levitation.

Magnetic suspension was not common even at the end of the twentieth century for several reasons: (a) real-time calculation speeds and peripheral functions in low cost digital processors were limited; (b) current regulators, i.e., power electronic inverters, were quite expensive; (c) controller structures required specialist design knowledge; (d) sensor devices for shaft displacement detection require space and are costly; (e) a large bearing space, inherent instability and specialist knowledge and experience are required to design, operate and repair such suspension systems; and (f) magnetic suspension needs both an electrical and a mechanical engineering background.

As we enter the twenty-first century, most of the problems described above have been solved. Fast A/D converters, 3-phase PWM functions and multipliers, as well as fast processing are available in the latest digital processors and field programmable gate arrays. Inverter costs have significantly reduced, thanks to package integration and mass production of power devices. Bearingless drives and generators integrate the magnetic bearings with the motor or generator, producing a compact size; while experiences in developing these systems are increasingly being reported in books and international conference proceedings.

The bearingless system is a key technology in the following power mechatronic applications: (a) high efficiency and compact systems with integrated magnetic suspension and high rotational shaft speed; (b) devices suspended by magnetic forces and operating in harsh environments; and (c) ultra-high-speed motor drives and generators with long flexible shafts requiring integrated vibration damping along the axial length.

The purpose of this book is to provide a fundamental understanding of bearingless drives for both mechanical and electrical engineers. The first part describes the basics of magnetic bearings including aspects of electromagnetics, controllers, practical problems, mechanical dynamics and power electronics with plenty of examples. During the first part, the reader will absorb the basic principles of magnetic suspension. The basic idea of 4-pole and 2-pole flux-wave combination machines is developed. Suspension force and current relationships are derived, and then the general controller design method is introduced. In the second part, the different bearingless machines are introduced; individual bearingless motors, i.e., permanent magnet, synchronous reluctance, induction, homopolar, consequent-pole permanent magnet and switched reluctance, are described and discussed. Some basic aspects of displacement sensors, controllers and power electronic circuits are also described. In the last chapter, some test machines and applications are reviewed. The book includes all the necessary material so that people will find that they have a thorough understanding of the different types of bearingless drives.

The authors of this book work in different institutions and have different backgrounds. However, they have come together to produce this book.

Professor Tadashi Fukao was with the Tokyo Institute of Technology. He says, "power electronics has introduced a freedom of frequency in electrical power technology". He has been studying high frequency power converters and high-speed motors and generators for many years. He has supervised research projects on synchronous reluctance, homopolar and disk-type bearingless drives. Since 2001 he has been with the Musashi Institute of Technology. He was the President of the Institute of Electrical Engineers of Japan (IEEJ) during 2003–2004. He is responsible for the introduction chapter of this book.

Dr Osamu Ichikawa began working in the field of bearingless machines in 1992 with the Tokyo Institute of Technology. He studied synchronous reluctance and disk-type bearingless drives for his PhD thesis and homopolar bearingless drives for his post-doctoral studies. He is now with the Polytechnic University. He is responsible for the chapters on the synchronous reluctance, homopolar and hybrid types of machine, and also the mechanical structure and position regulation chapters.

Mr Masahide Oshima, a PhD candidate, started his permanent magnet bearingless project in 1992 at the Tokyo University of Science, Suwa. He has been studying surface-mounted permanent magnet and buried permanent magnet bearingless drives. He is responsible for the cylindrical PM, salient-pole PM and buried PM bearingless motor chapters.

Mr Masatsugu Takemoto, a PhD candidate, started studying switched reluctance type bearingless motors in the Tokyo University of Science in 1996. He continues the project, and has extended into 2-pole permanent magnet bearingless motors, in the Tokyo Institute of Technology. He is responsible for the switched reluctance machine and the controllers and power electronics chapters.

Dr David Dorrell is a senior lecturer with the University of Glasgow, UK. He joined a consequent pole bearingless project while he was on an International Invitation Program in the Tokyo University of Science in 2002. Since then he has been with the project and is responsible for editing all the chapters.

As the primary author, I started to work with high-speed synchronous reluctance drives in 1985 on a PhD program with the Tokyo Institute of Technology. In experiments, I tested some combinations of rotors and stators in the power range of 1–3 kW and up to a speed of 24 000 r/min. The changing of one rotor for another in the machine was not very easy. It usually took more than two weeks. The problem was associated with the mechanical ball bearings. In replacing the rotor, one end's housing and a ball bearing must be removed. After replacing the rotor, much attention should be paid to the mechanical alignment while fixing the housing with bolts and nuts. Despite careful attention, the mechanical loss at 10 000 r/min could increase by about 40 W from a correctly-fitted and run-in value of 20 W. Therefore the motor needs to be driven for several days to reduce the mechanical loss. The rotational speed can be increased by 1000 r/min per day and after two weeks the test machine can be operated up to the rated speed of 24 000 r/min with a reasonably low mechanical loss. During this period I wondered if there was any other way to suspend the rotor.

Let me explain how I began to think about bearingless motors. One day a test machine was excited by a dc current to adjust the origin of rotor rotational position. After the dc power was switched on, the rotor was supposed to align to the stator magneto motive force (MMF) direction, though it did not rotate. The rotor was positioned in the stator by a magnetic attractive force because the housing bolts were loose. If the dc current was decreased the rotor could be rotated by hand. It was also found that there was a significant magnetic attractive force between the stator and the rotor at the rated excitation. This experience led me to think about useful applications of magnetic force. In spring 1987, the basic concept of the bearingless motor was proposed. In 1988, a prototype bearingless motor was constructed in the Tokyo University of Science. A patent, including the general idea of bearingless motors for various electrical machines, was submitted in January 1989. A short one-page paper was also presented in the IEEJ national convention. Since that time I have been studying several types of bearingless drives and generators. I am responsible for the magnetic bearing chapters, the primitive bearingless chapters and several other chapters.

All the chapters are carefully selected so that both mechanical and electrical engineers can have a thorough understanding. Basic theories are described in detail with practical examples. Most bearingless machine variations are covered; however, some recent developments such as 2-pole permanent magnet, coreless, short pitch and bridge windings are not included. Advanced readers may go to the proceedings of international conferences for further details.

The authors are based in universities. They not only analyse and simulate bearingless drives but also construct test machines to find practical problems. One of the purposes of this book is to help with a smooth transfer of technology into industry.

Akira Chiba
Tokyo
September 2004

Acknowledgements

This book arises largely from the work of an academic group at the Tokyo University of Science, the Tokyo Institute of Technology and the Tokyo University of Science, Suwa, in the period from about 1988 to 2004. All the authors have worked closely in bearingless motor developments. We would like to thank Professor Tazumi Deido and Professor Koichi Ikeda of the Tokyo University of Science for their support during the early developments. We would like to thank Professor Fukuzo Nakamura of the Tokyo University of Science and Professor Satoru Miyazawa of the Tokyo University of Science, Suwa, for coordinating and supporting the joint bearingless project on permanent magnet drives between two academic organizations. We also like to thank Professor M. Azizur Rahman of the Memorial University of Newfoundland. He invited Dr Chiba to be an International Post-Doctoral Fellow supported by Natural Science and Engineering Council of Canada. A prototype induction bearingless drive was built and tested in the Memorial University. He suggested that we use "bearingless" as a technical term in 1990. He and the authors have published several joint papers since the beginning of the era of bearingless drive developments. His strong support and nice suggestions have been very much appreciated. He has stayed in Tokyo through joint bearingless projects several times.

We would like to thank Professor Hirofumi Akagi and Dr Koji Takahashi of the Tokyo Institute of Technology; and Professor Yoichi Hayashi of the Aoyama Gakuin University for their kind support and co-operation in bearingless researches. We also like to thank Mr Chikara Michioka, who was a research associate in the Tokyo Institute of Technology, for his developments of homopolar hybrid and synchronous reluctance type bearingless motors. We also acknowledge research co-operations and discussions with Dr Mikihiko Matsui and Dr Fang Z. Peng while they were with the Tokyo Institute of Technology. We would like to thank Dr Andres O. Salazar of the Federal University of Rio Grande do Norte and Dr Jose Andres Santisteban L. of the Fluminense Federal University for their joint projects and co-operation while they were with the Tokyo University of Science.

We would like to acknowledge Mr Masaru Ohsawa, Mr Satoshi Mori and Mr Tadashi Satoh of the Ebara Research Company, Mr Masato Oota of the Seiko Seiki Company, Mr Yoshiaki Konishi of the Nikkiso Company and Mr Kazunobu Ogawa of J. Morita Tokyo Mfg Co. for their co-operation and support. We would also like to acknowledge Professor Yohji Okada and Professor Toru Masuzawa of Ibaraki University, Dr Reto Schoeb and Dr Thomas Gempp of Levitronix GmbH, Dr Paul Meuter of Sulzer Pumps Ltd and Dr Hideki Kanebako of Sankyo Seiki Mfg Co. for their permissions to include their developments as well as valuable discussions on bearingless projects.

We appreciate the valuable discussions on bearingless drives with Dr Tetsuo Ohishi, Dr Ken Matsuda, Dr Satoshi Ueno of Ibaraki University, Professor Daiki Ebihara, Dr Masaya Watada of the Musashi Institute of Technology, Professor Jorg Hugel, Dr Natale Barletta, Dr Christian Redemann of the Swiss Federal University, Professor Wolfgang Amrhein, Dr Siegfried Silber, Dr Klaus Nenninger of the Yohannes Kepler University of Linz, Professor Wilfried Hofmann of the Technical University of Chemnitz, Professor Hannes Bleuler of EPFL, Professor Richard M. Stephan of COPPE-UFRJ, Professor Paul E. Allaire in the University of Virginia and Dr Lyndon S. Stephen in the University of Kentucky.

We would like to express our appreciation to Professor T. J. E. Miller of the University of Glasgow for his suggestion and encouragement to write a book on magnetic bearings and bearingless motors. We also appreciate the efforts and patience of Mr Matthew Deans and Ms Jodi Burton of Newnes Publishers in Butterworth-Heinemann.

We would like to thank all the students who have contributed in bearingless machines developments through projects. My special thanks to Master students Messrs K. Chida, T. Hatsuda, S. Nomura, Y. Takamoto, R. Miyatake, R. Furuichi, Y. Yokotani, Y. Aikawa, K. Shimada, S. Hara, K. Hiraguri, H. Hosoda, E. Itho, K. Yoshida, Y. Shima, T. Gotou, N. Sugitani, M. Yamashita, K. Muronoi, K. Inagaki, K. Yasuda, T. Suzuki, N. Fujie, K. Kobayashi, R. Yoshimatsu, Y. Sakata, K. Yoshizue, T. Kuwajima, R. Hanawa, K. Kiryu, T. Nomoto, K. Yamashita, Y. Kubota, S. Aoyagi, T. Emori, T. Nobe, T. Fujishiro, S. Ogawa, S. Orita, T. Aoyagi, T. Kurokawa, T. Takenaga, K. Ebara, H. Saitoh, J. Amemiya, Y. Sakata, Y. Kurono, S. Itou and T. Hachiya; and Bachelor students Messrs N. Chino, Y. Kasahara, K. Yamada, T. Yokoyama, S. Onoya, T. Kikuchi, M. Yanagida, A. Suzuki, A. Tanaka, T. Ebihara, T. Ichikawa, K. Ninomiya, D. Yamaguchi, G. Kimura, R. Tsukakoshi, K. Ishiwaka (now Yoshida), S. Iwao, R. Yasuda, T. Tanimoto, T. Nakajima, T. Iida, S. Nogami, K. Maruyama, N. Andoh, T. Suzuki, A. Yajima, T. Takahashi, T. Sekiyama, T. Tera, M. Sakagami, T. Kobayashi, K. Asami, M. Nakagawa, D. Akamatsu, T. Katou, D. Fujimoto, R. Imaoka, N. Fujita of the Tokyo University of Science. We also would like to thank all the students in the Tokyo Institute of Technology who have engaged in bearingless projects in their graduate researches – Messrs Y. Miwa, H. Kawamura, M. Hanazawa, G. Fukuda, T. Sakamoto, J. Q. Li, H. Takita, I. Tomita, Y. Toyoshima, C. Kijja, T. Amakasu,

F. Saito, T. Sugawara, I. Surya, T. Kikuchi and M. Uyama – and undergraduate researches – Messr K. Furuta, T. Tsumura, K. Shimbo, M. Haris, K. Yakushi, R. Sato, W. S. Ho, H. Nozawa, R. Matsumoto, Y. Matsumoto, H. Nakakoji, T. Matsumoto, K. Kitahara and T. Mantyu. We would also like to thank the students Messer's M. Yamana (now Oshima), O. Uchikawa, M. Miyazawa, H. Muroga, K. Tanzawa, Y. Kikuchi, K. Sakai, H. Takahashi, H. Hashiba, M. Hasebe, T. Nagao, N. Kuroiwa, Y. Takizawa, T. Maejima and Y. Ito of the Science University of Tokyo, Suwa College. We would like to thank the students Messrs K. Miyagawa, D. Ishida, Y. Tanaka, N. Sugano, Y. Irino, Y. Kodama and D. Itou of the Musashi Institute of Technology. Special thanks are due to Mr Kazuya Kikuchi, Mr Takeshi Koshirakata and Ms Sae Ishii of Fukao's Laboratory.

I thank Mr Tetsuya Nobe, a graduate student in the Tokyo University of Science, for his figure and table preparation for this book. Finally I thank my wife Hiromi for her kind encouragement.

Akira Chiba
Tokyo
September 2004

1

Introduction

Tadashi Fukao and Akira Chiba

In this chapter, an overview of bearingless drives and magnetic bearings is presented; the principle of radial force generation is discussed and a typical motor drive with magnetic bearings is introduced and compared with the bearingless drive. In addition, a definition of the bearingless motor is given and the related technologies and early developments are reviewed. Typical application structures, winding configurations, radial force and torque comparisons and applications are also included.

1.1 Magnetic bearing and motor drive

A new bearingless concept was introduced into ac drive technology in the late 1980s. Since then, the theory and background of the concept has been studied, with many test drives developed to gain experience of the operation and behaviour of a variety of bearingless ac drives.

Ac motor drive technology has been developed and applied in a wide range of applications since the 1970s because of their advantages over dc motor drives, such as high performance, compactness, lighter weight, use of low maintenance motors and lower motor cost. The increase in power and rotational speed of ac drives has widened the application area. One maintenance task that still remains with an ac drive is bearing lubrication and renewal. In some applications, bearing maintenance is still a significant problem. For example, the bearings can present a major problem in motor drive applications in outer space, and also in harsh environments with radiation and poisonous substances. In addition, lubrication oil cannot be used in high vacuum, ultra high and low temperature atmospheres and food and pharmacy processes. Hence motor drives with magnetic suspension can enlarge the possible application areas of motor drives.

Figure 1.1 shows the principles of rotor radial force generation in both the magnetic bearing and bearingless motors. A rotating shaft is surrounded by the stator core. The rotor and stator are magnetized with four poles in a north, south, north, south sequence. There are strong magnetic attractive forces under

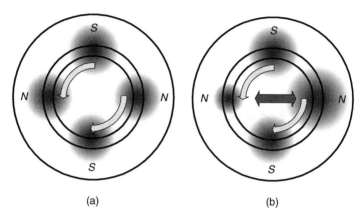

(a)   (b)

Figure 1.1 Radial force generation by unbalanced airgap flux density: (a) balanced airgap flux density; (b) unbalanced airgap flux density

these magnetic poles between the rotor and the stator cores. For example, a magnetic force of 40 N is generated in 1 cm^2 with an airgap flux density of 1 T. In Figure 1.1(a), these four magnetic poles have equal flux density and hence equal attractive force magnitudes. Thus, a vector sum of the four radial forces is zero. However, in Figure 1.1(b), one north pole is stronger than the other three poles so that the net attractive force is strong. Hence the unbalanced airgap flux density distribution results in radial magnetic force acting on the rotor. In this case, rotor radial force on the rotor is generated in the right-hand direction. In both radial magnetic bearing and bearingless motors, rotor radial force is generated by an unbalanced magnetic field; i.e., the rotor radial force is generated by the difference of radial forces between the magnetic poles. The attractive force is an inherently unstable force as it is stronger if the rotor moves in the force direction. The zero-radial-force point at the centre of the stator bore is an unstable point so that negative feedback is necessary.

In the Japanese Railways (JR) magnetically levitated train, the magnetic suspension force is generated by the interaction between the superconductor coil flux and an induced coil current. This means that a negative feedback controller is not required. In some flywheel applications, superconductor materials are used as magnetic bearings. These superconductor coils or materials provide stable magnetic suspension. However, the cost is quite high because a refrigerator and thermal insulation are required. These days, attractive magnetic force production requires simple, light-weight and cost-effective solutions for the magnetic suspension. Hence this book focuses on magnetic attractive forces.

Figure 1.2 shows the typical structure of a motor drive system equipped with magnetic bearings. The motor is located between the two radial magnetic bearings. Each radial magnetic bearing generates radial forces in two perpendicular radial axes. The radial forces are controlled by negative feedback control systems so that the radial shaft position is regulated to the centre of the stator bore. The left-hand magnetic bearing is regulated in two radial axis coordinates x_1 and y_1.

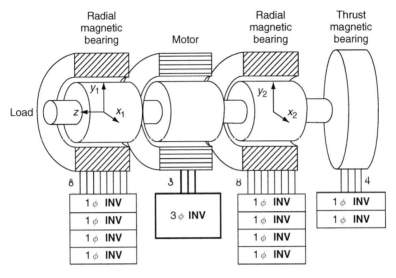

Figure 1.2 Motor with magnetic bearings

The right-hand radial magnetic bearing is regulated in radial axis coordinates x_2 and y_2. The thrust position on the z-axis, i.e., in the shaft direction, is regulated by axial forces generated by a thrust magnetic bearing. In total there are the five axes, x_1, y_1, x_2, y_2 and z, which are regulated by magnetic bearing systems.

Each radial magnetic bearing has four coils in a stator. Two coils are arranged on the x-axis and another two coils are on the y-axis. With a current in one coil, a magnetic attractive force is generated. A radial force in the x-axis direction is generated due to a difference in the magnetic attractive forces generated by the x-axis coils.

Coil currents in magnetic bearings are regulated by power-electronic circuits. In most cases, single-phase voltage-source inverters are utilized. A single-phase inverter can regulate one coil current so four single-phase inverters with eight output wires are necessary in a radial magnetic bearing.

In the thrust magnetic bearing, there are two coils so two single-phase inverters are connected to regulate the coil currents and generate radial force in axial direction.

The motor is responsible for generating torque around the shaft or z-axis. The rotational speed of the shaft is controlled by the motor torque and described by the torque equation of the system. A 3-phase inverter is connected to the motor terminals through three wires, with the motor windings connected in a wye or delta form (assuming 3-phase operation). The inverter supplies variable frequency and variable voltage based upon the shaft rotational speed and torque control requirements. The inverter frequency is proportional to the speed for most motors and the voltage/frequency ratio is usually almost constant up to the field weakening region.

1.2 Bearingless drives

Figure 1.3 shows the structure of a bearingless drive. Two bearingless units are constructed on a single shaft. Each bearingless unit generates radial forces as well as rotational torque. The left-hand bearingless unit is responsible for x_1 and y_1 radial positioning while the right-hand unit is responsible for x_2 and y_2 positioning. The total drive torque is twice the rated torque of the bearingless unit because the units share the torque production. Each bearingless unit has three terminals for the suspension windings and another three terminals for the motor windings. The respective motor phase windings of each unit are connected in series as shown in Figure 1.3 so that a wye connection is formed with two series-connected phase windings in each phase leg. It is important that the series-connected phase windings and rotor are correctly aligned in both units so that if the motor current lies on the rotor q-axis in one unit then it also lies on the rotor q-axis of the other. A single 3-phase inverter is connected to the series-connected motor windings and supplies variable voltage and frequency for the motor drive. At the suspension winding terminals, two independent 3-phase inverters are connected to supply the required levitation currents in order to generate radial forces in four axes as dictated by negative feedback controllers and radial shaft position sensing.

One can see the following advantages of bearingless drives:

Compactness – The shaft length is short in a bearingless drive, resulting in high critical speeds and more stable operation.

Low cost – The number of wires is less in the bearingless drive. The number of inverters is also less. Low-cost standard 3-phase inverters are also employed.

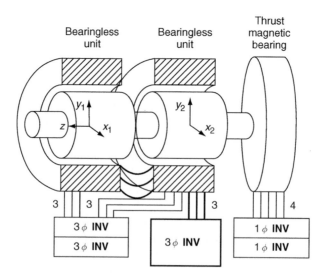

Figure 1.3 Bearingless drive

High power – The bearingless drive can generate increased power if the shaft length is the same.

These advantages are realized as a result of the integration of a radial magnetic bearing and a motor.

1.3 Definition and related technologies

Table 1.1 gives two alternative definitions for the bearingless motor. The term "magnetically integrated" is very important. The drive shown in Figure 1.2 has an integrated magnetic bearing; however, it is only structurally integrated and not magnetically integrated. The drive in Figure 1.3 has magnetically integrated bearings.

Electrical engineers may feel familiar with the first definition because they may be familiar with motors. On the other hand mechanical engineers may feel familiar with the second definition. Therefore bearingless technology lies between electrical and mechanical engineering. It can also be noted that the bearingless generator can be defined. Most of the electrical motors can be used as generators.

Where does the term "bearingless" come from? Several researchers had used the term "bearingless" in their papers independently up to 1994. The origin of the word may be a simple modification of "brushless dc motors". It is well known that "brushless dc motor" is used to describe an inverter-driven permanent magnet synchronous motor supplied with a quasi-square wave phase current (a sinusoidal phase current fed to the motor is a "brushless ac motor"). These have electrical characteristics similar to standard commutator-type dc motors, for example the rotational speed is proportional to the applied voltage. In commutator-type dc motors, the brushes are electro-mechanical contacts and can be a cause of troubles such as noise and sparking; they also have a limited lifetime and require maintenance. For dc motor users, the elimination of the brushes is very attractive. In a brushless dc motor the mechanical switching action of the commutator is replaced by electronic switching of the phase windings. Thus, inverter-driven synchronous motors are often called "brushless dc motors".

These days, most of the maintenance requirement in an industrial drive is related to mechanical bearings. Lubrication oil should be replaced periodically. The bearings should also be replaced periodically, which requires the opening of the motor frame. If the shaft is suspended by a magnetic force, these maintenance

Table 1.1   Definitions of a bearingless motor

1. A motor with a magnetically integrated bearing function.
2. A magnetic bearing with a magnetically integrated motor function.

Note: "Motor" can be replaced by "generator".

tasks are not required. Hence going "bearingless" has many attractions for the motor user.

Figure 1.4 shows the related technologies of bearingless drives. The following significant developments have aided magnetic bearing technology:

1. Industrial experience and acceptance.
2. Radial positioning strategies such as inertial rotational centre alignment.
3. Low-noise and fast-response radial displacement sensors.

Power electronic technology has made a great contribution to the development of both magnetic bearing and bearingless motors. Instantaneous current control is possible with high switching-frequency IGBTs and MOSFETs. These power devices are now integrated into one power module. The power module includes six power devices forming a 3-phase inverter arrangement, as well as gate drive circuits and protection circuits, making the current regulation more reliable. Thanks to recent technological improvements, the cost of the current regulation system has decreased significantly and the cost of the power module has also decreased because mass-produced 3-phase voltage-source inverters are now widely used in industry and domestic appliances in such systems as variable-speed induction-motor and permanent magnet-motor drives. Most of the air-conditioners sold in Japan are driven by high-efficiency inverters. The benefits of low cost 3-phase inverters will further aid the development of the bearingless motor.

Digital signal processing has been improving over a period of time so the calculation speed has increased immensely while the cost has reduced. It should be noted that magnetic suspension requires relatively short sampling times

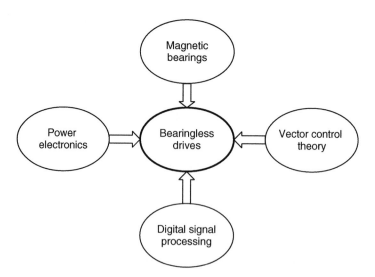

Figure 1.4 Related technologies

compared with motor drives. This is because magnetic suspension is inherently unstable so that derivative or phase-lead controllers are necessary to realize stable magnetic suspension. In order to obtain phase margin at a crossover frequency, fast sampling is necessary. The requirement for the sampling frequency depends on mechanical inertia and stiffness design. Thanks to recent signal processing developments, such as digital signal processors and one-chip microprocessors, there is now enough calculation power to include fast calculation in a magnetic suspension system.

Bearingless motors often take advantage of the magnetic field set up by the motor winding currents. Therefore it is very important to have information about this revolving magnetic field. Controllers based on vector control theory provide instantaneous torque regulation as well as revolving magnetic field regulation. Hence the rotational position and amplitude of the magnetic field can be regulated. Based on the angular position and the amplitude of the motor magnetic field, radial forces are generated by generating additional magnetic fields using the suspension winding current. Therefore it can be said that bearingless technology stands on vector control theory. This leads to the fact that bearingless motors could not be realized before vector control theory (or, as it is sometimes called, field oriented control theory) was developed in the 1980s.

We can now see that bearingless motor technology was developed only after 1994 since it relies on four technologies that have only recently matured.

1.4 Early developments

Table 1.2 summarizes some notable developments of bearingless motors. In the middle of 1970s, a primitive electromagnet with stator windings having pole numbers of p and $(p+2)$ was proposed by Hermann [1–2]. This electromagnet was proposed as a motor which has a radial magnetic bearing function. Moreover, a split winding motor was proposed by Meinke [3]. However, development was limited since there was little knowledge of inverters, digital signal processors and field-oriented control theories at that time.

In 1985 a stepping motor, which was magnetically combined with a magnetic bearing, was proposed by Higuchi [4]. It included an inherently decoupled structure for torque and radial force, while taking an advantage of the motor exciting current.

In 1988, a stator winding structure with different pole combination was proposed for static force generation in order to support a rotor weight against gravity force [5]. In the same year, a disc type motor with axial force generation produced by adjusting the motor exciting current was proposed [6]. To the authors' best knowledge, the word "bearingless" was used for the first time.

In 1989, the primary authors proposed a general concept for the bearingless motor [7]. Based on field-oriented theory, they concluded that most electrical machines can be used in a bearingless drive with additional suspension windings driven by a 3-phase inverter. The concept has been developed theoretically

Table 1.2 Notable inventions

Year	Topics	Ref. No.	Researchers	Reference
1974	p-Pole and $(p+2)$-pole windings combination	[1–2]	Hermann	UK patent
1976	Split winding structure	[3]	Meinke	US patent
1985	Magnetically combined stepping motor	[4]	Higuchi	Japan patent
1986	Static force generation for gravity force	[5]	Williamson	UK patent
1988	Disc type axial force bearingless motor	[6]	Bosch	ICEM
1989–	General bearingless concept, etc.	[7–8]	Chiba and Fukao	Japan patent, IEEE
1990–	PM bearingless machine, etc.	[9–10]	Bichsel and Hugel	Dissertations
1990–	Split winding for induction machine, etc.	[11]	Salazar and Stephan	IEEE MAG
1992–	PM bearingless machine, etc.	[12]	Okada and Ohishi	US patent

with strong support of Rahman [8]. Since then, the concept has been applied to synchronous reluctance, induction, permanent magnet and disc-type bearingless drives, as well as homopolar, hybrid and consequent-pole bearingless drives.

In 1990, Bichsel [9] finished his dissertation under supervision of Hugel of ETH, Zurich, Switzerland. They proposed a 6-pole permanent magnet rotor with 4-pole suspension windings as well as digital signal processor–controlled inverters. In the Zurich group, the continuous development in induction machine theory has led to the direct type of field-oriented control theory by Schoeb [10], Gempp and Redemann. Two-axis slice bearingless drives have also been developed by Barletta. The developments in ETH have been steadily spreading by co-operation with Amrhein and Silber.

It is also noted that radial force generation using split motor windings has been investigated by Salazar, Stephan and Watanabe [11].

The basic pole combination has been investigated by Okada and Ohishi [12] in Ibaraki University. Continuous developments for blood pump applications by Matsuda, Ueno and Masuzawa are also noted.

Since the middle of 1990s, development of the bearingless machine has been taking place in Switzerland, Austria, Germany, UK, France, Canada, USA, China, Korea and other places. Some authors refer to the bearingless motor in their own ways. Examples of alternative names are "bearing motor", "bearing and drive", "levitated motor", "floating actuator", "lateral force motor", "combined motor-bearing", "self bearing motor" and "integrated motor bearing". These names provide an interesting selection of alternative names for the bearingless motor. However, most technical papers use "bearingless motor". Hence it is recommended to use "bearingless motor" in technical writing for universal understanding.

1.5 Bearingless structures

In this section, several typical structures for the bearingless motor and generator are described. Two-axis active magnetic suspension, five-axis suspension and combinations of conventional magnetic bearings and mechanical bearings are included in the description.

Figure 1.5(a,b) shows two-axis active magnetic suspension. In Figure 1.5(a), a shaft is inserted into a rotor core. Two-axis magnetic suspension is realized by magnetic forces between the rotor and the stator. At the bottom of the shaft, a pivot bearing is located for axial and radial positioning of the shaft end. This structure is suitable for vertical shaft machines. On the other hand, in Figure 1.5(b), the shaft is removed. The two-axis active positioning provides passive magnetic suspension in tilting and axial motion. Hence there is a restriction in axial core length to realize passive suspension. However, with correct design, compact and low-cost bearingless drives of this form have been realized.

Figure 1.6(a–d) shows cross-sections of several arrangements for five-axis active suspension. Two bearingless units are required in order to generate radial forces in four axes. There is also a thrust magnetic bearing for active axial positioning on the fifth axis. In Figure 1.6(a), there are two rotors on the shaft acting in tandem. The rotor and the shaft are rotating inside two stator cores. A load machine such as a pump and compressor impeller can be attached to one end of the shaft. In Figure 1.6(b), the rotor is on the outside of the two stators. This structure is suitable for flywheel drives as well as digital video disk (DVD) and hard disk drives. The arrangement in Figure 1.6(c) is a modification of Figure 1.6(a). The shaft is hollow, allowing flow down the centre, and the thrust bearing is located between the bearingless units for full five-axis suspension. This structure is suitable for flow meters, canned pumps, spindles, etc. The hollow shaft can also be used for locating a wheel. In Figure 1.6(a–c), a thrust magnetic bearing is used. However, it is not necessary in some cases, where axial force is low or precise axial

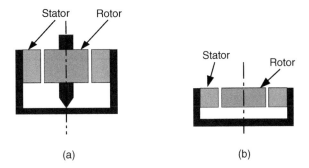

(a)　　　　　　　　　　　(b)

Figure 1.5 Two-axis active suspension: (a) pivot bearing at the bottom; (b) no contacts

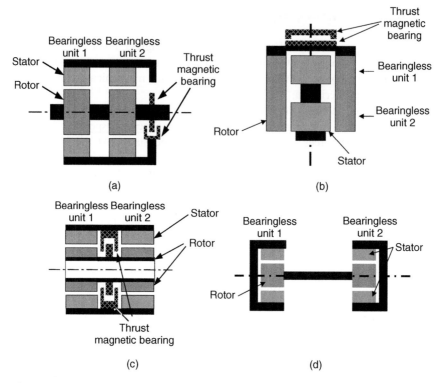

Figure 1.6   Five-axis active suspension variations: (a) inner rotor; (b) outer rotor; (c) hollow rotor; (d) load machine space on the shaft centre, 4-axis

positioning is not required. In these cases, the axial positioning can be realized by passive positioning as shown in Figure 1.6(d), where the rotors in the bearingless units are naturally drawn to their axial centres by magnetic forces. Since the bearingless units generate considerable magnetic flux, the shaft experiences enough spring force under axial movement to keep the drive axially stable.

Figure 1.7(a–c) shows the cross-sections of different bearingless units when they are combined with conventional, mechanical or magnetic bearings. In Figure 1.7(a), the left-hand unit is a conventional radial magnetic bearing. Only the right-hand unit is a bearingless unit. This arrangement is suitable for a high radial force load connected to the left shaft end. In Figure 1.7(b), mechanical bearings are located at both ends of a bearingless unit. When the rotational speed is extremely high, the shaft has critical speeds caused by bending of the flexible shaft. At these speeds vibrations are produced. The bearingless unit can suppress these vibrations and support the shaft weight. In Figure 1.7(c), a load is driven by a conventional motor with a long shaft. The bearingless unit is located near the middle of the shaft in order to suppress shaft vibrations.

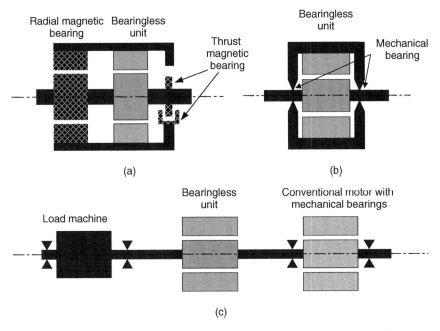

Figure 1.7 Combinations with conventional magnetic and mechanical bearings: (a) with conventional magnetic bearing; (b) with conventional mechanical bearing; (c) with long shaft

1.6 Comparisons

Table 1.3 compares several aspects of the different types of bearingless motor. The characteristics and performance criteria are divided into two parts, i.e., torque and force generation. Let us inspect the torque generation criteria. The motor efficiency is a ratio of shaft output and electrical input power. The efficiency is generally high in permanent magnet motors. They are also compact thanks to the recently developed rare-earth permanent magnets. Induction motors have a robust cylindrical rotor and smooth torque. Synchronous and switched reluctance motors have the advantage of low material costs so are suitable for mass production.

From the point of view of radial force generation, induction and permanent magnet motors are temperature dependent. In induction motors, the rotor resistance is a function of the rotor winding or cage temperature. The conductivity of the copper or aluminium can vary considerably over the operating temperature range of the machine, resulting in misalignment of the calculated field orientation in some cases. This misalignment of the field orientation sometimes results in shaft suspension problems. A similar problem may also occur in permanent magnet motors because the remanent magnetism (or strength) of permanent magnets decreases at high temperature. The variation is dependent on the magnet material. The discrepancy in magnet strength results in a field orientation calculation error. However, homopolar, switched reluctance and synchronous reluctance

Table 1.3 Comparisons of bearingless motors

Criteria	Induction	Permanent magnet	Homo hybrid consq.	Synchronous reluctance	Switched reluctance
Torque					
Efficiency	4	5	4	4	4
Compactness	4	5	3–5	4	4
Rotor robustness	3–5	3–5	4	3–5	4
Torque smoothness	5	4	4	4	3
Motor cost	4	3	4	5	5
Radial force					
Temperature independence	3	4	4–5	5	5
Indirect control possibility	3	4	5	4	4
Inherent field decoupling	4	4	5	4	4
Force/current ratio	4	3–5	5	5	5
Motor INV fault independence	3	5	3–5	3	3

5 – Excellent, 4 – Very good, 3 – Good, 2 – fair, 1 – Poor

motors are robust in temperature variations since they do not have an electric circuit or magnets on the rotor.

When considering the possibility of indirect control, it should be remembered that the machine parameters depend on the level of magnetic saturation in synchronous and switched reluctance bearingless machines and, to a lesser extent, in interior and buried permanent magnet bearingless machines. As already mentioned, permanent magnet machine parameters also depend on temperature.

The decoupling of torque and radial force generation is excellent in homo-polar, hybrid and consequent-pole bearingless motors. In these motors, the direction and amplitude of generated radial forces do not depend on torque generation (to any great extent). The radial force is mainly generated by an interaction of flux produced by suspension winding current and rotor magnets.

Suspension force produced by suspension current depends on the permeance of the paths that the generated flux follows. The relative permeability of permanent magnet is low. It is usually close to unity depending on the magnet's material. Hence, permanent magnets can often be considered as airgap so that the magnets and airgap can be lumped together to give low permeance flux paths, producing high reluctance in the flux paths. This results in a low force/current ratio. This ratio is also low in squirrel-cage induction motors since the flux produced by the suspension winding current is damped by current in the rotor cage. This is because the suspension-current flux will induce electro motive forces (EMFs) into the cage, producing current and hence flux that will oppose the primary flux. This problem can be overcome by using pole-specific rotor bar connections in the rotor circuit. When using this sort of connection, most of the suspension winding current effectively generates suspension force so that the force/current ratio is as high as in reluctance motors. In synchronous and switched reluctance motors, the radial force/current ratio is high because of the high permeance of the levitation flux paths (in the aligned position in the case of a reluctance machine and on the d-axis in the case of a salient-pole synchronous machine). In addition, since the airgap between the rotor and stator is usually designed to be small in these types of motor, high permeance also exists for the main motor flux paths. In homopolar, hybrid and consequent-pole machines, the permeance of the flux paths can be designed to be high if the airgap length is as small as in reluctance machines, leading to a high force/current ratio.

It can be noted that the force/current ratio is easily improved with an increase in the number of series-connected conductors in the suspension windings. Therefore the force/current ratio is only meaningful if discussed under the condition of equal number of series turns.

The inverter fault tolerance is also an important aspect of bearingless motors. In some cases, motor current may be stopped because of a motor inverter fault. Even in this situation, magnetic suspension should continue in some applications to prevent further damage. In permanent magnet, hybrid and consequent-pole bearingless motors, magnetic suspension operates independently of the motor winding current by using the permanent magnet flux.

1.7 Winding structures

Table 1.4 summarizes various stator winding configurations. Conventional motors have 3-phase, 2-phase or single-phase windings. It is necessary to add 2- or 3-phase windings to generate radial force and various winding configurations have been proposed. The "4-pole and 2-pole winding" indicates a bearingless motor with a 4-pole motor winding set for torque generation as well as a 2-pole winding set for radial force generation. For example, 4-pole 3-phase windings and 2-pole 3-phase windings can be wound in a stator with a 4-pole permanent magnet rotor. This winding configuration applies in general to the theory of bearingless motors so it is applicable to various motor types such as induction and synchronous reluctance motors and permanent magnet motor types including surface-mounted, interior, inset and buried-spoke rotor configurations. The motor and the suspension windings are quite separate and the suspension winding operates using the differential principle.

The "4-pole and 2-pole winding" can also indicate a 2-pole motor winding and a 4-pole suspension winding, the functions of the winding sets are simply exchanged. A 2-pole revolving magnetic field is generated by the motor winding,

Table 1.4   Stator winding configurations

Windings and operation	Winding combinations
Separated and differential	4-Pole and 2-pole windings
Separated and differential	p-Pole and $(p \pm 2)$-pole windings
Separated and differential	2-Pole winding with homopolar rotor
Separated and differential	Short-pitch differential winding around stator poles
Separated and differential	Single-phase motor and 2-phase suspension winding
Motor common and current unbalance	Split motor windings

so a 2-pole permanent magnet rotor is required. This winding strategy is well suited to magnetically cylindrical rotors, such as induction motors and surface-mounted permanent magnet motors. In salient-pole motors like synchronous reluctance or interior permanent magnet motors, the radial force is also a function of rotor angular position, so additional compensation is necessary. These winding structures are extensively described in this book.

The "2-pole winding with homopolar rotor" indicates a homopolar, a hybrid or, possibly but not necessarily, a consequent-pole bearingless machine. These machines have rotor poles excited in one direction only, i.e., the flux will cross the airgap once and return via another route. Thus, a 2-pole suspension winding Magneto Motive Force (MMF) is needed to generate an unbalanced magnetic field. Radial force is generated in the direction of the MMF. With the correct selection of the rotor magnetic pole number, a radial force is generated which is independent of the rotor angular position. The details are described in Chapter 14.

The "short-pitch differential winding" has short-pitch windings on each stator pole. This winding configuration is particular to switched reluctance motors although it is also used in compact permanent magnet motors with high stator-slot fill and magnetic bearings. Differential windings can be wound over the motor windings so that the MMF can be made unsymmetrical in a controlled manner to produce an unbalanced flux distribution and hence generate a net radial magnetic force between the stator and rotor. As already mentioned, differential windings are applicable in permanent magnet motors; however, these windings are described in the chapter on switched reluctance type bearingless motors Chapter. Another way to generate radial force is to regulate the winding currents independently as illustrated in the split windings description in Chapter 16. Other winding structures are also described in Chapter 16.

1.8 Applications

There are several suitable applications of magnetically suspended motors and generators. Table 1.5 summarizes some applications. High-speed drives and generators require frequent bearing maintenance if mechanical bearings are fitted. However, magnetic suspension provides maintenance-free high-speed rotation. A flywheel drive and generator used in energy storage needs very low friction suspension. A satellite reaction wheel needs a rotating wheel for attitude

Table 1.5 Applications

- High speed drives and generators.
- Flywheel drives and generators, satellite reaction wheels, momentum wheels.
- Food processes, pharmacy processes, harsh environments such as low temperature, high temperature, vacuum and poisonous gas atmospheres.
- Swinging motors.
- Medical equipment, implantable blood pumps.
- Information storage drives.

regulation. In food and pharmacy processes, oil leaks caused by broken seals on mechanical bearings should be avoided. In harsh environments, such as in extremely low and high temperatures, as well as in vacuum, shaft suspension is always a problem. Pumps and blowers for poisonous, explosive and acidic gases and fluids also have problems with the mechanical seals although stainless steel cans with carbon bearings are usually employed. Magnetic suspension provides a long lifecycle and is maintenance-free.

A swinging motor also has a problem with bearings. If the swinging motor operates 24 hours a day without a break, ball-bearing grease is dispersed, so the bearing lifetime is not guaranteed. Therefore magnetic suspension is an effective solution. More information on applications is given in the last chapter of this book.

References

[1] P. K. Hermann, "A Radial Active Magnetic Bearing", London Patent No. 1 478 868, 20 November 1973.
[2] P. K. Hermann, "A Radial Active Magnetic Bearing Having a Rotating Drive", London Patent No. 1 500 809, February 9, 1974.
[3] P. Meinke and G. Flachenecker, "Electromagnetic Drive Assembly for Rotary Bodies using a Magnetically Mounted Rotor", United States Patent No. 3 988 658, July 29, 1974.
[4] T. Higuchi, "Magnetically Floating Actuator Having Angular Positioning Function", United States Patent No. 4 683 391, March 12, 1985.
[5] S. Williamson, "Construction of Electrical Machine", United States Patent No. 4 792 710, February 20, 1987, UK.
[6] R. Bosch, "Development of a Bearingless Electric Motor", ICEM, pp. 373–375, 1988.
[7] A. Chiba and T. Fukao, "Electric Rotating Machinery with Radial Position Control Windings and its Rotor Radial Position Controller", Japan Patent No. 2835522, January 1989.
[8] A. Chiba, T. Deido, T. Fukao and M. A. Rahman, "An Analysis of Bearingless AC Motors", IEEE Transaction on Energy Conversion, Vol. 9, No. 1, March, 1994, pp. 61–68.
[9] J. Bichsel, "Contributions on Bearingless Electric Motors", ETH Thesis No. 9303, 1990.
[10] R. Schoeb, "Contributions on Bearingless Asynchronous Machines", ETH Thesis No. 10417, 1993.
[11] A. O. Salazar, W. Dunford, R. Stephan and E. Watanabe, "A Magnetic Bearing System using Capacitive Sensors for Position Measurement", IEEE Transanction on Magnetics, Vol. 26, No. 5, September 1990, pp. 2541–2543.
[12] T. Ohishi, "Magnetic Bearing Device with a Rotating Magnetic Field", United States Patent No. 5 237 229, April 1992.

2

Electro-magnetics and mathematical model of magnetic bearings

Akira Chiba

In this chapter, basic structures, analyses and mathematical models of magnetic bearings and related actuators are presented in addition to the following. The principles of a feedback control strategy for magnetic bearings are introduced, and a simple electromagnetic actuator is developed in order to understand the electrical equivalent circuits. An analysis is carried out that calculates the inductance, flux densities, stored magnetic energy and magnetic forces. Typical structures for the radial magnetic bearing and related actuators are described. As a result of the analysis, a simple representation of the magnetic bearing is introduced. The maximum radial force is also derived based upon the saturated flux density in the magnetic circuits. A block diagram is drawn in order to develop a controller.

2.1 Electro-mechanical structure and operating principles

Figure 2.1 shows the basic structure of an electromagnet with a feedback controller for one-axis magnetic suspension. Excitation of the winding produces a magnetic force to suspend the rectangular iron object. The object is free only in the vertical axis. The current i produces a flux ψ. The flux path is shown by the dotted line and it crosses the airgap twice in the vertical direction. The attractive force between the suspended object and C-core is a function of i, and is proportional to the square of i if the core is not saturated. Under steady-state conditions, the generated attractive force is adjusted to be exactly equal to the

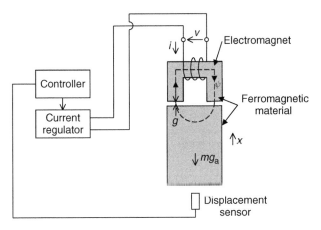

Figure 2.1 Magnetic suspension system

product of the object weight m and gravity acceleration g_a to satisfy the force equilibrium.

The displacement sensor detects the vertical position of the suspended object. The sensor output voltage is the input to a controller. A magnetic force command is generated to stably suspend the object. The force command is the sum of damping- and spring-force commands. The damping-force command is proportional to the speed of the suspended object. The spring-force command is proportional to the displacement of the suspended object. These commands are in the opposite direction to the speed and displacement for negative feedback. The controller generates a current command so that the generated force follows the force command. The current regulator controls the current by applying a voltage to the electrical terminals.

The current i excites a series-wound coil. Let us suppose that the number of turns in the winding is N so that a magneto motive force (MMF) Ni is produced. Since the permeability is high in ferromagnetic materials, the flux follows the path shown. The flux crosses the airgap twice. Note that only one flux path is shown; however, the flux is distributed in the airgap. The maximum flux density in the airgap determines the force capability of the electromagnet. A high flux density results in high magnetic force. However, the maximum flux density is limited to 1.7 to 2 T in conventional silicon steel. It is also important to make the airgap length as small as possible, which will reduce the current and losses.

2.2 Electric equivalent circuit and inductance

Figure 2.2 shows an electromagnet used to suspend an I-shaped core with a magnetic force. The C-core has a width w with a stack length l. The main flux path is indicated by the dotted line. The lengths of the flux path in the C-core are defined by l_1 and l_2. The flux path length in the I-shaped core is l_3. The winding

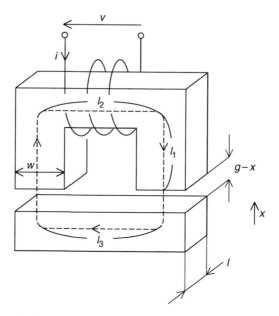

Figure 2.2   C- and I-shaped cores with a winding

has N turns. The instantaneous current value is i, so that the MMF is Ni. The airgap length is g at the nominal position. A coordinate position x is defined in the I-shaped core position so that the airgap length is $(g - x)$.

The reluctance of the magnetic circuit is defined as

$$R = \frac{l_{fp}}{\mu_{mt} S} \tag{2.1}$$

where:

l_{fp} = flux path length
μ_{mt} = permeability in the material
S = cross-section area of flux path.

A permeance is the inverse function of magnetic reluctance, i.e., for a permeance

$$P_a = \frac{\mu_{mt} S}{l_{fp}} \tag{2.2}$$

Figure 2.3 shows an "electrical" equivalent circuit for the magnetic circuit of the electromagnet. In terms of MMF (voltage), flux (current) and reluctance (resistance), a constant (dc) magnetic circuit can be treated in the same way as an electric circuit. The main difference is that magnetic reluctance is an energy storage component rather than a loss component. The "dc voltage" source Ni represents the MMF generated by the winding current. R_c and R_I are the magnetic

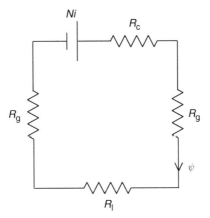

Figure 2.3 Equivalent magnetic circuit

reluctances in the C- and I-cores, respectively, and R_g represents the magnetic reluctance at the airgap. These magnetic reluctances are written as

$$R_g = \frac{g - x}{\mu_0 wl} \tag{2.3}$$

$$R_c = \frac{2l_1 + l_2}{\mu_0 \mu_r wl} \tag{2.4}$$

$$R_I - \frac{l_3}{\mu_0 \mu_r wl} \tag{2.5}$$

where μ_0 is the permeability of free space ($\mu_0 = 4\pi \times 10^{-7}$ H/m) and μ_r is the relative permeability ($\mu_{mt} = \mu_r \mu_0$). The value of μ_r for iron is typically in the range of 1000–10 000. The relative permeability of air is approximately equal to 1.0. In most cases, the airgap reluctance is significantly larger than the iron reluctance, so that the magnetic reluctance in the iron can be neglected in the following calculation. Therefore the equivalent electrical circuit is simplified. The flux ψ is then

$$\psi = \frac{Ni}{2R_g} = \frac{Ni}{2} \frac{\mu_0 wl}{g - x} \tag{2.6}$$

The flux linkage λ_1 of the coil is defined as the number of turns N multiplied by the flux passing through the coil:

$$\lambda_1 = \frac{N^2 i}{2} \frac{\mu_0 wl}{g - x} \tag{2.7}$$

But inductance is defined as flux linkage divided by the current value ($L = \lambda_1/i$), giving

$$L = \frac{N^2 \mu_0 wl}{2(g - x)} \tag{2.8}$$

If the displacement x is small compared to the airgap length, the following series expansion is applicable

$$\frac{1}{g-x} = \frac{1}{g}\frac{1}{\left(1-\dfrac{x}{g}\right)} = \frac{1}{g}\left(1+\frac{x}{g}+\frac{x^2}{g^2}+\frac{x^3}{g^3}+\cdots\right) \tag{2.9}$$

and if only the first and the second terms are considered, the inductance can be approximated to

$$L = L_0\left(1+\frac{x}{g}\right) \tag{2.10}$$

where L_0 is defined as the nominal inductance

$$L_0 = \frac{N^2\mu_0 wl}{2g} \tag{2.11}$$

In addition, the flux density B in the airgap can be derived as

$$B = \frac{\psi}{wl} = \frac{N\mu_0 i}{2(g-x)} \tag{2.12}$$

Example 2.1   Calculate the magnetic reluctances when $x = 0$, assuming the following dimensions: $w = 1\,\mathrm{cm}$, $l = 1\,\mathrm{cm}$, $l_1 = 1\,\mathrm{cm}$, $l_2 = 3\,\mathrm{cm}$, $l_3 = 4\,\mathrm{cm}$ and $g = 1\,\mathrm{mm}$. Assume the relative magnetic permeability in the iron $\mu_r = 1000$.

Answer
From the dimensions given above, the magnetic reluctances can be calculated: $R_g = 8 \times 10^6\,\mathrm{A/Wb}$, $R_c = 5.6 \times 10^5\,\mathrm{A/Wb}$, $R_I = 3.1 \times 10^5\,\mathrm{A/Wb}$. The airgap reluctance is dominant since $R_g \gg R_c + R_I$.

Example 2.2   Calculate the flux density B_0, flux ψ and inductance L_0 when $x = 0$. Assume that $N = 100$ turns and $i = 16\,\mathrm{A}$.

Answer
The values are calculated to be $B_0 = 1\,\mathrm{T}$, $\psi = 0.1\,\mathrm{mWb}$, $L_0 = 0.628\,\mathrm{mH}$.

2.3 Stored magnetic energy and force

In this section, the magnetic force is derived at an arbitrary displacement by consideration of the stored magnetic energy in the system. The concept of co-energy is also introduced and several simple examples of actuators are shown.

Figure 2.4 shows a relationship between the flux linkage λ and current i in magnetic circuits with ferromagnetic components such as the cores in Figure 2.2.

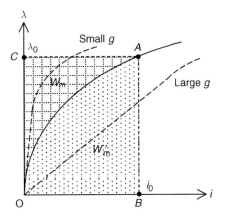

Figure 2.4 Magnetic energy and co-energy

The flux linkage is proportional to the current only at low current values. At high current, the ferromagnetic cores become saturated, producing a nonlinear characteristic. In conventional silicon steel, magnetic saturation occurs at a flux density between 1.2 and 1.8 T. The exact saturation curve depends on the type of silicon steel. Such information is provided by the steel manufacturer.

The magnetization curves also depend on the airgap length between the C- and I-cores. With a small airgap length, the λ/i curve is quite nonlinear because the magnetic circuit is dominated by the saturable C- and I-core reluctances at high current. A wide airgap length results in a more linear characteristic because the circuit reluctance is now dominated by the airgap reluctance component even at high current, which is non-saturable.

Let us suppose that the operating point is A in Figure 2.4, with current i_0 and flux linkage λ_0. The stored magnetic energy W_{m} in a magnetic system is obtained from

$$W_{\mathrm{m}} = \int_0^{\lambda_0} i \, \mathrm{d}\lambda \qquad (2.13)$$

The integration corresponds to the area bounded by points O, C and A. In addition to the magnetic energy, we can introduce the magnetic co-energy W'_{m}. The magnetic co-energy is defined as

$$W'_{\mathrm{m}} = \int_0^{i_0} \lambda \, \mathrm{d}i \qquad (2.14)$$

This definition indicates that the magnetic co-energy is represented by the area bounded by points O, B and A. The sum of W_{m} and W'_{m} is equal to a product of current i_0 and flux linkage λ_0. W_{m} is the stored magnetic energy in the system whereas the co-energy is a component we introduce to aid analysis but has no physical reality.

The independent variables in a magnetic suspension system are normally the winding current and object displacement. If the system is moved by δx then it can be shown that the work done is equal to the change in co-energy of the system. Since the work done is the force $F \times \delta x$, the electromagnetic force F is given as the partial derivative of the magnetic co-energy

$$F = \frac{\partial W'_m}{\partial x} \tag{2.15}$$

This equation provides the expression for the electromagnetic force between the C- and I-cores with nonlinear magnetizing characteristics.

If the magnetizing characteristic curve is linear (i.e., no saturation) then the flux linkage and current relationships are as shown in Figure 2.5. It is seen that magnetic energy and magnetic co-energy are equal so that

$$W_m = W'_m \tag{2.16}$$

In this instance, the magnetic force can be also written as

$$F = \frac{\partial W_m}{\partial x} \tag{2.17}$$

Note that this equation is valid only for a magnetically linear system.

Assuming a linear system, where the self-inductance L is constant and $Li = \lambda$, the magnetic co-energy W'_m is derived as

$$W'_m = \int_0^i Li\,di = \frac{1}{2}Li^2 \tag{2.18}$$

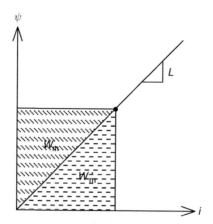

Figure 2.5 Magnetically linear relationship

The electromagnetic force F is

$$F = \frac{\partial W'_m}{\partial x} = \frac{\partial L}{\partial x} \frac{i^2}{2} \qquad (2.19)$$

From (2.10), the partial derivative of L with respect to x is

$$\frac{\partial L}{\partial x} = \frac{L_0}{g} \qquad (2.20)$$

Substituting (2.20) into (2.19) yields

$$F = \frac{L_0}{g} \frac{i^2}{2} \qquad (2.21)$$

Example 2.3   Using equation (2.21), derive the magnetic force acting on the body shown in Figure 2.2 for an airgap of 1 mm.

Answer
Using equation (2.21), the magnetic force of Figure 2.2 can be derived. Substituting $L_0 = 0.628$ mH and current of 16 A, F is 80 N. Note that 80 N is equivalent to 8.2 kgf with a gravity acceleration of 9.81 m/s^2.

2.3.1 Flux density and electromagnetic force

In equation (2.21), the magnetic force is expressed as a function of current. Another useful expression obtains the force from the flux density in the airgap. L_0 in equation (2.21) is substituted by (2.11) so that

$$F = \frac{N^2 i^2 \mu_0 wl}{4g^2} \qquad (2.22)$$

Assuming that $x = 0$, solving (2.12) for Ni yields

$$Ni = \frac{2gB_0}{\mu_0} \qquad (2.23)$$

Therefore substituting (2.23) into (2.22) produces

$$F = \frac{B_0^2}{2\mu_0} \times 2wl \qquad (2.24)$$

But $2wl$ is the area of airgap. Let us define the airgap area S as $2wl$ so that the magnetic force is expressed as in the well-known Maxwell stress equation where

$$F = \frac{B_0^2}{2\mu_0} S \qquad (2.25)$$

This equation gives us a straightforward insight into magnetic force generation. It can be said that the magnetic force is proportional to the square of flux density in the airgap. The force is also proportional to the airgap area. The following example provides a practical way to estimate the magnetic force.

Example 2.4 Consider the arrangement in Figure 2.6 and answer the following questions:

1. At flux density of 1 T, derive the magnetic force assuming $S = 1\,\text{cm}^2$.
2. At the same flux density, derive the magnetic force at $S = 1\,\text{m}^2$.
3. Suppose that the saturated flux density in silicon steels is 1.7 T. How much radial force is generated for $S = 1\,\text{cm}^2$?

Answers
1. $F = 1^2/(2 \times 4\pi \times 10^{-7}) \times 100 \times 10^{-6} = 40\,\text{N}(\cong 4\,\text{kgf})$. Remember that the flux density level in the airgap is usually less than 1 T so that a magnetic force density of $40\,\text{N/cm}^2$ (approximately $4\,\text{kgf/cm}^2$) gives the highest practical value. Figure 2.6 illustrates this. There is an airgap between two iron poles. The flux density is 1 T and the area under the poles is $1\,\text{cm}^2$ so that there is an attractive force of 40 N between these two poles.
2. $400\,\text{kN}\,(\cong 40\,\text{t})$.
3. $116\,\text{N}\,(\cong 12\,\text{kgf})$. If the flux density in the airgap is increased to 1.7 T then the magnetic force is three times the value at 1 T. To achieve 1.7 T, an extremely high MMF is usually required. Increasing the MMF may possibly require bulky coil windings and external cooling. Engineers usually find that it is more practical to increase the airgap area rather than flux density.

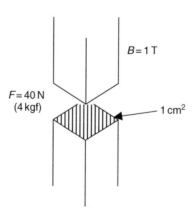

Figure 2.6 Flux density and attractive magnetic force

2.3.2 Nonlinear magnetic characteristics

So far we have considered the magnetic force by assuming that iron cores have linear characteristics, i.e., the permeability is constant. It is also assumed that the iron permeability is high so that the iron core reluctance can be neglected. However, in some cases, the core reluctance should be considered. For example, a magnetic circuit with a very short airgap length but long flux path in the iron core has high per-unit iron reluctance. In this case, the relationship between flux linkage and current is highly nonlinear.

Figure 2.7 shows magnetization characteristics of a C- and I-core combination under three different conditions. Without an airgap, the magnetization curve is determined by the nonlinear magnetic characteristics of the iron core. If a small airgap is introduced, the flux linkage is decreased for a given current because the airgap results in an increase in magnetic reluctance. If the airgap is further increased, the magnetization curve becomes straight because the magnetic reluctance is mainly due to the airgap reluctance. In this case, we can assume a linear characteristic; however, it has to be noted that more current is required to achieve the same flux linkage level. For example, to achieve λ_0, the currents i_0, i_1 and i_2 are needed for $g = 0$, g_1 and g_2, respectively. In order to reduce copper losses and coil dimensions, the airgap length is usually made as small as possible.

2.3.3 Force calculation in nonlinear magnetization curve

We have derived two force equations. One is based on the current and the other is based on the flux density in the airgap. These are (2.21) and (2.25), and are repeated here:

$$F = \frac{L_0}{g}\frac{i^2}{2}$$

$$F = \frac{B_0^2}{2\mu_0}S$$

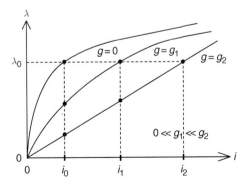

Figure 2.7 Magnetic linearity and airgap length

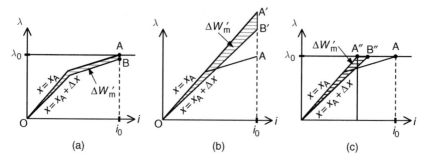

Figure 2.8 Linear approximation errors: (a) nonlinear system co-energy; (b) magnetic co-energy in a magnetically linear model based on current value; (c) magnetic co-energy in a magnetically linear model based on flux linkage value

Let us see how accurate these equations are under magnetically saturated conditions.

Figure 2.8(a) shows nonlinear magnetization curves of a C- and I-core with a small airgap using a piece-wise linear approximation. At low current, the relationship between the flux linkage and current is linear. The derivative of flux linkage with respect to the current is the self-inductance L. At high current, the core is saturated. Suppose the operating point is A with λ_0, and i_0 at the airgap length of x_A. Let us suppose that the airgap length is increased by Δx so that the magnetization curve is moved to line OB for $x_A + \Delta x$. The area OAB bounded by the magnetizing curves at $x = x_A$ and $x_A + \Delta x$ is the co-energy variation $\Delta W'_m$. A product of the magnetic force and the displacement Δx is equal to the area $\Delta W'_m$. Hence the shaded area is proportional to the magnetic force.

In Figure 2.8(b) the co-energy obtained from the current equation is shown. The equation assumes a linear relationship even at high current. At the current i_0, the flux linkage is Li_0 so the operating point is A'. The shaded area $OA'B'$ is the co-energy $\Delta W'_m$. We can see that the calculated $\Delta W'_m$ is too high. Therefore the magnetic force calculated from (2.21) is higher than the actual values under saturated conditions.

In Figure 2.8(c) the co-energy based on the flux density equation is shown. Again, the magnetization curve is the extension of linear relationship, i.e., $\lambda = Li$. Since the flux density is proportional to the flux linkage, even in the nonlinear conditions, the operating point is A'' with flux linkage of λ_0. Therefore the shaded area $OA''B''$ is the co-energy variation. One can see that $\Delta W'_m$ is smaller than that in Figure 2.8(b). Hence the overestimation of magnetic force can be avoided under saturated conditions. Equation (2.25) is widely used in estimating the achievable magnetic force to avoid overestimation.

2.3.4 Differential actuator

Figure 2.9 shows a differential type of magnetic actuator. The cylindrical object can move in the x direction. It is suspended in the air by a controlled magnetic

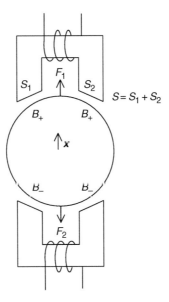

Figure 2.9 Differential actuator

force. Two C-shaped actuators are used as illustrated. Suppose that the flux distribution in the airgap is uniform with flux densities of B_+ and B_- in the upper and lower airgaps respectively. The airgap area is S with a sum of S_1 and S_2. The magnetic forces F_1 and F_2 are applied to the cylindrical object. These forces are $F_1 = B_+^2 S/(2\mu_0)$ and $F_2 = B_-^2 S/(2\mu_0)$ so the sum of these magnetic forces F is written as

$$F = \frac{S}{2\mu_0}\left(B_+^2 - B_-^2\right) \tag{2.26}$$

It can be seen that the magnetic force is proportional to the difference of flux density squared.

Example 2.5   Suppose that $B_+ = B_0 + \Delta B$ and $B_- = B_0 - \Delta B$, i.e., these flux densities are controlled by the bias flux density B_0 and differential flux density ΔB. The object weight is supported by a magnetic force. Derive $B_+, B_-, \Delta B, F_1$ and F_2 assuming that $S = 2\,\mathrm{cm}^2$, $B_0 = 0.5\,\mathrm{T}$ and the weight of the cylindrical object is 1 kg.

Answer
Substituting $B_+ = B_0 + \Delta B$ and $B_- = B_0 - \Delta B$ into (2.26), one can derive

$$F = \frac{2S}{\mu_0}B_0\Delta B$$

Substituting $F = 9.8\,\mathrm{N}$ and the given values we get $\Delta B = 0.062\,\mathrm{T}$. Hence $B_+ = 0.562\,\mathrm{T}$, $B_- = 0.438\,\mathrm{T}$, $F_1 = 25.14\,\mathrm{N}$ and $F_2 = 15.27\,\mathrm{N}$.

2.3.5 Surface-mounted permanent magnet motor (SPM motor)

Figure 2.10 shows a permanent magnet motor. On the rotor, a cylindrical-ring permanent magnet is mounted on the shaft. The permanent magnet is pre-magnetized with two poles. On the stator, there are four salient poles. Each pole has a winding; however, only one phase winding is shown in the Figure 2.10. Motor fluxes are produced by the permanent magnets. Therefore attractive magnetic forces exist between the rotor surface and stator poles. Normally, the sum of these forces is zero if the rotor is centred, and the currents in opposite coils are equal or zero.

Example 2.6   A machine has a rotor with a diameter of 50 mm with an axial length of 50 mm. Let us suppose that the flux density distribution in the airgap between the rotor and stator poles is uniform with a peak of 0.7 T. The rotor is made of NdFeB permanent magnet and silicon steel. The mass density of these materials can be averaged to 7.65 g/cm^3.

1. Find airgap area of one stator pole.
2. Find magnetic force between one stator pole and the rotor.
3. What is the ratio of magnetic force to the rotor weight?
4. Find total magnetic force acting on the rotor.
5. Suppose that the flux density in the airgap of stator poles 1 and 3 are actively controlled as in the example shown in Figure 2.9. Find the ΔB necessary to support a rotor.

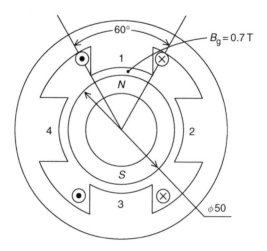

Figure 2.10   Surface mounted permanent magnet motor

Answer

1. The area $S = \pi \times 5\,\text{cm} \times \left(\dfrac{60}{360}\right) \times 5\,\text{cm} = 13.1\,\text{cm}^2$.

2. The magnetic force is $F = 4\,\text{kgf/cm}^2 \times (0.7\,\text{T})^2 \times 13.1\,\text{cm}^2 = 25.7\,\text{kgf}$.

3. The rotor volume is $V = \pi \times \left(\dfrac{5}{2}\,\text{cm}\right)^2 \times 5\,\text{cm} = 98.1\,\text{cm}^3$. The density of silicon steel and permanent magnet is approximated to $7.65\,\text{g/cm}^3$, therefore the weight is $0.750\,\text{kg}$. This means that the magnetic force at one stator pole is about 34 times the force exerted by the rotor weight.

4. Total magnetic force acting on the rotor is zero because radial forces under stator poles 1 and 3 cancel each other out. The forces under poles 2 and 4 also cancel.

5. Substituting $F = 0.750 \times 9.8\,\text{N}$, $S = 13.1 \times 10^{-4}\,\text{cm}^2$ and $B_0 = 0.7\,\text{T}$ into $F = \dfrac{2S}{\mu_0} B_0 \Delta B$ gives $\Delta B = 0.005\,\text{T}$.

In this example, we can see that significant attractive forces are generated between the rotor and the stator salient poles. There are disadvantages and advantages with these radial forces as described below.

If an electrical motor shaft is misaligned from the rotor centre position so that it no longer rotates on its own axis, the four radial forces are unbalanced so that total magnetic force is no longer zero. The generated magnetic force rotates with the rotor. This rotating radial force results in vibrations and acoustic noise and it also shortens the bearing life.

If the radial magnetic forces in the four stator poles are actively controlled then a rotor shaft can be supported by magnetic forces. A slight unbalance of flux density of $0.005\,\text{T}$ generates a magnetic force which will suspend the rotor weight. Even if the rotor experiences a centripetal force of $9\,\text{G}$, a flux density variation of only $0.05\,\text{T}$ is sufficient to suspend the rotor.

2.3.6 E- and I-cores with a permanent magnet

Figure 2.11 shows E- and I-shaped iron cores with a permanent magnet. The permanent magnet is characterized by the residual flux density B_r, which is in the range 1–$1.45\,\text{T}$ for rare-earth permanent magnets employing Neodymium or

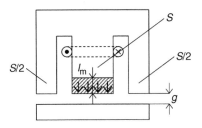

Figure 2.11   E- and I-cores with permanent magnet

Samarium, and about 0.5 T (or less) for low-cost ferrite permanent magnets. The flux generates an attractive force between the E- and I-cores. The permeability (called the recoil permeability μ_{rec}) of a permanent magnet depends on the magnet's material but is generally slightly greater than one for rare-earth magnets.

Example 2.7   Derive the magnetic force using the following procedures and answer the following questions:

1. Assume a one-turn coil in the E-core as shown. Obtain the coil current which generates the same flux as the permanent magnet. Assume that the residual flux density of the magnet is B_r, the magnet thickness is l_m, the flux path area is S and the airgap length is g. Also assume that the iron reluctance and leakage and fringing fluxes are negligible and that the magnet recoil permeability is one.
2. Derive the self-inductance of the one turn coil.
3. Derive the magnetic force.
4. Find the flux density in the airgap.
5. How can we increase the magnetic force?
6. Suppose that both a permanent magnet and a winding are used in this electromagnet. A static force is provided by the permanent magnet MMF and, in addition, a dynamic force is provided by the winding current MMF. Find the electromagnetic force. If the permanent magnet is very thin, the force/current is low because of low flux density in the airgap. If the permanent magnet is thick, the force/current is low because of the high magnetic reluctance of the permanent magnet. Find the condition for maximum force/current.
7. Find the condition for maximum force/displacement.

Apply the numerical parameters of $S = 1\,\mathrm{cm^2}$, $B_r = 1\,\mathrm{T}$, $l_m = 1\,\mathrm{mm}$ and $g = 0.5\,\mathrm{mm}$.

Answer
1. Figure 2.12 shows a magnetic equivalent circuit of the E- and I-cores. The airgap reluctances are $g/(\mu_0 S)$ and $2g/(\mu_0 S)$ while $l_m/(\mu_0 S)$ is the permanent magnet reluctance. The permeability of the permanent magnets is not exactly equal to that of free space, e.g., $\mu_r = 1.05$ for NeFeB; however, it can be approximated to unity. The current of the supply is also the MMF of the one-turn coil with current i.

   Let us suppose that the airgap length is zero. The equivalent circuit is then simplified as shown in Figure 2.13. The permanent magnet is short-circuited and the flux ψ is obtained from $\psi = \mu_0 Si/l_m$. Hence the flux density B in the permanent magnet is given as $B = \psi/S = \mu_0 i/l_m$. This flux density is equal to the residual flux density B_r so that

$$i = \frac{B_r l_m}{\mu_0} \tag{2.27}$$

Substituting with the given numerical parameters gives $i = 800\,\mathrm{A}$.

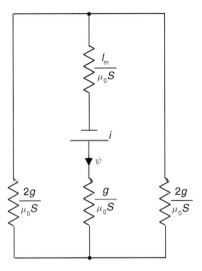

Figure 2.12 Equivalent magnetic circuit including permanent magnet

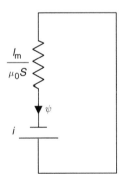

Figure 2.13 Short-circuited permanent magnet

2. In Figure 2.12, the flux is written as

$$\psi = \frac{i}{\dfrac{l_{m}}{\mu_{0}S} + \dfrac{2g}{\mu_{0}S}} \tag{2.28}$$

The self-inductance is derived from the flux per unit current so that

$$L = \frac{\mu_{0}S}{l_{m} + 2g} \tag{2.29}$$

3. From the inductance L and current i, the magnetic energy is equal to $Li^{2}/2$. Substituting this into (2.29) yields

$$W_{m} = \frac{1}{2}\frac{\mu_{0}S}{l_{m} + 2g}i^{2} \tag{2.30}$$

Assuming a magnetically linear flux path, the magnetic force F is the partial derivative of the energy with respect to the airgap length $F = \delta W_m / \delta g$. Thus

$$F = \frac{\mu_0 S}{(l_m + 2g)^2} i^2 \tag{2.31}$$

Note that the direction convention of the magnetic force is opposite to the coordinate of displacement g so we change the polarity. The current i is substituted from (2.27), giving

$$F = \frac{B_r^2 S}{\mu_0} \frac{l_m^2}{(l_m + 2g)^2} \tag{2.32}$$

Using the given parameters, $F = 19.9\,\mathrm{N}$.

4. The flux density is obtained from the flux in (2.28) divided by the area S, and substituting into (2.27) gives $B = [l_m/(l_m + 2g)] \times B_r$. The flux density is 0.5 T.

5. One way to increase the magnetic force is to increase the permanent magnet thickness. An increase in l_m results in an increase in the airgap flux density. In the flux density equation, the coefficient is 0.5; however, this value can be increased to nearly 1.0 by an increase in l_m. It can also be noticed that a decrease in the airgap length also increases the airgap flux density. Another way to increase the magnetic force is to use a stronger permanent magnet with high residual flux density. In this example, $B_r = 1\,\mathrm{T}$; however, magnets with B_r up to 1.45 T are available on the market.

6. Let us define the current in the one-turn coil as i_c. Let us also define the equivalent current caused by the permanent magnet as I_p. Then, the coil current is a sum of the actual coil current and the equivalent permanent magnet current so that $i = I_p + i_c = B_r l_m / \mu_0 + i_c$ (from 2.27). Substituting this current into (2.31):

$$F = \frac{S}{\mu_0} \left(\frac{B_r l_m + \mu_0 i_c}{l_m + 2g} \right)^2 \tag{2.33}$$

This equation shows that the force is not a linear function of coil current. In most cases, linearization is carried out near the operating point so that the maximum radial force for the current i_c can be obtained from a partial derivative of F with respect to i_c. The partial derivative of F can be derived as

$$\frac{\partial F}{\partial i_c} = \frac{2S(B_r l_m + \mu_0 i_c)}{(l_m + 2g)^2} \tag{2.34}$$

The optimum permanent magnet thickness which maximizes the force/current is given when the partial derivative of the above equation is zero.

$$\frac{\partial}{\partial l_m} \left(\frac{\partial F}{\partial i_c} \right) = 0$$

The above equation gives several solutions for permanent magnet thickness l_m; however, only one reasonable solution is obtained. Assuming that coil current is much smaller than the equivalent permanent magnet current, the equation simplifies to

$$l_m = 2g \tag{2.35}$$

The result of this equation indicates that the maximum radial force for a given winding current is obtained if the permanent magnet thickness is twice the airgap length. This fact gives us a simple criterion when designing a permanent magnet-assisted magnetic suspension system such as a zero power magnetic suspension system.

7. Let us define x as a displacement from the reference airgap between the E- and I-cores. The airgap length is expressed as $g - x$. With no coil current the magnetic force is dependent on x so that (2.31) can be re-written as

$$F = \frac{\mu_0 S I_p^2}{(l_m + 2(g - x))^2} \tag{2.36}$$

The force and displacement have a nonlinear relationship. Thus, the partial derivative of the force with respect to x gives force/displacement around the operating point. Solving the derivative and substituting I_p into i in (2.27) yields

$$\frac{\partial F}{\partial x} = \frac{4 S B_r^2}{\mu_0} \frac{l_m^2}{(l_m + 2(g - x))^3} \tag{2.37}$$

Solving the equation

$$\frac{\partial}{\partial l_m} \left(\frac{\partial F}{\partial x} \right) = 0$$

the condition at $x = 0$ can be obtained, where

$$l_m = 4g \tag{2.38}$$

Hence the force/displacement is maximum when the permanent magnet thickness is four times the nominal airgap length. In some cases, increasing the force/displacement is important because of the wide frequency response requirement for the magnetic suspension system.

2.4 Radial magnetic bearing

In this section, the principles, basic analyses, examples and linearization of a typical 8-pole radial magnetic bearing are described and discussed.

Figure 2.14 shows the cross section of a typical radial magnetic bearing. The rotor is cylindrical in shape with a centre shaft surrounded by ferromagnetic material such as laminated silicon steel. The stator surrounds the rotor and has eight poles. The areas between the stator poles are the slots and contain the windings. The stator yoke completes the magnetic paths of the eight stator poles. The width of the stator yoke is designed to be wide enough to avoid magnetic saturation and produce high mechanical stiffness in order to avoid vibration caused by radial magnetic forces.

The eight poles are divided into four electromagnets, i.e., magnets numbered 1 to 4. Windings are shown only for magnets 1 and 3. In magnet 1, two short-pitch coils are wound around two stator poles. These coils are connected in series so that only two terminals are required for one magnet. With a current i_1 in a coil, MMF, flux and attractive radial force F_1 are generated as previously shown in Figure 2.2. Magnet 1 generates radial force in the x-axis direction whereas magnet 3 generates radial force in the opposite ($-x$-axis) direction. Therefore magnets 1 and 3 are working in differential mode, as previously shown in Figure 2.9. Magnets 2 and 4 produce y-axis radial force also in differential mode.

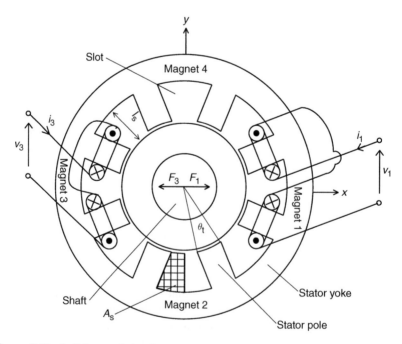

Figure 2.14 Radial magnetic bearing

In the radial magnetic bearing, two perpendicular radial forces, i.e. the x-axis and y-axis forces, are generated. As already mentioned, the four magnets are operating in differential mode with the currents in the four magnets being regulated independently. In total, eight wires are necessary for connection between the radial magnetic bearing and the four current drivers.

Let us define the following parameters:

$D =$ rotor outer diameter (m)
$l\ \ =$ stack length of a rotor core, i.e., rotor axial length (m)
$\theta_t =$ stator pole arc angle (deg)

This means that the area S of one stator pole in airgap is

$$S = l \left(\pi D \times \frac{\theta_t}{360} \right) \tag{2.39}$$

The radial force F_1 generated by two stator poles is derived from (2.25). Since the poles have an angular position of 22.5 deg, the force is:

$$F_1 = \frac{B^2 S}{\mu_0} \cos \left(\frac{\pi}{8} \right) \tag{2.40}$$

The self-inductance of one electromagnet with a nominal airgap length of g is

$$L_0 = \frac{N^2 \mu_0 S}{2g} \tag{2.41}$$

where N is the sum of the number of turns of two short-pitch coils. The radial force can also be obtained from (2.21):

$$F_1 = \frac{L_0}{2g} \cos \left(\frac{\pi}{8} \right) i^2 \tag{2.42}$$

Example 2.8 Derive the maximum radial force produced by magnet 1 for the given parameters. Find the rotor weight and compare the maximum radial force with the rotor weight. Also find the radial force density for the shadow area of the rotor. The shadow area is defined as the product of the rotor diameter D and the rotor axial length l. The parameters are $D = 50\,\text{mm}$, $l = 50\,\text{mm}$, $\theta_t = 25\,\text{deg}$ and the maximum flux density is 1.7 T.

Answer
From (2.39) and (2.40), $S = 5.45 \times 10^{-4}\,\text{m}^2$, so $F = 1158\,\text{N}$. The mass densities of iron and silicon steel are $7.86\,\text{g/cm}^3$ and $7.65\,\text{g/cm}^3$ respectively. Here, the density is approximated to $7.9\,\text{g/cm}^3$ for simplicity. Thus, the rotor weight W, including the shaft, is $\pi \times (5/2)^2 \times 5 \times 7.9 = 0.775\,\text{kg}$ so that the maximum radial force is 152 times the rotor weight. The force density is $1158/(9.8 \times 5 \times 5) = 4.72\,\text{kgf/cm}^2$.

Example 2.9 In Figure 2.14 the short-pitch coil on each stator pole has 50 series-connected turns. One turn is composed of two parallel wires, each with a diameter of 0.8 mm. The rated current density in a conductor is 8 A/mm². The airgap length between the stator pole and rotor surface is 0.4 mm and the slot depth l_s is 17 mm. The slot fill factor is defined as the ratio of conductor area to slot area. Find the slot fill factor, the rated current, airgap flux density and radial force of magnet 1 at the rated current. Also find the value of the ratios: radial force/Dl and radial force/rotor weight.

Answer
The half slot area A_s shown in Figure 2.14 is derived by approximating the area as a trapezoid:

$$A_s = \frac{1}{2}\left[\frac{1}{2}\left(\pi(D+2g)\frac{45}{360} - t_w\right) + \frac{1}{2}\left(\pi(D+2g+2l_s)\frac{45}{360} - t_w\right)\right]l_s$$

$$= 132\,\text{mm}^2 \tag{2.43}$$

where t_w is a tooth width which is given by

$$t_w = \pi(D+2g)\frac{\theta_t}{360}$$

The cross-sectional area of the 0.8 mm diameter wire is $\pi \times (0.8/2)^2 = 0.5\,\text{mm}^2$. Therefore the conductor area in a half slot is 100 times the cross-sectional area, i.e., $50\,\text{mm}^2$. The slot fill factor S_{fl} is $50/132 = 0.38$. The slot fill factor of this short-pitched winding is usually in the range 0.3–0.42 using normal fabrication processes.

The self-inductance is obtained from (2.41). $N = 100$ turns since there are two coils connected in series, and the inductance is 8.56 mH. The rated current I_{rt} is the product of rated current density, the number of parallel conductors and the conductor cross-sectional area so that $I_r = 8\,\text{A/mm}^2 \times 0.5\,\text{mm}^2 \times 2 = 8\,\text{A}$. The flux density at the rated current is $B_{rt} = L_0 I_{rt}/(NS) = 1.26\,\text{T}$ and the radial force generated by magnet 1 is given by (2.40) which is calculated to be 632 N.

The radial force density for the rotor shadow area is

$$\frac{F}{Dl} = \frac{\pi B^2}{\mu_0}\frac{\theta_t}{360}\cos(22.5°)\,\text{N/m}^2$$

$$= 1.6B^2\,\text{kgf/cm}^2$$

At $B = 1.26\,\text{T}$, $F/(Dl) = 2.6\,\text{kgf/cm}^2$. The radial force/rotor weight ratio is given by

$$\frac{F}{W} = \frac{4B^2}{\mu_0 D \times 7.9 \times 10^3 \times 9.8} \times \frac{\theta_t}{360}\cos(22.5°)$$

$$= 261\frac{B^2}{D}\,\text{kgf/cm}^2$$

At $D = 5\,\text{cm}$ and $B = 1.26\,\text{T}$, $F/W = 83$. Note that F/W increases for a small diameter rotor. A small diameter is suited for high acceleration applications.

The derived radial force in this example is about half the value of that in Example 2.8. In Example 2.8, only the magnetic design was considered. However, in Example 2.9, both the magnetic and winding designs are included. For practical radial force limitation and operation, the differential-mode magnetic force should be considered, as shown in the next example.

2.4.1 Linearization

In equations (2.40) and (2.42), radial force is expressed as a function of flux density and current respectively. In order to regulate the radial force the flux density or current should be regulated. Detecting current rather than flux density is advantageous for the following reasons:

1. Current detectors are low cost. These sensors can be installed in current drivers.
2. Flux detection is not easy and can be expensive. One flux detection device is the Hall sensor. The Hall sensor must be extremely thin to be installed in the airgap. Thin Hall sensors are very expensive and mechanically weak. Wiring from the Hall sensor to the controller can also be a problem. Another way to detect the flux is to use search coils. However, dc flux components cannot be detected with search coils.

In most cases, instantaneous current is controlled for radial force regulation. One can see that the relationship between the current and radial force is nonlinear. Without the effect of saturation, the radial force is proportional to the square of the current. In practice, radial force is not proportional to i^2, it is rather proportional to $i^{1.6}$. Let us express the radial forces as

$$F_1 = \frac{k_i'}{4} i_1^2 \tag{2.44}$$

$$F_3 = \frac{k_i'}{4} i_3^2 \tag{2.45}$$

where

$$k_i' = \frac{2L_0 \cos(\pi/8)}{g}$$

To realize the linear relationships between the radial force and current components, the winding currents in electromagnets 1 and 3 are divided into two current components, i.e., bias current I_b and force regulating current i_b:

$$i_1 = I_b + i_b \tag{2.46}$$

$$i_3 = I_b - i_b \tag{2.47}$$

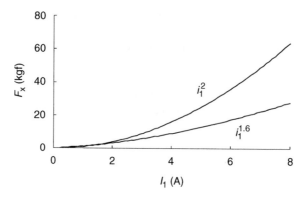

Figure 2.15 Radial force and current relationships

Note that the currents i_1 and i_3 are positive values. Hence i_b should be less than I_b. The radial force acting on the shaft in x-axis direction is

$$F_x = F_1 - F_3 \tag{2.48}$$

Substituting (2.44) and (2.45) yields a simple radial force equation

$$F_x = k_i' I_b i_b \tag{2.49}$$

It is now seen that the radial force is proportional to the force regulating current i_b when the bias current I_b is kept constant.

Figure 2.15 shows the non-linear characteristics between radial force and winding current for two cases, one is proportional to i^2 and another is proportional to $i^{1.6}$.

Figure 2.16 shows the relationships between the radial force regulating current component i_b and radial force for the above two cases. This confirms that the radial force and i_b have a linear relationship as indicated in (2.49). In addition,

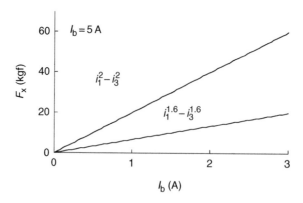

Figure 2.16 Linear relationship with bias current

an almost linear relationship is also obtained for the case where radial force is proportional to $i^{1.6}$. This numerical result shows the effectiveness of this current component control strategy. One can see that radial force is expressed as

$$F_x = k_i i_b \tag{2.50}$$

where $k_i = k_i' I_b$ and k_i is referred to as a force-current factor [1].

Figure 2.17 shows a block diagram for the current regulation. In the controller, a radial force command F_x^* and a regulating current component i_b^*, which is proportional to the force command, are generated. Then the current command is added or subtracted to a constant bias current command I_b^* based on (2.46) and (2.47). Winding current commands i_1^* and i_3^* are fed to the current regulators which regulate winding currents to follow these commands. Therefore the sum of generated radial force in electromagnets 1 and 3 will follow the radial force reference F_x^*.

Example 2.10 Find the force-current factor in the 8-pole radial magnetic bearing for the bias current $I_b = 4\,\text{A}$. Find the radial forces generated by electromagnets 1 and 3 for $i_b = 4\,\text{A}$ at $x = 0$ and also find the total radial force in the x direction.

Answer
The force-current factor is given by

$$k_i = \frac{2L_0 \cos(\pi/8)}{g} I_b = 158\,\text{N/A}$$

The total radial force is given as a product of k_i and I_b, which yields 632 N. The radial forces produced by magnets 1 and 3 are

$$F_1 = \frac{k_i(I_b + i_b)^2}{4I_b} = 632\,\text{N}$$

$$F_3 = \frac{k_i(I_b - i_b)^2}{4I_b} = 0\,\text{N}$$

The calculated results indicate the fact that magnet 1 is generating an x-direction radial force of 632 N, while magnet 3 is generating a radial force of 0 N.

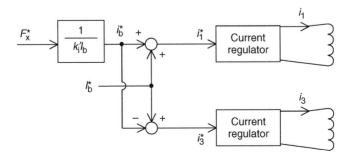

Figure 2.17 Current regulating strategy

2.5 Unbalance pull force

The principles, analysis and examples of unbalance pull force (sometimes called unbalanced magnetic pull) are put forward and discussed in this section. The unbalance pull force is important because it is the cause of inherently unstable characteristics for magnetic suspension systems. The principles are illustrated using a simple example, and then the analysis, including derivation of the unbalance pull force, is introduced. The result is applied to the 8-pole radial force magnetic bearing from the previous section.

2.5.1 Principles

Figure 2.18(a,b) shows simple C- and I-cores with windings. Let us suppose that the winding currents are constant. Note that only the airgap length is different in these figures. In Figure 2.18(a) the nominal airgap length is g, however, in Figure 2.18(b) the airgap length is smaller, i.e., $g - x$. Let us define B_a and B_b as airgap flux densities in Figure 2.18(a,b) respectively. The flux density in Figure 2.18(b) is high so that $B_a < B_b$. Therefore the attractive magnetic force between C- and I-cores is high in Figure 2.18(b) since $F_a < F_b$. Hence the magnetic force is increased when the airgap length is decreased. This characteristic causes the unstable mechanism described below.

If the I-core can be moved in the x direction then it is attracted to the C-core due to the magnetic force. Hence the airgap length decreases. However, this causes the magnetic force to increase. Further movement will only be limited by the two cores becoming clamped together by the force. Therefore the above mechanism is an inherently unstable negative spring system. Hence the magnetic force generated as a function of x is an unstable force. In a successful magnetic suspension system the unstable force should be cancelled by a negative feedback force.

In the case of the differential actuator shown in Figures 2.9 and 2.14, the unstable magnetic force is balanced at the centre position. If a suspended object

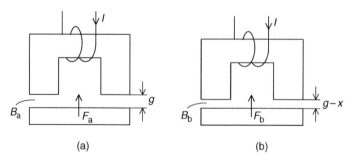

(a)   (b)

Figure 2.18   Magnetic attractive force and airgap length: (a) wide airgap length; (b) decreased airgap length

is moved from the centre position, an unstable magnetic force is generated. The unstable force is a function of eccentric position of a suspended object, so the magnetic force is called an unbalance pull force.

2.5.2 Analysis in C- and I-cores

The principles of unstable force suggest that it is not only a function of current but also a function of displacement x. In the previous sections radial force was expressed as a function of current only and it was assumed that $x = 0$. However, it is important to include the influence of x.

In section 2.2 an approximation was applied in a mathematical expansion. However, $1/(g - x)$ can be expanded to include further terms

$$\frac{1}{g-x} = \frac{1}{g} \frac{1}{\left(1 - \dfrac{x}{g}\right)} = \frac{1}{g}\left[1 + \frac{x}{g} + \left(\frac{x}{g}\right)^2 + \left(\frac{x}{g}\right)^3 + \cdots\right] \qquad (2.51)$$

In sections 2.2 to 2.4, only the first and the second terms in the brackets were considered. The unstable force originates from the third or higher terms. Let us consider up to the third bracketed term. The self-inductance L is

$$L = L_0\left[1 + \frac{x}{g} + \left(\frac{x}{g}\right)^2\right] \qquad (2.52)$$

Therefore the stored magnetic energy is written as

$$W_{\mathrm{m}} = \frac{1}{2}i^2 L_0\left[1 + \frac{x}{g} + \left(\frac{x}{g}\right)^2\right] \qquad (2.53)$$

Hence the radial force is obtained from the partial derivative of the stored magnetic energy, assuming magnetically linear material, so that

$$F = \frac{L_0}{2g}i^2 + \frac{L_0}{g^2}i^2 x \qquad (2.54)$$

The second term is the unstable force which is a function of displacement x.

2.5.3 Analysis in radial magnetic bearing

Let us define x as a displacement of the rotor from the stator centre position. The unstable force generated by electromagnet 1 in Figure 2.14 is

$$F = \frac{L_0}{g^2}i^2 x \cos\left(\frac{\pi}{8}\right) \qquad (2.55)$$

For electromagnet 3, the same amount of radial force is also generated; however, the definition of displacement and direction of radial force is opposite, so the total unstable force in x-axis is twice that of (2.55) since the force increases for one electromagnet and decreases for the other:

$$F = \frac{2L_0}{g^2}i^2 x \cos\left(\frac{\pi}{8}\right) \tag{2.56}$$

If the winding current is almost equal to the bias current I_b then

$$F = k_x x \tag{2.57}$$

where

$$k_x = \frac{2L_0 \cos(\pi/8)}{g^2}I_b^2 \tag{2.58}$$

The coefficient k_x, as previously mentioned, is called as a force-displacement factor [1]. In radial magnetic bearings, the force-displacement factor is positive. Hence this force makes the magnetic bearing inherently unstable. Therefore there is a requirement to provide enough negative position feedback to cancel the effect of this coefficient; this is shown in a later section.

Example 2.11   For the 8-pole radial magnetic bearing, find the force-displacement factor for $I_b = 4\,\text{A}$. Also find the radial force at a displacement of $10\,\mu\text{m}$.

Answer
From (2.58) k_x is $1.58 \times 10^6\,\text{N/m}$, i.e., $1.58\,\text{N}/\mu\text{m}$. At a displacement of $10\,\mu\text{m}$, the unstable force is $15.8\,\text{N}$. This force is about twice the rotor weight.

2.6 Block diagram and mechanical system

In this section, a magnetic bearing is represented in a block diagram as a linear actuator. In addition, the mechanical system, including the mass of a rotor, is represented.

2.6.1 Radial force

Using the analyses in sections 2.4 and 2.5 the radial force can be derived for the radial magnetic bearing as a function of both the current i_b and rotor radial displacement x. The radial force F_x is the sum of these forces

$$F_x = k_i i_{bx} + k_x x \tag{2.59}$$

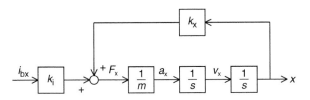

Figure 2.19 Block diagram of magnetic suspension system

where i_{bx} is the force regulating current in the x-axis. The force-current and force-displacement factors are given by

$$k_i = 2L_0 \left(\frac{I_b}{g} \right) \cos \left(\frac{\pi}{8} \right) \tag{2.60}$$

$$k_x = 2L_0 \left(\frac{I_b}{g} \right)^2 \cos \left(\frac{\pi}{8} \right) \tag{2.61}$$

Figure 2.19 shows a block diagram of (2.59) and the mechanical system. In the block diagram, F_x is a sum of $k_i i_{bx}$ and $k_x x$. The radial force is divided by mass m, so that the output of the block is the acceleration a_x, which is the input of an integration block. s is the Laplace operator, hence a block $1/s$ indicates the integration of the input. The integration of acceleration is the radial rotor speed v_x. The integration of the speed is the radial displacement x. The mass m is the mass of the suspended object. It can be seen that k_x provides a positive feedback loop, which makes the transfer function unstable. A similar block diagram can be drawn for the y-axis variables.

2.6.2 Block diagram

Figure 2.20 shows a simplified block diagram. Now, it is seen that the transfer function is unstable because the denominator $(ms^2 - k_x)$ is missing a first-order term and includes a negative term $-k_x$. The characteristic equation $ms^2 - k_x = 0$ gives

$$s = \pm \sqrt{\frac{k_x}{m}}$$

There is a pole in the right half plane so that the transfer function from radial force current component to the radial rotor displacement is unstable. In order

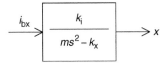

Figure 2.20 Simplification of block diagram

to have stable rotor suspension, negative feedback controllers are necessary. In the next chapter, designs of controllers are discussed assuming that the magnetic bearing behaviour is characterized by equation (2.59).

Reference

[1] G. Schweitzer, H. Bleuler and A. Traxler, "Active Magnetic Bearings", Hochschulverlag AG an der ETH Zurich, ISBN 3 7281 21320.

3

Magnetic bearing controllers

Akira Chiba

In the previous chapter, electromagnetic designs and mathematical expressions for magnetic bearings and related actuators were described. The magnetic bearings were modelled as two simple force constants, which were defined with respect to displacement and current. The transfer function is inherently unstable so that the controller design is important.

In this chapter, the basic controller design strategy of a magnetic suspension system is described and discussed in detail. Based on a simple one-axis magnetic suspension system, the design of a proportional-derivative (PD) controller is explained in order to realize simple magnetic suspension. Designs with good frequency and time domain responses are illustrated and disturbance force suppression is also discussed. The one-axis model is extended to a two-axis model. The displacement variations caused by mechanical unbalance and also by eccentric misalignment of a sensor target and a magnetic bearing rotor are included. Finally the application of synchronizing disturbance components is introduced.

3.1 Design principles in one-axis magnetic suspension

Figure 3.1 shows the block diagram of a magnetic bearing and its controller for one-axis magnetic suspension. The position x of a suspended object is detected and amplified by a displacement sensor with gain k_{sn} and then compared with the position reference x^*. The error is amplified by a controller G_c. A current i_b is fed to the one-axis magnetic bearing. In the magnetic bearing block, m is the object mass, k_i is the force-current factor and k_x is the force-displacement factor.

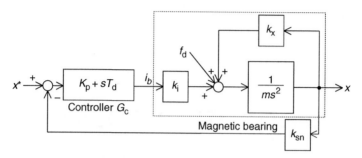

Figure 3.1 Magnetic suspension system

The transfer function of the magnetic bearing is not stable so that the controller must stabilize it using a feedback loop.

A simple controller suitable for a magnetic bearing is a PD controller. The ideal transfer function of a PD controller is

$$G_c = K_p + sT_d \qquad (3.1)$$

where K_p is the constant gain of the proportional controller and T_d is the time constant of the derivative controller. The loop transfer function can be written as

$$G_L = k_{sn}(K_p + sT_d) \times \frac{k_i}{ms^2 - k_x} \qquad (3.2)$$

Hence the transfer function from position reference to the displacement is derived to be

$$\frac{x}{x^*} = \frac{(K_p + sT_d)k_i k_{sn}}{(K_p k_i k_{sn} - k_x) + T_d k_i k_{sn} s + ms^2} \qquad (3.3)$$

Equating the denominator to zero provides a characteristic equation. Solving the equation for s gives

$$s = \frac{1}{2m}\left[-T_d k_i k_{sn} \pm \sqrt{(T_d k_i k_{sn})^2 - 4m(K_p k_i k_{sn} - k_x)}\right] \qquad (3.4)$$

This equation provides some conditions for the magnetic suspension stability:

(a) If T_d is zero and $K_p = 0$ then the value inside the brackets is

$$[\ ] = \pm\sqrt{4mk_x} \qquad (3.5)$$

Hence one pole is located in the right half plane. This system is unstable.
(b) If T_d is zero and $K_p k_i k_{sn} - k_x > 0$ then

$$[\ ] = \pm j\sqrt{4m(K_p k_i k_{sn} - k_x)} \qquad (3.6)$$

Hence two poles are located on the imaginary axis. Theoretically speaking, this condition is stable; however, in practice, it is not useful since it is marginally stable.

(c) If T_d is positive and $K_p k_i k_{sn} - k_x > 0$ then

$$[\ ] = -T_d k_i k_{sn} \pm \sqrt{(T_d k_i k_{sn})^2 - 4m(K_p k_i k_{sn} - k_x)} \qquad (3.7)$$

Since two poles are located in the left half plane, the feedback system is stable.

This shows that a derivative controller is necessary to make the magnetic suspension feedback loop stable. In addition, there is a minimum proportional gain for the controller, which is given by $K_p > k_x/(k_i k_{sn})$. This minimum gain condition is obtained as follows: the radial force is generated by $k_x x$, i.e., a product of a force/displacement constant and displacement. This force makes the magnetic suspension system unstable. Therefore $k_x x$ should be cancelled by the negative feedback force in order to make the system stable. The radial force feedback is given by $-k_i k_{sn}(K_p + sT_d)x$. In steady-state conditions, sT_d can be neglected so that the radial force feedback is $-K_p k_i k_{sn} x$. Hence the sum of the radial force is $k_x x - K_p k_i k_{sn} x$. This is negative when $K_p > k_x/(k_i k_{sn})$ and this condition is necessary for a stable feedback loop.

3.1.1 Equivalent spring–mass–damper system

Figure 3.2 shows a one-axis mechanical system. A mass m [kg] is suspended by a spring and damper. In a steady-state condition, the length of the spring is $X_s + x$. The spring generates a force $k_s x$, which is proportional to the spring displacement x so that

$$mg_a = k_s x \qquad (3.8)$$

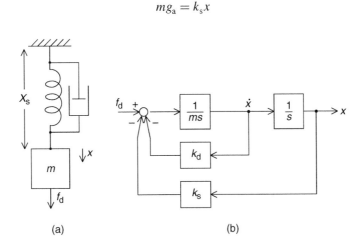

(a)   (b)

Figure 3.2   Spring–mass–damper system: (a) structure; (b) block diagram

The damper generates a damping force which is proportional to the speed of displacement of x, i.e., k_d (dx/dt). An external force f_d is also applied to the mass so the motion is governed by the dynamic equation as:

$$m\frac{d^2x}{dt^2} = f_d - k_d\frac{dx}{dt} - k_s x \qquad (3.9)$$

Using this equation, a block diagram can be drawn as shown in Figure 3.2(b). It can be seen that the spring and damper force blocks k_d and k_s are negative feedback forces. The transfer function is

$$\frac{x}{f_d} = \frac{1}{ms^2 + k_d s + k_s} \qquad (3.10)$$

This transfer function is the typical second-order lag system; the standard form is

$$\frac{x}{f_d} = \frac{\dfrac{1}{k_s}}{1 + 2\zeta\dfrac{s}{\omega_n} + \left(\dfrac{s}{\omega_n}\right)^2} \qquad (3.11)$$

where

$$\zeta = \frac{k_d}{2}\sqrt{\frac{1}{mk_s}} \qquad (3.12)$$

$$\omega_n = \sqrt{\frac{k_s}{m}} \qquad (3.13)$$

From these equations the following can be concluded:

(a) The speed of response to the external disturbance force is determined by the natural angular frequency ω_n. Faster response is obtained at higher ω_n. It can be seen that fast response is realized with a light mass and stiff spring constant.

(b) The damping constant ζ is proportional to the damping constant k_d. A strong damper with high k_d results in high ζ. For a step input, the response overshoot is not seen if $\zeta > 1$. An estimate gives overshoots of 25 and 50% for $\zeta = 0.4$ and $\zeta = 0.2$ respectively so that an increase in damper constant results in oscillation suppression.

(c) The static displacement is proportional to the spring constant k_s. The displacement is small if a stiff spring is employed.

Next, let us look at the similarity between the spring–mass–damper system and magnetic suspension system. Figure 3.3 shows a modified block diagram of the magnetic bearing system previously shown in Figure 3.1. In the figure, a disturbance force f_d is applied to the suspended object. Two other forces caused by the object displacement and speed are added. It is assumed that the position

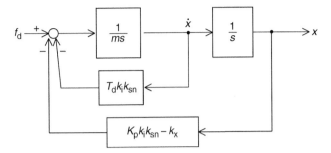

Figure 3.3 Magnetic suspension system equivalent to spring–mass–damper system

reference x^* is zero. The similarities between Figures 3.3 and 3.2 can be seen. The constants of the damper and spring are replaced by magnetic suspension system constants such that

$$k_d = T_d k_i k_{sn} \tag{3.14}$$

$$k_s = K_p k_i k_{sn} - k_x \tag{3.15}$$

This shows that the time constant T_d in the derivative controller is equivalent to the damper constant. In addition, the proportional controller gain K_p is equivalent to the spring stiffness. The damper and spring constants are adjustable in the magnetic suspension system and the force/displacement constant k_x results in a negative influence on the spring stiffness. The condition of $K_p > k_x/(k_i k_{sn})$ produces positive spring stiffness. In practice, k_x usually has variations and fluctuations caused by bias current and nonlinearity. Therefore $K_p k_i k_{sn}$ should be carefully selected to be significantly higher than k_x.

From (3.11) to (3.13), the magnetic suspension system is expressed as

$$\frac{x}{f_d} = \frac{1}{K_p k_i k_{sn} - k_x} \times \frac{1}{1 + 2\zeta \left(\dfrac{s}{\omega_n}\right) + \left(\dfrac{s}{\omega_n}\right)^2} \tag{3.16}$$

where

$$\zeta = \frac{T_d k_i k_{sn}}{2} \sqrt{\frac{1}{(K_p k_i k_{sn} - k_x)m}} \tag{3.17}$$

$$\omega_n = \sqrt{\frac{K_p k_i k_{sn} - k_x}{m}} \tag{3.18}$$

Hence one can see the following characteristics in the magnetic suspension:

(a) An increase in derivative controller gain results in better damping in response.
(b) An increase in proportional controller gain results in fast response and high stiffness.

Example 3.1   Now let us see how we can obtain a stable system with PD controllers. In a practical system, a displacement sensor is used to detect the instantaneous displacement of the suspended object. A displacement of 1 mm gives a 5 V signal, hence the sensor gain k_{sn} is 5000 V/m. Suppose $k_i = 158\,\text{N/A}$ and $k_x = 1.58\,\text{N/}\mu\text{m}$ in the radial magnetic bearing example in the previous chapter. Also suppose that $m = 3.14\,\text{kg}$ and includes the shaft mass. These values are used throughout this chapter unless otherwise specified. Obtain the following:

(a) Draw the bode plot of G_p, i.e., a transfer function of sensor output by input current.
(b) Draw the bode plot for a PD controller with $G_c = 10(0.001s + 1)$.
(c) Find the loop gain. Also draw the bode plot. Specify the phase margin.
(d) Draw a step response for x.

Answer
(a) The G_p transfer function is $G_p = k_{sn}k_i/(ms^2 - k_x)$. Substituting the values yields

$$G_p = \frac{0.25 \times 10^6}{s^2 - 0.5 \times 10^6} \tag{3.19}$$

The bode plots are drawn in Figure 3.4(a). At low frequency the gain is $(k_ik_{sn}/k_x) = -6\,\text{dB}$ with a phase delay of 180 deg. At an angular frequency of $\omega_c = \sqrt{k_x/m} = 707\,\text{rad/s}$, the gain decreases at a rate of $-40\,\text{dB/dec}$.
(b) Figure 3.4(b) shows the bode plots for the transfer function G_c of the PD controller. At low frequency, the gain is 20 dB with a phase lead of 0 deg. In the line approximation, the gain increases at a rate of 20 dB/dec. The numerically calculated characteristics are 3 dB higher than the approximated line at the crossover angular frequency of 1 krad/s. A line approximation is also drawn in the phase characteristics. The line links the three points. One is 45 deg at 1 krad/s. The other two points are 0 and 90 deg at 100 rad/s and 10 krad/s. The numerical calculation shows a good correspondence of the simple line approximation. The important characteristic is that a phase lead is provided in the PD controller. This phase lead gives a phase margin to make the system stable as shown in the next bode plots.
(c) The loop transfer function G_L is the product of G_c and G_p (plant and controller transfer functions). The product becomes a sum of the gains when the gain characteristic is on a logarithmic scale. At low frequency the gain is $20 - 6\,\text{dB} = 14\,\text{dB}$. At 707 rad/s, the plant gain decreases at 40 dB/dec, although the controller gain increases at 20 dB/dec from 1 krad/s so that the loop gain decreases at 40 dB/dec and then at 20 dB/dec. As a result of numerical calculation, the loop gain is 0 dB at a gain crossover angular frequency ω_{cg} of 2300 rad/s.

At this angular frequency, the phase angle is slightly advanced from $-180\,\text{deg}$ due to the phase-lead angle of the PD controller.

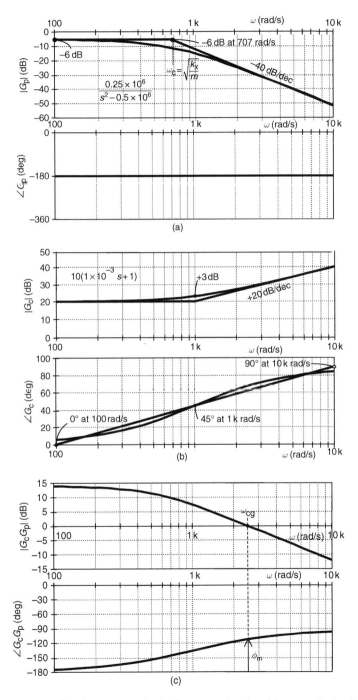

Figure 3.4 Controller design principle in frequency domain: (a) magnetic bearing G_p; (b) controller G_c; (c) loop transfer function $G_p G_c$

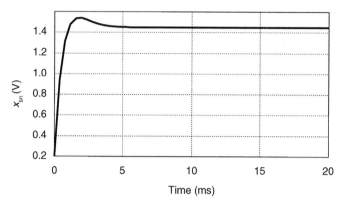

Figure 3.5 Step response of position x_{sn}

The advanced angle generates the phase margin ϕ_m and this margin must be positive for stable suspension. If a loop transfer function can be approximated as a second-order lag system, the relationship between the damping constant and phase margin is well known. The damping constant ζ is approximately 0.2, 0.4 and 1 at ϕ_m angles of 25, 45 and 75 deg respectively so that larger ϕ_m results in better damping of the response.

(d) Figure 3.5 shows the calculated response of the sensor output x_{sn} with a unit step input in x^*. A 7% overshoot results at 1.8 ms. It is quite important to see the correspondence between the loop transfer function and this step response. The response time can be roughly estimated from the loop transfer characteristics. The gain-crossover angular frequency ω_{cg} is about 2.3 krad/s in Figure 3.4(c) so that $\pi/(2.3 \times 10^3) = 1.4$ ms. In this case, the estimated time of 1.4 ms is close to 1.8 ms. It is usual to find that π/ω_{cg} provides an estimated time when overshoot occurs. Strictly speaking, the denominator should be corrected for a damping constant and the second-order lag approximation should be applicable.

As for the overshoot, a rough estimation is also possible from loop frequency response. The phase margin is about 70 deg in Figure 3.4(c). This value is slightly smaller than 75 deg, which corresponds to $\zeta = 1$ when no overshoot exits. A phase margin of 45 deg corresponds to $\zeta = 0.4$ with an overshoot of 25%. About 10% overshoot is reasonable.

3.2 Adjustment of PID gains

In this section, gain optimization in the PD controller is discussed and practical derivative and integral controllers are introduced. The disturbance force suppression characteristics are examined with adjustment of the proportional-integral-derivative (PID) controller gains.

In a magnetic suspension system, the displacement reference is usually set to zero at the nominal position. For example, the position reference is always the centre position in a magnetic bearing system. The displacement response with respect to the reference is not usually discussed; rather it is the displacement response to a disturbance force that is addressed, although improvements in the disturbance force suppression also result in a better response to the displacement reference. Therefore the displacement response with respect to the disturbance force is examined in this section.

From Figure 3.3, the transfer function from the disturbance force to the displacement, i.e., the dynamic stiffness, is

$$\frac{x}{f_d} = \frac{1}{ms^2 + T_d k_i k_{sn} s + K_p k_i k_{sn} - k_x} \tag{3.20}$$

Note that the denominator is the same as the transfer function of x/x^*, as shown in (3.3). Therefore these transfer functions have the same pole loci. This fact indicates that a better response can be obtained in x/x^* if the response of x/f_d is improved, illustrating the statement made above. In steady-state conditions the static stiffness is obtained by substituting $s = 0$ so that

$$\frac{x}{f_d} = \frac{1}{K_p k_i k_{sn} - k_x} \tag{3.21}$$

This static stiffness provides information about the static displacement from the nominal position. Let us examine the static stiffness. When a mass of m is levitated by the electromagnet, the displacement is X_s. When an additional mass of m' is added the displacement increases to $X_s + x$. An increase of x results from the additional external force of $m' g_a$. Based on (3.21), $X_s = m g_a / (K_p k_i k_{sn} - k_x)$ and $X_s + x = (m + m') g_a / (K_p k_i k_{sn} - k_x)$. If $(K_p k_i k_{sn} - k_x)$ is high, a small displacement and hence high stiffness are obtained.

Example 3.2 Find the static stiffness in the system given in Example 3.1. Also find the displacement if the mass is doubled.

Answer
The static stiffness is $x/f_d = 0.158\,\mu m/N$. For a mass of 3.14 kg, the mass force $f_d = 3.14 \times 2 \times 9.8\,N$ so that $x = 10\,\mu m$.

Example 3.3 Suppose that a step disturbance force of 1 N is applied to the suspended object given by the parameters in the previous example. Construct a block diagram for digital simulation. Find the responses at half and twice of the proportional gain. Also find the responses at half and twice the derivative gain. Show the bode plots for these cases. Sweep the disturbance frequency and observe the response.

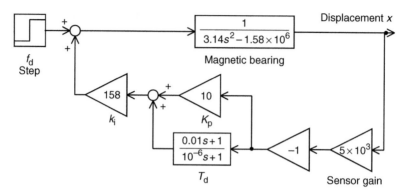

Figure 3.6 Digital simulation block diagram

Answer

Figure 3.6 shows a block diagram for digital simulation. A step disturbance is applied to the magnetic bearing. The displacement is amplified by a negative sensor gain for negative feedback. K_p and T_d are the proportional and derivative gains respectively. In the digital simulation, a transfer function of $(0.01s + 1)$ is multiplied by the first-order lag element of $(10^{-6}s + 1)^{-1}$ with a high cut-off frequency of 10^6 rad/s. This lag element is necessary even in digital simulation. The lower cut-off frequency is usual for practical implementation. The outputs of these controllers are added and it is assumed that the amplitude of the current driver is unity. Therefore the output of k_i is a negative feedback force.

Figure 3.7(a) shows the displacement responses for the given proportional gains. It is seen that the steady-state displacement is smaller for a high proportional gain. The final values are small for high K_p because the static stiffness is high. Figure 3.7(b) shows the displacement responses for the given derivative gains. Improved damping is seen for high T_d but the final values are not dependent on T_d.

Figure 3.7(c,d) shows the bode plots for the dynamic stiffness x/f_d. The gain is about −140 dB at low frequency. This gain corresponds to a static stiffness of 0.158 μm/N since $20 \log_{10}(0.158 \times 10^{-6}) = -136$ dB. A discrepancy of stiffness for proportional gains is seen only at low frequency. The magnitude of x/f_d is low for high K_p, resulting in better disturbance force suppression. At high frequency, the stiffness does not depend on the PD controller gains. This is represented by a line which decreases with frequency. In this region, the mechanical inertia prevents vibration.

The phase angle of x/f_d is also important in testing the magnetic suspension system. In frequency response measurement, an electrical disturbance signal is injected into the magnetic suspension system. The frequency of the injected signal is varied from low frequency to high frequency to observe a wide frequency transfer function. In such a test, the phase angle of the displacement response is in phase at low frequency; however, the phase angle is gradually

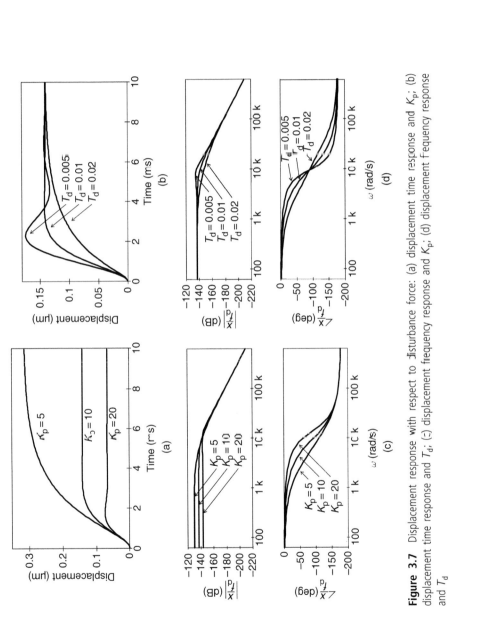

Figure 3.7 Displacement response with respect to disturbance force: (a) displacement time response and K_p; (b) displacement time response and T_d; (c) displacement frequency response and K_p; (d) displacement frequency response and T_d

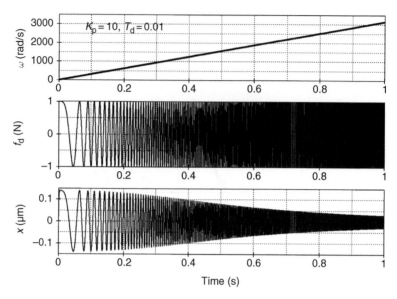

Figure 3.8 Disturbance frequency sweep

delayed with increasing frequency. This phase delay is shown in the figure as the phase angle.

Figure 3.8 shows the disturbance frequency, disturbance force and displacement. The frequency of the disturbance force is increased linearly while the amplitude is kept a constant value of 1 N. As the frequency increases, the displacement decreases as expected from the bode plot. The phase angle is difficult to see from Figure 3.8; however, the displacement phase angle is gradually delayed with respect to f_d as the frequency increases.

3.2.1 PD gain adjustment

Better dynamic response of the disturbance force suppression can be obtained in a system with improved damping with fast response. Adjusting the derivative and proportional gains individually does not lead to the required response in a straightforward manner. It is important to increase both the derivative and proportional gains whilst maintaining a certain relationship. This is based on equations for the proportional and derivative gains that are given for a required natural frequency and damping factor. In the previous section, the natural angular frequency and the damping factor are expressed in (3.17) and (3.18) as

$$\zeta = \frac{T_d k_i k_{sn}}{2} \sqrt{\frac{1}{(K_p k_i k_{sn} - k_x)m}} \tag{3.22}$$

$$\omega_n = \sqrt{\frac{K_p k_i k_{sn} - k_x}{m}} \tag{3.23}$$

From these equations, the proportional and derivative gains can be expressed as

$$K_{\mathrm{p}} = \frac{m\omega_{\mathrm{n}}^2 + k_{\mathrm{x}}}{k_{\mathrm{i}}k_{\mathrm{sn}}} \tag{3.24}$$

$$T_{\mathrm{d}} = \frac{2m\omega_{\mathrm{n}}}{k_{\mathrm{i}}k_{\mathrm{sn}}}\zeta \tag{3.25}$$

The following observation can be made from these equations:

(a) For a fast response, the natural angular frequency ω_{n} should be high. To increase ω_{n}, the proportional gain should be increased as the square of ω_{n}.
(b) The derivative gain should be increased in proportion to ω_{n} and ζ

Example 3.4 Find the proportional and derivative gains as a function of the natural angular frequency for a damping factor of 0.9. For the cases of $T_{\mathrm{d}} = 0.01$ and 0.02, find the corresponding proportional gain and natural angular frequency. Show the response of the displacement for a step disturbance force of 1 N.

Answer
From (3.24) and (3.25), K_{p} and T_{d} are calculated as shown in Figure 3.9(a,b). From this figure, it can be seen that the natural frequencies $\omega_{\mathrm{n}} = 1500$ and 2900 correspond to $T_{\mathrm{d}} = 0.01$ and 0.02 respectively. The corresponding proportional gains are 10 and 35. For a step disturbance, the displacement responses are shown in Figure 3.10. High controller gains result in a fast response with small displacement.

3.2.2 Practical derivative controller

The transfer function of a derivative controller is proportional to s in the Laplace domain. In the frequency domain, this is replaced by $j\omega$, so that the magnitude is proportional to the frequency. In practice, the output of an ideal derivative

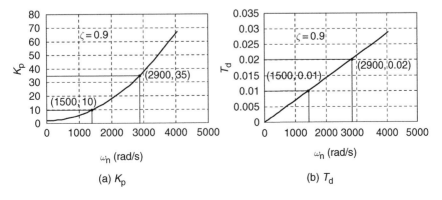

(a) K_{p} (b) T_{d}

Figure 3.9 Proportional and derivative gains

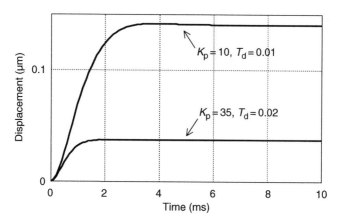

Figure 3.10 Displacement response for optimized PD gains

element unfortunately includes considerable noise. High-frequency noise at the input terminals results in significant amplification at the output terminals so that the ideal derivative should be avoided in practice.

Figure 3.11 shows ideal and practical derivative controllers. In the practical controller, the transfer function is known as a phase-lead compensator. This transfer function acts as a derivative function in the angular frequency range of $1/T_d \ll \omega$. The denominator determines the high frequency limit, which is $\omega = 1/T_f$. Thus, the practical derivative block executes a derivative function in the range $1/T_d \ll \omega \ll 1/T_f$. The low frequency gain is 0 dB and the high frequency gain is limited to T_d/T_f. Therefore the engineer can determine T_d/T_f from the signal conditions. Note that, in practice, additional high-cut filters are usually employed above $1/T_f$ to aid rapid cut-off. The proportional term in the numerator can also be eliminated when a proportional controller is connected in parallel.

Figure 3.12 shows the bode plots drawn with line approximation. The magnitude is 0 dB at low frequency; however, it increases at $\omega = 1/T_d$ by 20 dB/dec up to $\omega = 1/T_f$ where the magnitude levels out to a constant of T_d/T_f. For the phase angle, phase advancement starts at $\omega = 1/(10T_d)$, with 45 deg lead at $\omega = 1/T_d$. If $1/(10T_f)$ is low compared with $10/T_d$, as found in practice, the phase lead is flat in the frequency range of $1/(10T_f)$ to $10/T_d$, then it begins to decrease. The phase-lead angle is 45 deg at $\omega = 1/T_f$, then becomes zero at $\omega = 10/T_f$.

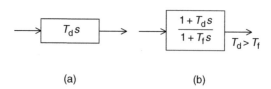

(a) (b)

Figure 3.11 Derivative controllers: (a) ideal derivative; (b) practical derivative

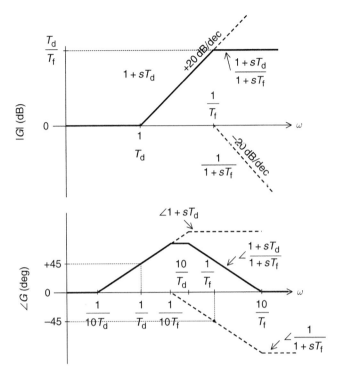

Figure 3.12   Bode diagrams of a practical derivative transfer function

---

Example 3.5
(a) Draw the bode plots for a practical derivative controller $K_p + (1 + sT_d)/$
    $(1 + sT_f)$ for the cases of $T_d = 0.01$ and $T_f = 10^{-6}$, 10^{-4} and 10^{-3}.
(b) Apply the above controls to the previous example of the one-axis magnetic
    suspension. Draw the bode plots and show the phase margins.
(c) Show the displacement response for a step disturbance force. Discuss the
    T_f selection.

Answer
(a) Figure 3.13(a) shows the magnitude and phase characteristics for the
    given T_f. The case of $T_f = 10^{-6}$ is close to the ideal case for the drawn
    frequency range. The magnitude increases at a rate of 20 dB/dec at an
    angular frequency of 1 krad/s. At this angular frequency, the derivative
    block gain is equal to the proportional gain so that the phase-lead angle is
    about 45 deg. The phase-lead angle is close to 90 deg at high frequency.
    For the case of $T_f = 10^{-4}$, the magnitude is 20 dB/dec for 1 dec from 1 to
    10 krad/s. The high frequency limit of 10 krad/s corresponds to $1/T_f$. The
    phase-lead angle decreases when the frequency is more than 1 krad/s. For
    the case of $T_f = 10^{-3}$, the phase-lead angle is only 20 deg.

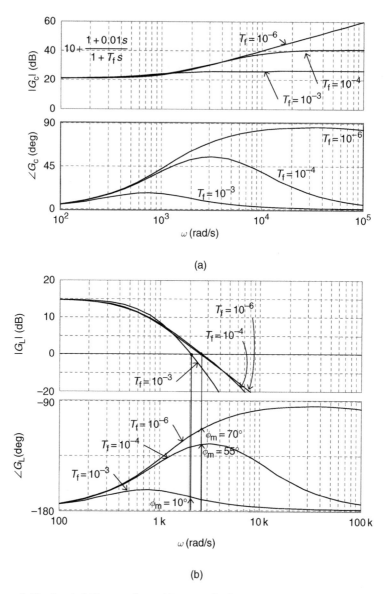

Figure 3.13   Practical PD controller and loop transfer function variations with T_f: (a) practical derivative controller; (b) loop transfer function variation with T_f

(b) Figure 3.13(b) shows the bode plot of the loop transfer function. It is seen that, for the cases of $T_f = 10^{-4}$ and 10^{-6}, the characteristics are similar, although there is a slight decrease in phase margin. However, for the case of $T_f = 10^{-3}$, the phase margin is only 10 deg, which is a significant decrease when compared to the cases of 55 and 70 deg for $T_f = 10^{-4}$

and 10^{-6}. A phase margin of 10 deg indicates insufficient damping. If we draw the $x/f_{\rm d}$ bode plot, $T_{\rm f} = 10^{-3}$ has a peak, which indicates vibration at a peak frequency of 2000 rad/s (318 Hz).

(c) The figure is not shown for the displacement time responses for a step disturbance force applied to the suspended object. However, for the case of $T_{\rm f} = 10^{-3}$, poorly damped vibration would be seen. The vibration term is about 3 ms, which corresponds to 318 Hz. For the other two cases, well-damped responses would be seen. One can conclude that $T_{\rm f} = 10^{-4}$ is the best choice from these three cases, because the noise is less than for the $T_{\rm f} = 10^{-6}$ case, while a similar response is obtained.

3.2.3 Steady-state position error and integrator

With only a PD controller, the airgap length between the suspended object and the electromagnetic actuator is increased when external static force is applied due to the steady-state error. In many applications, an integral controller is often employed in addition to a PD controller to prevent this.

Figure 3.14(a) shows the block diagram for a PID controller. In addition to a PD controller, an integrator $K_{\rm in}/s$ is added. The input, i.e., a position error e_x, is integrated and the output is added to that of the PD controller so that the force command F_x^* is generated.

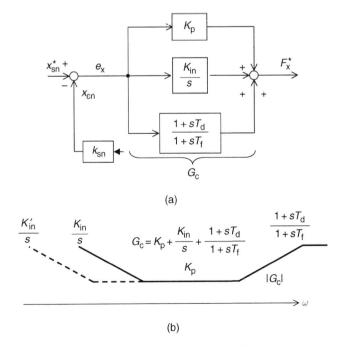

(a)

(b)

Figure 3.14 PID controller: (a) block diagram; (b) frequency characteristics

The function of the integrator can be explained by Figure 3.15. This figure shows output waveforms with a step input in position reference. At 5 ms, a transient response of the PD controller has almost finished. x_{sn} has overshoot with respect to reference x_{sn}^* so that the position error is negative. However the output of the integrator is negative and levels out with time. But an x-axis force is generated in the negative direction. The force tries to move the object in the opposite direction to the x-axis so that the position x_{sn} gradually decreases. The integrator output decreases until the position is on the reference. At a time of 30 ms x_{sn}^* is equal to x_{sn} and the integrator output is constant (which is called the integrator constant). Hence F_x^* is not zero, but is a constant value, which is necessary for force equilibrium for zero position error.

Figure 3.14(b) shows a simplified bode plot for the PID controller. In the mid-frequency range the proportional gain is dominant; however, at low and high frequency ranges, the integral and derivative controllers play important roles. The dotted line indicates a case with low integral controller gain. It can be seen that the magnitude of controller gain G_c is increased at low frequency ranges, thanks to the integral controller. This fact indicates that the loop gain is increased at low frequency so that the negative feedback is improved.

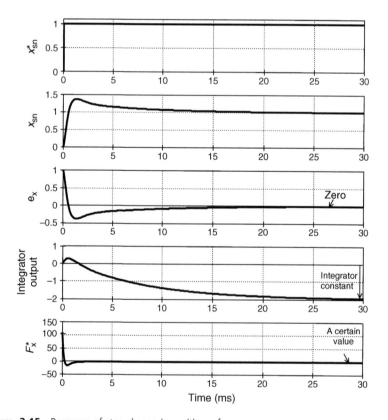

Figure 3.15 Response of step change in position reference

Let us examine the displacement error in the steady-state condition. The displacement error/position command and the displacement/disturbance force can be written as a function of the controller gain G_c, where

$$\frac{e_x}{x_{sn}^*} = \frac{ms^2 - k_x}{ms^2 + G_c k_i k_{sn} - k_x} \tag{3.26}$$

$$\frac{x}{f_d} = \frac{1}{ms^2 + G_c k_i k_{sn} - k_x} \tag{3.27}$$

Let us substitute for G_c using the following equation:

$$G_c = K_p + \frac{K_{in}}{s} + sT_d \tag{3.28}$$

Further, calculation of (3.26) and (3.27) can be executed. The results are not shown here; however, the results have the operator s in the numerator so that there is a zero at the origin. For a step response the final values are

$$x_\infty = \lim_{s \to 0} s \left(\frac{e_x}{x_{sn}^*} \frac{1}{s} \right) \tag{3.29}$$

$$x_\infty = \lim_{s \to 0} s \left(\frac{x}{f_d} \frac{1}{s} \right) \tag{3.30}$$

These values are zero. Hence the steady-state position error is eliminated.

Example 3.6 Draw the bode plot for a PID controller with the cases of $K_{in} = 0$, 100, 1000 and 10 000 and the parameters $K_p = 10$, $T_d = 0.01$ and $T_f = 10^{-4}$. Also draw the bode plot of the loop transfer function and dynamic stiffness with $k_{sn} = 5 \times 10^3$ and $k_i = 158\,\text{N/A}$. Show the step response of the disturbance force suppression.

Answer
Figure 3.16(a) shows the bode plots of the PID controller. The low frequency characteristics depend on the integrator gain K_{in}. When K_{in} is zero, i.e., equivalent to the PD controller, the magnitude is a constant value, i.e., about 20 dB, which is mostly determined by the proportional controller gain of 10. However, the magnitude increases as the frequency decreases with other K_{in} values. When $K_{in} = 100$ the magnitude increases at an angular frequency of less than 10 rad/s. The flat magnitude frequency range is 10 to 1 krad/s. In the case of $K_{in} = 1000$, it increases when the frequency is less than 100 rad/s. It can be said that the latter case is more effective for low frequency disturbance suppression. In the case of $K_{in} = 10000$, one cannot observe the flat frequency response. This condition results in a decrease in phase-lead angle at a frequency of around 1 krad/s.

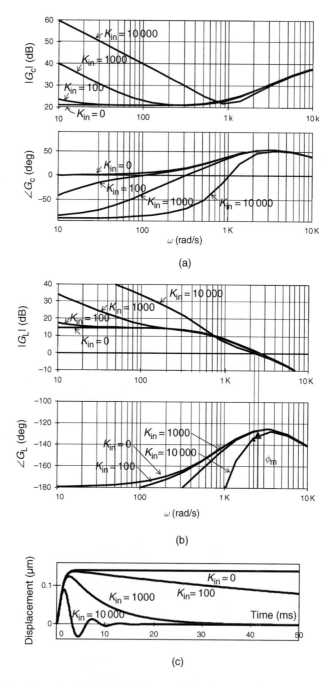

Figure 3.16 Effect of an integrator: (a) frequency characteristics of a PID controller; (b) loop transfer function with several integrator gains; (c) displacement response

Figure 3.16(b) shows the bode plots of the loop transfer function. One can see that high loop gain is obtained with high K_{in} at low frequency. In addition, a decrease in phase margin is seen for the case of $K_{in} = 10000$. Therefore among these four cases, $K_{in} = 1000$ is the best choice due to the following reasons: (a) high loop gain in the low frequency range and (b) no decrease in phase margin.

Figure 3.16(c) shows the displacement response with a step disturbance force. For the case of $K_{in} = 10000$, poor damping can be seen. When $K_{in} = 100$, the displacement response is slow. When $K_{in} = 0$, the displacement does not converge to zero. Therefore the best response is obtained at $K_{in} = 1000$.

The bode plots for x/f_d are not shown here; however, one would see that x/f_d is lower, i.e., high stiffness is obtained, for high K_{in}. For the case of $K_{in} = 10000$, a peak would be observed at 1 krad/s. It can be seen that a peak in stiffness should be avoided to enable damped responses.

3.2.4 Influence of delay

In an actual magnetic suspension system there are many causes of possible delay. The delay may be due to the following reasons:

(a) Iron losses in the actuator iron core.
(b) Flux delay with respect to current, caused by eddy current.
(c) Voltage saturation in a current driver.
(d) Limited frequency response of a current driver.
(e) Limited sensor frequency response.
(f) The sampling period of a digital controller.
(g) An analogue or digital filter at an A/D converter input.
(h) The first-order lag element of a winding resistance and an inductance.

Depending on the application, some of these points may cause serious problems. Therefore it is important to see the influence of delay in a magnetic suspension system. In this section, an example is given to show how the delay could ruin the stability in a magnetic suspension system.

Example 3.7 Let us suppose that a first-order lag element, representing the delay, exists in a magnetic suspension system. Let us assume that the cut-off frequency of the lag element is 500 Hz so that a transfer function of $[1 + s/(2\pi 500)]^{-1}$ is connected in series with the magnetic bearing. Compare the frequency responses of the loop transfer function and dynamic stiffness with and without the lag element. Consider the loop gain adjustment and show the displacement variation when a 100 N radial force is applied with a frequency range between 0 and 500 Hz. Also compare the step disturbance force suppression.

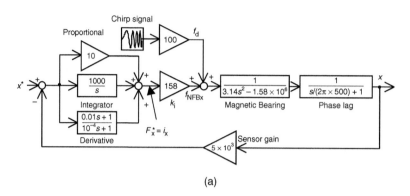

(a)

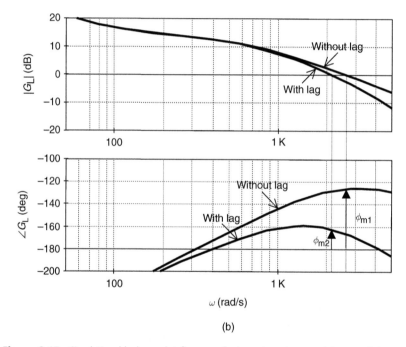

ω (rad/s)

(b)

Figure 3.17 Simulation blocks and influence of phase lag element: (a) one-axis magnetic suspension system with a delay element; (b) loop transfer function and phase margin

Answer

Figure 3.17(a) shows a block diagram for digital simulation. The first-order lag element is connected just after the magnetic bearing block. The PID controller gains are adjusted using the parameters quoted.

Figure 3.17(b) shows the bode plots of the loop transfer functions with and without the phase-lag elements. It can be seen that there is significant discrepancy in the phase angle characteristics. The phase margin decreases from 50 to 20 deg. In addition, the phase angle is less than -180 deg at high frequency. This fact suggests that this is one of the important practical aspects in loop gain adjustment. If the loop gain is increased, for example, by 10 dB, then the gain crossover frequency is 4.2 krad/s, where the phase angle is less than −180 deg. Therefore the magnetic suspension loop would not be stable. On the other hand, if loop gain is decreased by 12 dB, then the gain crossover frequency is 400 rad/s, where the phase angle is again less than −180 deg. Hence, in this example, the magnetic suspension is stable only when the loop gain variation is within the range from −12 to +10 dB if there is a lag. In practice, it is usually found that the gain adjustment is difficult to set until a stable point is obtained. If the integrator is eliminated, the low frequency restriction disappears and low loop gain is acceptable. Therefore adjusting the controller gains becomes easier.

If the dynamic stiffness x/f_d characteristic is drawn then there is a peak in the magnitude at an angular frequency of 2200 rad/s with a phase lag (which corresponds to 350 Hz).

Waveforms with a sinusoidal sweep in disturbance force input can also be drawn. If the disturbance force has an amplitude of 100 N with a frequency range from 1 to 500 Hz linearly, the displacement x is about 15 μm at low frequency. However, a peak exists at 350 Hz, after which the displacement decreases significantly. These characteristics correspond to the magnitude and frequency characteristics of the dynamic stiffness. The current also has a sinusoidal waveform with peak amplitude at 350 Hz.

A displacement response characteristic can also be drawn when a step disturbance force of 100 N is applied to the suspended object (though not drawn here). It would be seen that the damping is inferior in response when the phase-lag element is included. In addition, a significant fluctuation in feedback force f_{NFBx} would be observed.

In this example, it is shown that the first-order lag element causes a serious problem. In this case the cut-off frequency is 500 Hz; however, it is obvious that the lower cut-off frequency results in instability. For practical implementation, attention should be paid to the elimination of slow response due to the influences listed in (a–h) earlier.

3.3 Interference in two perpendicular axes

The generated radial forces are aligned on two perpendicular axes which usually coincide with the x- and y-axis displacements. However, there are some causes of radial force misalignment, i.e., when they may not be aligned with the x- and y-axes. Suppose that an electromagnet of a radial magnetic bearing is constructed with a misalignment at an angle with respect to the radial displacement sensors. The direction of the generated radial force has an angular position error, which results in interference of x- and y-axis force components. There are several other possible causes of interference, as listed below:

1. Flux due to eddy currents can generate a delay in the radial force. At high rotational speed and with a solid rotor, eddy currents flow on a rotor surface, which generates phase-delayed components in the flux wave distribution with respect to the rotor. This phase lag in the flux results in a direction error for the generated radial force.
2. The gyroscopic effect is apparent with short-axial-length machines with large-radius rotors. The gyroscopic effect generates interference radial force.
3. In bearingless motors, the radial force direction includes errors if the controller has errors in the estimated or detected angular positioning of the revolving magnetic field. Also direction errors can be caused by space harmonics of MMF and permeance distribution.

The interference between the x- and y-axis radial forces can cause a serious problem. In this section radial interference and its influence in the feedback system are examined.

Figure 3.18 shows two perpendicular axes x and y as well as the feedback radial forces f_{NFBx} and f_{NFBy} with slightly delayed angular position. The rotor is rotating in a counter-clockwise direction. The direction angle error is defined as θ_{er} and the interference radial forces can be written as:

$$f_{\text{dmy}} = K_{\text{mx}} f_{\text{NFBx}} \tag{3.31}$$

$$f_{\text{dmx}} = K_{\text{my}} f_{\text{NFBy}} \tag{3.32}$$

where

$$K_{\text{mx}} = -\sin\theta_{\text{er}}, \quad K_{\text{my}} = \sin\theta_{\text{er}}$$

Figure 3.19 shows a block diagram that includes the interference. The outputs are the radial positions and the inputs are the references. The upper and lower

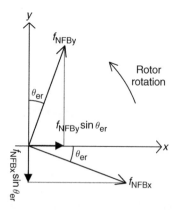

Figure 3.18 Angular position error

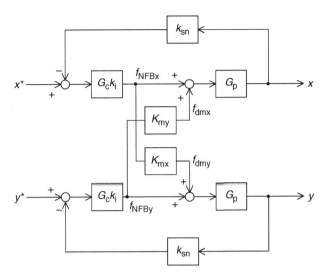

Figure 3.19 Interference in two-axis magnetic suspension

blocks are for x- and y-axis dynamic models. The interference is generated by K_{mx} and K_{my}. The decrease in radial force due to $\cos(\theta_{er})$ is neglected because θ_{er} is small.

Let us apply a step reference in x^* and see the response of displacement x. Figure 3.20(a) shows the step responses. The parameter K_m is set as $K_{mx} = -K_m$ and $K_{my} = K_m$. A serious damping problem is seen when K_m is increased. If K_m is increased to more than 0.3, which corresponds to a 17 deg direction error, the negative feedback has difficulty in damping the system.

Figure 3.20(b) shows the loop transfer function of the x-axis system. At an angular frequency of 2200 rad/s, a large phase delay is seen which increases with K_m. In an acceleration test, with $K_m = 0.3$, high vibrations are observed, again, at a speed of around 2200 rad/s. Also the dynamic stiffness is low around this angular frequency.

To see the results of digital simulation, it is also important to analyse the influence of the interference. In Figure 3.19, the radial position reference x^* is compared with radial displacement x, then amplified by PID controller gain G_c and k_i. A disturbance force f_{dmx} is added and the sum of radial force is applied to the magnetic bearing system G_p, so that the radial displacement x is given. The same structure exists for the y-axis.

Suppose that the y-axis position command y^* is zero; in practice, this assumption is valid in most cases so that the block diagram of the interference from f_{NFBx} to f_{dmx} is simplified as shown in Figure 3.21. It can be seen that the interference is represented by two series blocks that have inputs f_{NFBx} and f_{dmx}. The block transfer function is

$$\frac{f_{dmx}}{f_{NFBx}} = -K_{mx}K_{my}\frac{G_p G_c k_{sn} k_i}{1 + G_p G_c k_{sn} k_i} \tag{3.33}$$

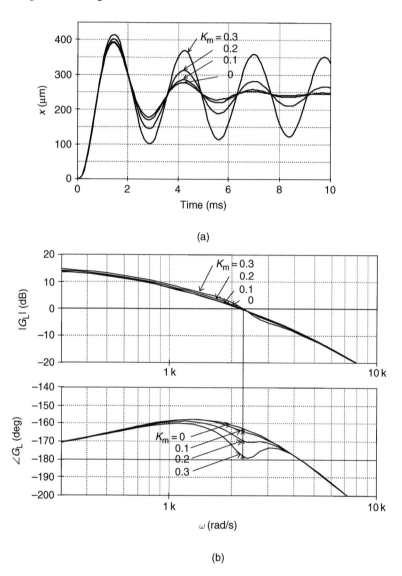

Figure 3.20 Step response and frequency characteristics: (a) displacement response with interference gain variations; (b) loop transfer function and phase margin

It is important to examine the transfer function of $f_{\mathrm{dmx}}/f_{\mathrm{NFBx}}$. Let us define $G_i = f_{\mathrm{dmx}}/f_{\mathrm{NFBx}}$. G_i is the transfer function of interference. If $K_m = 0$ and $K_{\mathrm{mx}} = K_{\mathrm{my}} = 0$ then $G_i = 0$, hence there is no interference. The fraction part is quite interesting: $G_p G_c k_{\mathrm{sn}} k_i$ is the loop transfer function of the radial magnetic bearing without interference so that $G_p G_c k_{\mathrm{sn}} k_i$ is high at low frequency. However, it decreases at a frequency determined by the mass and

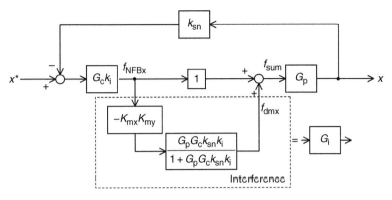

Figure 3.21 The x-axis block diagram with interference

spring constants as previously shown in section 3.1. The fraction part transfer function is near unity at low frequency.

Figure 3.22(a) shows the transfer function of G_i for $K_m = 0.1$ to 0.3. It can be seen that the low frequency gain is proportional to K_m^2. At 2200 rad/s the magnitude of G_i exhibits a peak, then it decreases in the high frequency range.

The interference force f_{dmx} is added to the normal feedback force f_{NFBx} in parallel. Therefore the transfer function from f_{NFBx} to the sum of radial force f_{sum} is written as

$$\frac{f_{sum}}{f_{NFBx}} = 1 + (-K_{mx}K_{my})\frac{G_p G_c k_{sn} k_i}{1 + G_p G_c k_{sn} k_i} \tag{3.34}$$

Figure 3.22(b) shows a transfer function of f_{sum}/f_{NFBx}. The magnitude is close to 0 dB. The important point is that there is a phase lag at around 2200 rad/s. The phase lag characteristic of f_{sum}/f_{NFBx} is not a common shape. This is explained by (3.34). The 1st proportional term is dominant across most of the frequency range so that the phase angle is 0 deg both in the low frequency and high frequency ranges. However, the second term has influence at a certain frequency, i.e., 2200 rad/s in this case, because it has a peak. At around the peak frequency, the phase angle rotates significantly. Hence the phase angle of f_{sum}/f_{NFBx} is delayed only at 2200 rad/s. This angular frequency is always close to the gain-crossover angular frequency because the fraction part in the second term is the closed-loop gain of the magnetic bearing. Therefore the interference can be a serious problem in a two-axis magnetic suspension system. Solutions are dependent on the causes of the interference. Steps should be taken to try to eliminate the misalignment of radial forces and x- and y-axis coordinates when designing a system.

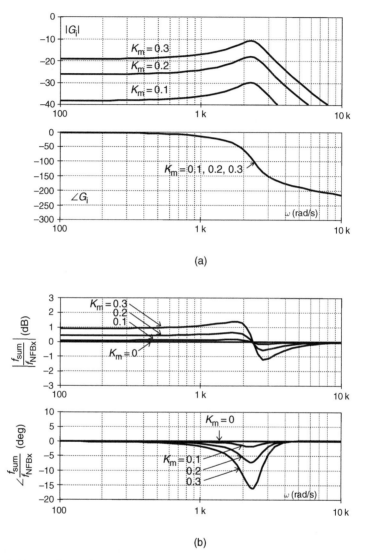

Figure 3.22 Interference frequency characteristics: (a) interference transfer function G_i; (b) bode diagrams of interference

3.4 Unbalance force and elimination

Vibration is caused by mechanical unbalance in a rotating system. In this section, the principles and mathematical representation of unbalanced mechanical force are introduced and the influence on the magnetic bearings is shown. An elimination method for unbalanced disturbance is included.

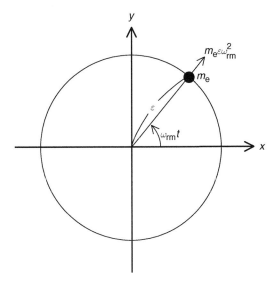

Figure 3.23 Unbalance mass

Figure 3.23 shows the principle of unbalance force generation. The x- and y-axes are perpendicular axes in a stationary frame. The shaft is perfectly symmetrical but there is an additional mass of m_e at a radius of ε in the direction of $\omega_{rm}t$, where ω_{rm} is the rotational angular speed of the shaft. The radial forces in the two perpendicular axes are defined by

$$F_x = m_e \varepsilon \omega_{rm}^2 \cos \omega_{rm} t \tag{3.35}$$

$$F_y = m_e \varepsilon \omega_{rm}^2 \sin \omega_{rm} t \tag{3.36}$$

These radial forces are synchronized to the shaft rotation. The amplitude is proportional to the square of rotational speed. The synchronous vibration is caused by an unbalanced force in the two-axis magnetic bearing.

Example 3.8 Figure 3.24 shows a simplified block diagram. The signal generator generates unbalanced radial forces for the case of $m_e \varepsilon = 10^{-6}$. These forces are added, as disturbance forces, to the negative feedback forces f_{NFBx} and f_{NFBy}. A similar system exists for the y-axis.

Figure 3.25(a) shows the waveforms for a rotational angular speed ω_{rm}, disturbance forces f_{dx} and radial position displacement x. The rotational speed increases as a ramp function. At a time of 1 s, the rotational speed is 6282 rad/s, i.e., 1000 r/s or 60 000 r/min. The following facts are obvious:

(a) At low rotational speed, the radial displacement increases with speed because the unbalanced force increases with speed.

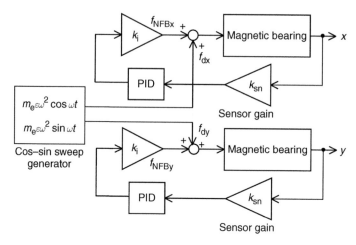

Figure 3.24   Unbalance force

(b) At a time of 0.37 s, the displacement is at the maximum value. This maximum vibration occurs at a rotational speed of about 2300 rad/s. This is the critical speed.
(c) At high rotational speed, the displacement decreases even though the amplitudes of the unbalanced forces are increasing. This decrease is caused by the shaft mass.

Figure 3.25(b) shows enlarged waveforms at around the 0.51 s point. The displacement is delayed by about 210 deg with respect to the disturbance force at a speed of 3200 rad/s. It can be seen that the delayed angle of the displacement is speed dependent and this is for the reasons described below.

If a bode plot is drawn for the transfer function from the disturbance force to the displacement x/f_{dx} then the following correspondence can be seen between the bode plots and time domain responses:

(a) The magnitude of x/f_{dx} is high in the speed range of 100–2000 rad/s, with a peak at about 2000 rad/s.
(b) The phase angle of x/f_{dx} has strong frequency dependence. At low speed, the phase angle leads; however, there is a gradual phase decrease as the speed increases. At 500 and 3250 rad/s, the phase delay angles are 18 and 214 deg respectively. These phase delay angles correspond to the time domain waveforms in the enlarged waveforms.

3.4.1 Synchronous disturbance elimination

As a result of the unbalance force, radial vibrations occur; these are synchronized to the shaft rotation so that the frequency of the vibrations increases with

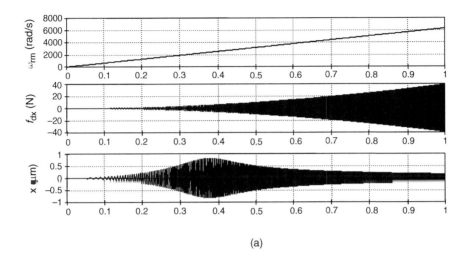

(a)

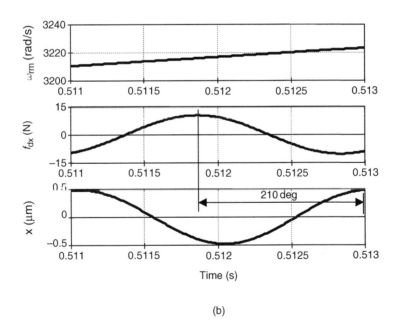

(b)

Figure 3.25 Waveforms with unbalance force: (a) acceleration and displacements; (b) enlarged

speed. Current drivers have to provide high frequency currents to magnetic bearings. The impedance of the magnetic bearing winding is approximated to ωL, where L is the inductance. The back electromotive force (EMF) is proportional to the rotational speed so that the voltage requirement increases. If the current driver cannot deliver enough voltage to provide current to the windings, the current driver becomes saturated so that the negative feedback is opened. Since the magnetic suspension system is unstable without the feedback, the system becomes unstable, which may result in shaft touchdown to the emergency bearings. Therefore it is important to eliminate the synchronous disturbance components from the negative feedback loops. In this section, a method of eliminating the synchronous disturbance is introduced. This method reduces shaft vibrations and the synchronized components in the current drivers.

The principles of synchronized disturbance elimination are explained below.

(a) Detect only the synchronized components from the detected radial positions.
(b) Subtract the detected components from the radial positions. Therefore only the synchronized components are eliminated from the radial positioning.

In power electronic applications, synchronizing a disturbance to eliminate it – i.e., stabilizing the voltage on a commercial power line – is often done by the controller. The same concept can be applied here.

A coordinate transformation from stationary x- and y-axes to rotational X_d- and Y_d-axes is given by

$$\begin{bmatrix} X_d \\ Y_d \end{bmatrix} = \begin{bmatrix} \cos \omega_{rm} t & \sin \omega_{rm} t \\ -\sin \omega_{rm} t & \cos \omega_{rm} t \end{bmatrix} \begin{bmatrix} x \\ y \end{bmatrix} \tag{3.37}$$

where $\omega_{rm} t$ is the detected shaft rotational position. If radial displacements x and y include the synchronized components then X_d and Y_d include dc components. Low pass filters can be used to detect only the dc components of X_d and Y_d. Let us define the dc components as X_d' and Y_d'. Therefore, the coordinate transformation from rotational axes to stationary axes of these dc components is

$$\begin{bmatrix} X_{sync} \\ Y_{sync} \end{bmatrix} = \begin{bmatrix} \cos \omega t & -\sin \omega t \\ \sin \omega t & \cos \omega t \end{bmatrix} \begin{bmatrix} X_d' \\ Y_d' \end{bmatrix} \tag{3.38}$$

where X_{sync} and Y_{sync} are the synchronized components of x and y.

Figure 3.26 shows a block diagram of a synchronizing component eliminator. From x and y, the equations (3.37) and (3.38) are calculated, hence X_{sync} and Y_{sync} are obtained. These are the detected synchronized components of x and y. These are subtracted from x and y so that x_{com} and y_{com} are obtained. These displacements do not include the synchronized components with angular frequency of ω_{rm}.

Figure 3.27 shows the effects of synchronized disturbance elimination. The shaft is rotating at a constant speed of 30 000 r/min. The synchronized disturbance elimination is switched on at a time of 0.4 sec. The disturbance force caused by unbalance is f_{dx}. Before the elimination function is switched on, the radial displacement x and feedback radial force f_{NFBx} are fluctuating. Once it is switched on the feedback radial force f_{NFBx} is significantly reduced. Since the winding current is proportional to the feedback radial force, the winding

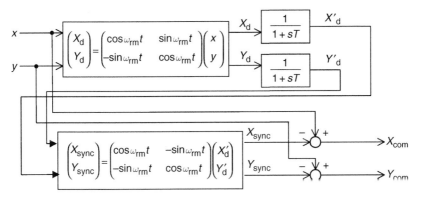

Figure 3.26 Synchronous disturbance elimination

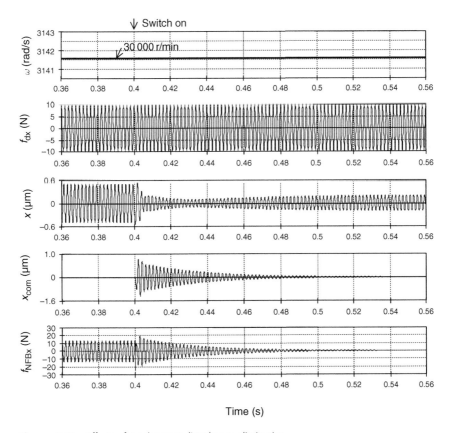

Figure 3.27 Effects of synchronous disturbance elimination

current is also decreased. In addition, the vibration of the radial displacement x is reduced.

Let us look at this in the detail. Before switch-on, the synchronized component X_{sync} is detected so that x_{com} is almost zero. After switch-on, a transient condition is started, lasting about 0.1 s, for which the time constant is mainly determined by the low pass filter. In this simulation, the cut-off frequency is set to 2 Hz. Once X_d approaches the final value, X_{sync} has constant amplitude, while x_{com} is almost zero.

Is the system stable if the synchronized components are eliminated from the negative feedback loops? The synchronized disturbance elimination is effective only when well over the critical speed. The eliminator is not effective, and is rather problematic, at low rotational speed and when the magnetic bearings need damping because the operation happens to be either under or slightly over the critical speed.

There are several other methods to reduce the current and displacement caused by the unbalance force. The disturbance forces are given in (3.35) and (3.36), and if identical electromagnetic forces are generated in the negative direction then the disturbance force is cancelled. The identification of disturbance forces in amplitude and direction can be realized using an observer, as described extensively in modern control theory. If the current caused by the disturbance force is identified then it can be subtracted in order to minimize the disturbance amplitude.

Another simple method is to identify the disturbance forces or currents as a function of rotational speed so that sinusoidal waves can be injected in a feed-forward strategy. In practice, a simple feed-forward controller is very useful because it does not influence the zero and pole assignment in the loop transfer function.

Adjusting the mechanical balance of the shaft is quite important. If there is less mechanical unbalance then there is less shaft vibration. Experience tells us that trying to reduce mechanical unbalance with magnetic force is not a good idea. It is better to balance the rotor accurately during the fabrication process.

3.5 Eccentric displacement

Figure 3.28 shows an overall view of a magnetic bearing arrangement. A magnetic bearing rotor is located on the shaft. In addition to the magnetic bearing rotor, there is a sensor target ring and two displacement sensors to detect the airgap lengths. Hence the centres of the target ring and the magnetic bearing rotor should be aligned to the shaft centre. However, mechanical fabrication will produce inaccuracies and tolerance variations. In this section, the practical problems associated with eccentric displacement are discussed.

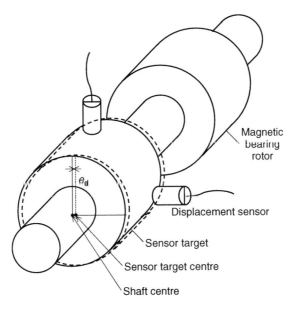

Figure 3.28 Sensor target

Figure 3.28 shows a definition of eccentric displacement. The sensor target centre is displaced from the shaft centre by e_d. The detected sensor outputs include position errors, so that

$$X_e = e_d \cos \omega_{rm} t \tag{3.39}$$

$$Y_e = e_d \sin \omega_{rm} t \tag{3.40}$$

In this situation, the shaft is suspended at the centre position but the target ring is not rotating about the shaft axis so that the sensor has a target eccentric displacement. Figure 3.29 shows the x-axis block diagram including the influence of the displacement errors for x-axis. At the output of the magnetic bearing displacement, X_e is added so that the sensor feedback signal includes the position error. A simplification of this block diagram provides a transfer function of x/X_e:

$$\frac{x}{X_e} = \frac{ms^2 - k_x}{ms^2 + k_i k_{sn} T_{da} s + k_i k_{sn} - k_x} \tag{3.41}$$

From this transfer function the following facts are obvious:

(a) At low rotational speed, the transfer function is $(-k_x)/(k_i k_{sn} - k_x)$. Hence the disturbance can be suppressed.

(b) At high rotational speed, the transfer function is unity. Therefore the sensor output includes the sinusoidal variation of the error (with amplitude e_d).

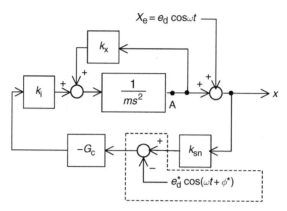

Figure 3.29 Position disturbance

If the magnetic bearing rotor has a misalignment from the shaft centre then eccentric displacement X_e is added to the point A in Figure 3.29. In this case, radial force is generated by the displacement-force factor.

Example 3.9 Draw the transfer function of x/X_e. Obtain the waveforms of the displacement and the negative feedback force for $e_d = 10\,\mu m$.

Answer

Figure 3.30(a) shows an example of the transfer function of x/X_e. The controller has an integral control so that the transfer function is quite low at low frequency. At about 1200 rad/s, the transfer function is more than 0 dB so that the sensor output is more than e_d. At 2400 rad/s it reaches a peak value of 10 dB then it decreases to 0 dB. The phase angle varies with frequency; it is 90 deg at low frequency then gradually delays to 360 deg at high frequency.

Figure 3.30(b) shows the result of acceleration. The shaft speed is increased from 0 to 60 000 r/min as a ramp function. The amplitude of X_e is constant at 10 μm. The displacement x is small at low speed; however, at about 2400 rad/s, the amplitude is about three times X_e, corresponding to 10 dB of the amplitude of x/X_e. The amplitude then gradually decreases to 10 μm, i.e., e_d. The current i_x characteristic is similar to the displacement x so that considerable high frequency current is required.

As shown in the previous section, higher frequency current is also eliminated by the synchronized disturbance elimination. In addition, a simple elimination method is shown by the dotted lines in Figure 3.29. If the eccentric displacement is known then a signal injection of $e_d \times \cos(\omega t + \phi^*)$ to the sensor output is effective. The amplitude e_d^* and the phase angle ϕ^* are adjusted to the actual values to eliminate the influence.

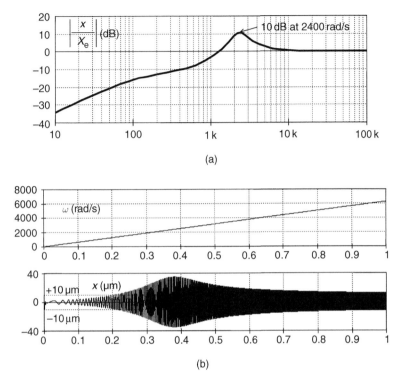

Figure 3.30   Sensor target eccentricity: (a) frequency characteristics; (b) acceleration test

3.6 Synchronized displacement suppression

In section 3.4, a synchronized disturbance elimination method was introduced. The elimination method detects disturbance components synchronized with rotor rotation and then subtracts the detected components from the radial displacement signals so that in the frequency characteristics the loop gain had a dip at the rotor rotational frequency where feedback forces and currents are significantly reduced. However, this does not suppress the radial displacements significantly.

In some applications, the shaft radial position should be forced to zero or to be as small as possible. If the frequency characteristics of the loop transfer function have a peak at the rotor rotational frequency then the rotor displacement can be reduced. Feedback forces and currents are required for displacement suppression. These have to be increased for precise positioning.

Figure 3.31 shows a block diagram for the required signal conditioning. The inputs are the radial displacements x and y. From these signals, synchronized components X_{sync} and Y_{sync} are detected as written in (3.37). These detected synchronized components are amplified by K_{ps} and then added to the radial displacements to obtain x_{com} and y_{com}. Note that synchronized components are not subtracted as in the elimination method, instead, these are added to the

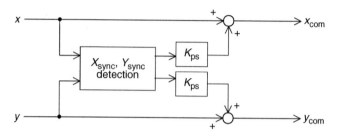

Figure 3.31 Synchronizing displacement suppression

displacements. If K_{ps} is increased then x_{com} and y_{com} include K_{ps}-scaled synchronized components. The gain increase at the rotor rotational frequency enhances the loop gain. The synchronized components are amplified in the PID controllers so that radial force commands are generated to reduce the radial displacements. Hence this method is valid at low speed range where PID controllers and current drivers can provide enough suppression current.

Example 3.10   Show the effectiveness of enhanced stiffness at a rotational speed of 1200 r/min. Consider the influence of both an unbalance of $m_e \varepsilon = 10^{-6}$ and a 1 µm misalignment of the sensor target ring. Then show the acceleration characteristics. Discuss the characteristics of the adding and subtracting methods for the synchronized components.

Answer
Figure 3.32 shows waveforms to illustrate the effectiveness of the synchronized displacement suppression. The rotor is rotating at a constant speed of 1200 r/min. Two causes of disturbance are included, one is an unbalance of $m_e \varepsilon = 10^{-6}$ and the other is a sensor target ring offset of 1 µm. The suppression is switched on at a time of 0.4 s with $K_{ps} = 10$. The waveforms are for an unbalance force f_{dx}, x-axis radial displacement, compensated radial displacement x_{com} and negative feedback force f_{NFBx}. Before switch-on, the radial displacement has fluctuations; however, the negative feedback force is small. After switch-on, x is significantly reduced and there is an increase in the negative feedback force. However, a large current is injected to generate high radial force in order to suppress the synchronized components of the radial displacements.

During acceleration tests the following are observed:

(a) The subtraction method in section 3.4 is effective only at high rotational speed above the critical speed. The advantage of this method is that the negative feedback current is reduced at high rotational speed.
(b) The adding method in section 3.6 is effective only at low rotational speed under the critical speed. The radial displacement is reduced although high current is required.

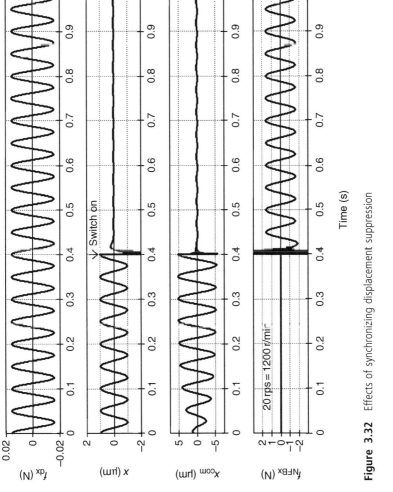

Figure 3.32 Effects of synchronizing displacement suppression

(c) Without the subtraction and the addition, the magnetic suspension is stable over the whole speed range. However, the displacement is high at low speed and high current is required at high speed.

It has been shown that synchronized component suppression and elimination should be adjusted depending on the rotor speed and critical speed to take advantage of these controllers. These controllers do not require information about the amplitude and direction of the unbalance and displacement of the sensor target ring and the magnetic bearing rotor. If these can be identified then compensation can be carried out by feed-forward controllers.

4

Mechanical dynamics

Akira Chiba

An object has six degrees of freedom, i.e., x-, y- and z-axis positions and rotational positions θ_x, θ_y, and θ_z around these three axes. In most magnetic bearing systems, one axis is assigned as the rotating shaft axis so that the rotating shaft should be suspended by active or passive regulation of the remaining five axis positions. Therefore understanding the dynamic constraints of the mechanical system is important.

In this chapter, the dynamic characteristics of a simple mechanical system are described. In the first section, the fundamental equations are described for a two-axis system including the gyroscopic effect. It is important to note that there is cross coupling between axes caused by the gyroscopic effect. Block diagrams provide easy understanding and computer simulations illustrate the system behaviour. In the second section, the two-axis system is extended to four- and five-axis systems. In the last section, a thrust magnetic bearing is described and the requirements for five-axis active suspension are shown.

4.1 Two-axis system

Figure 4.1(a) shows a shaft rotating at an angular speed of ω_{rm} around the k-axis. The three perpendicular x-, y- and z-axes are in the stationary coordinate reference frame. The other three perpendicular i-, j- and k-axes are in the rotational coordinate reference frame. The bottom of the shaft is fixed to the origin of these axes. At a length l_{rt} from the origin, a cylindrical magnetic bearing rotor is fixed to the shaft. Under normal conditions, the k- and z-axes are almost aligned; however, there is always a slight difference. Figure 4.1(b,c) shows the difference. In Figure 4.1(b), a view along the y-axis is shown. The k-axis is inclined by an angular position θ_y from the z-axis. A moment (or torque) N_y is applied around the y-axis. Note that the angular position and moment are defined around the y-axis using the right-hand rule. In Figure 4.1(c), a view along the

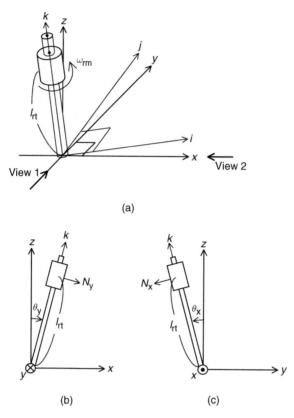

(a)

(b)   (c)

Figure 4.1   Coordinates: (a) Coordinate system; (b) view 1; (c) view 2

x-axis is shown. The angular position and the moment are also defined as θ_x and N_x respectively, again, based on the right-hand rule.

Let us define inertia I_i and I_k around i- and k-axis rotations respectively. The inertia on the j-axis is equal to that on the i-axis because of the symmetrical shaft structure. The product of the inertia and the second-order differential of the angular position with respect to time is equal to the moments so that

$$I_i\ddot{\theta}_x = -\omega_{rm}I_k\dot{\theta}_y + N_x \tag{4.1}$$

$$I_i\ddot{\theta}_y = \omega_{rm}I_k\dot{\theta}_x + N_y \tag{4.2}$$

The first terms on the right-hand side of these equations are moments originated from the gyroscopic effect. The gyroscopic moment is a product of a z-axis rotational speed ω_{rm}, the k-axis inertia and the angular speed of the perpendicular axis. These terms are effective at a high rotational speed and also with high k-axis inertia such as in a disc-shaped rotor.

Displacements from the cylindrical rotor centre alignment on the x- and y-axes can be obtained from Figure 4.1(b,c); assuming that the inclination of the rotating shaft is small, then

$$x = l_{rt} \sin \theta_y \simeq l_{rt} \theta_y \qquad (4.3)$$

$$y = -l_{rt} \sin \theta_x \simeq -l_{rt} \theta_x \qquad (4.4)$$

Note that x and y are proportional to each other's angular positions θ_y and θ_x. Solving these equations for θ_x and θ_y and substituting into (4.1) and (4.2) yields equations (4.5) and (4.6):

$$\ddot{y} = \frac{\omega_{rm} I_k}{I_i} \dot{x} - \frac{l_{rt}}{I_i} N_x \qquad (4.5)$$

$$\ddot{x} = -\frac{\omega_{rm} I_k}{I_i} \dot{y} + \frac{l_{rt}}{I_i} N_y \qquad (4.6)$$

Figure 4.2(a) shows a block diagram for these equations. Acceleration in y-axis is the sum of the gyroscopic term of x-axis speed and the moment term around the x-axis. The gyroscopic term is proportional to the ratio of inertias in the k- and i-axes. Therefore the gyroscopic effect is more substantial for the case of a disc-like shaft rather than a long and thin shaft.

Next, let us examine the external moment. Suppose that the centre of gravity of a magnetic bearing rotor has a height of h and a mass of m (kg). Note that l_{rt} is equal to h in the ideal case; however, in practice, the sensor target and shaft weight will also have to be considered. Let us define the gravity acceleration as g_a. Therefore the moments N_{xg} and N_{yg}, around the x- and y-axes, due to the mass, are written as

$$N_{xg} = mg_a h \sin \theta_x \simeq mg_a h \theta_x = -mg_a h \frac{y}{l_{rt}} \qquad (4.7)$$

$$N_{yg} = mg_a h \sin \theta_y \simeq mg_a h \theta_y = mg_a h \frac{x}{l_{rt}} \qquad (4.8)$$

Let us define the current-driven radial forces, generated in the magnetic bearing on the x- and y-axes, as F_x and F_y. Therefore the moments N_{xi} and N_{yi} are generated by currents and given by $N_{xi} = -F_y l_{rt}$ and $N_{yi} = F_x l_{rt}$. In addition to the current-driven radial forces, displacement-caused radial forces are also produced. The radial force in y-axis is written as $k_x y$ with a force-displacement factor k_x and y-axis shaft displacement y. The displacement-originated moment N_{xd} around the x-axis is written as $N_{xd} = -k_x y l_{rt}$ and the moment N_{yd} around the y-axis is given as $N_{yd} = k_x x l_{rt}$.

The sum of moments around the x- and y-axes are

$$N_x = N_{xg} + N_{xi} + N_{xd} = -\frac{mg_a h}{l_{rt}} y - F_y l_{rt} - k_x y l_{rt} \qquad (4.9)$$

$$N_y = N_{yg} + N_{yi} + N_{yd} = \frac{mg_a h}{l_{rt}} x + F_x l_{rt} + k_x x l_{rt} \qquad (4.10)$$

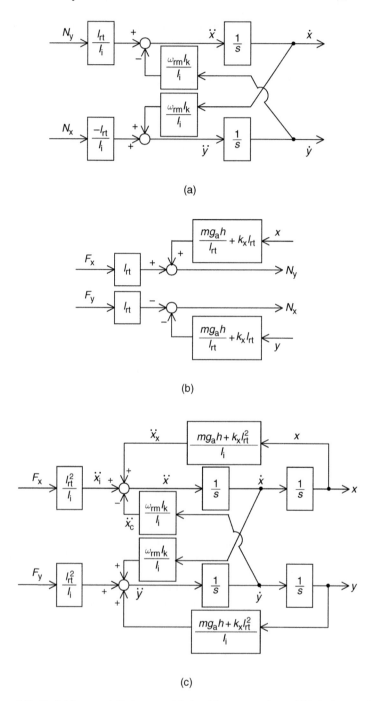

Figure 4.2 Radial force and displacement blocks: (a) torque and speed block diagram (b) the moments N_x and N_y around x- and y-axes (c) radial force and displacement blocks

Figure 4.2(b) shows a block diagram for the moments. The input variables are current-driven radial forces from the magnetic bearing and also forces from the rotor radial displacement. Radial forces caused by gravity and displacement are generated as a function of the rotor radial displacement. The outputs are the moments around the x- and y-axes. This block diagram can be merged with the previous block diagram.

The moment equations above are substituted into (4.5) and (4.6) to obtain dynamic motion equations such that

$$\ddot{y} = \frac{\omega_{\text{rm}} I_k}{I_i} \dot{x} + \frac{mg_a h + k_x l_{\text{rt}}^2}{I_i} y + \frac{l_{\text{rt}}^2}{I_i} F_y \tag{4.11}$$

$$\ddot{x} = -\frac{\omega_{\text{rm}} I_k}{I_i} \dot{y} + \frac{mg_a h + k_x l_{\text{rt}}^2}{I_i} x + \frac{l_{\text{rt}}^2}{I_i} F_x \tag{4.12}$$

Figure 4.2(c) shows the block diagram. The inputs are the suspension radial forces F_x and F_y and the outputs are the x- and y-axis shaft displacements. There are cross-coupling blocks caused by gyroscopic effects and positive feedback

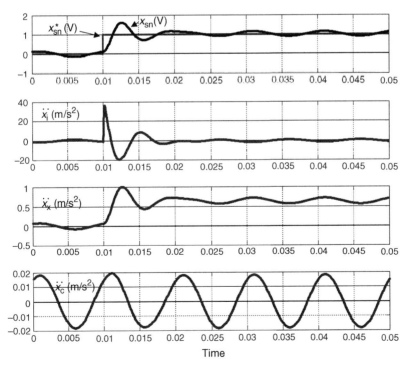

$m = 1.3\,\text{kg},\ l_{rt} = 0.09\,\text{m},\ h = 0.15\,\text{m},\ k_x = 1.7 \times 10^5\,\text{N/m},\ I_i = 1.6 \times 10^{-2}\,\text{kgm}^2,\ I_k = 2.3 \times 10^{-4}\,\text{kgm}^2$

Figure 4.3 Step change in x-axis radial position command

loops are caused by weight and the force-displacement factor. This block diagram provides the mechanical system dynamic response.

Figure 4.3 shows the waveforms from a computer simulation of the block diagram where a step change in displacement reference in x-axis is applied. Feedback controller blocks have been added and a mechanical unbalance force is considered. The rotational part is made up of a cylindrical rotor with a diameter of 5 cm and a thickness of 3 cm and an additional cylindrical iron part is also included to model the sensor target. The shaft length is about 20 cm long and it is assumed to be of a small diameter and long axial length. The figure shows the x-axis position command x_{sn}^*, the radial sensor output x_{sn}, the current-originated acceleration $\ddot{x}_i$, the displacement-originated acceleration $\ddot{x}_x$ and the gyroscopic acceleration $\ddot{x}_c$. As seen from the figure, the current-originated acceleration is dominant, which is generated by the error between x_{sn}^* and x_{sn}. Also, the displacement-originated acceleration is similar in shape to the x_{sn} characteristic, although the peak amplitude is small. The gyroscopic acceleration is almost a sinusoidal function. This is because the radial speed has a variation caused by the mechanical shaft unbalance. In this example, the amplitude is negligible at a speed of 6000 r/min; however, the influence increases with rotational speed.

4.2 Four-axis and five-axis systems

In the previous section, the dynamic behaviour of a magnetically suspended machine with two axes of freedom is explained. Based on the equations and block diagrams, dynamic representations of systems with four-axis and five-axis degrees of freedom are described in this section. A simple rigid rotor is considered with a symmetrical structure with respect to the centre of gravity so that the interference between translational and inclination forces can be neglected. Details on rigid and elastic rotor dynamics can be found in References [1,2].

4.2.1 Equation of motion

Figure 4.4(a) shows a shaft with a five-axis active suspension system. There are two radial magnetic bearings and one thrust magnetic bearing. The first radial magnetic bearing, numbered 1, generates radial forces in x_1- and y_1-axis directions. The second radial magnetic bearing, numbered 2, generates radial forces in x_2- and y_2-axis directions. A thrust magnetic bearing generates a z-axis suspension force.

The radial shaft movements can be expressed by translational and inclination movements. Figure 4.4(b,c) shows these shaft movements. In the translational displacement, the shaft rotational axis is moved in parallel to the stator centre

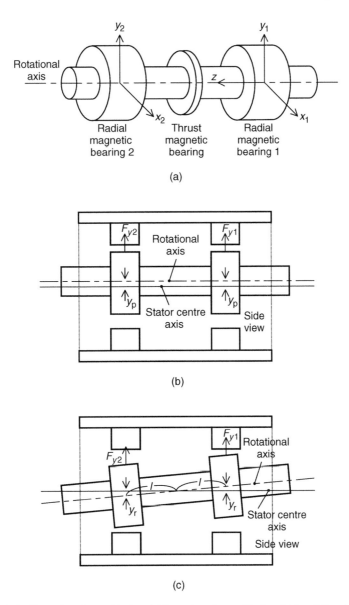

Figure 4.4 Five-axis active positioning: (a) five-axis active positioning; (b) translational displacement; (c) inclination displacement

axis. In the example figure, both the rotors are moved by y_p in the y-axis direction. In the Figure 4.4(c), a rotor shaft is rotated on the y-axis and rotors 1 and 2 are displaced by y_r and $-y_r$, respectively. In the inclination displacement, the shaft rotational axis is rotated with respect to the stator axial centre.

The rotor radial movement is expressed by translational and inclination displacements so that

$$\left.\begin{array}{l} x_1 = x_p + x_r \\ y_1 = y_p + y_r \\ x_2 = x_p - x_r \\ y_2 = y_p - y_r \end{array}\right\} \tag{4.13}$$

The translational and inclination forces can be expressed in terms of the radial forces of the magnetic bearings as

$$\left.\begin{array}{l} F_{xp} = F_{x1} + F_{x2} \\ F_{yp} = F_{y1} + F_{y2} \\ F_{xr} = F_{x1} - F_{x2} \\ F_{yr} = F_{y1} - F_{y2} \end{array}\right\} \tag{4.14}$$

The dynamic motion equations for the translational movement have to consider the fact that the radial force is generated by both magnetic bearings. Hence

$$m\ddot{x}_p = F_{xp} + 2k_x x_p \tag{4.15}$$

$$m\ddot{y}_p = F_{yp} + 2k_x y_p - mg_a \tag{4.16}$$

Note that the gravity force of the shaft weight m is applied in the negative direction of the y-axis. The y_p block surrounded by the dotted lines in Figure 4.5 shows the above relationship on the y-axis. The input and output variables are the translational radial force and the displacement.

For inclination movement, the dynamic equations in the previous section illustrate that a shaft inclined at an angle given by y_r/l_{rt} has an inclination angular acceleration given by $\ddot{y}_r/l_{rt}$. The external moment is a sum of current- and displacement-originated terms, i.e., $F_{yr}l_{rt}$ and $(2k_x y_r)l_{rt}$. Therefore the dynamic motion equations are

$$\ddot{y}_r = \frac{\omega_{rm} I_k}{I_i}\dot{x}_r + \frac{2k_x l_{rt}^2}{I_i}y_r + \frac{l_{rt}^2}{I_i}F_{yr} \tag{4.17}$$

$$\ddot{x}_r = -\frac{\omega_{rm} I_k}{I_i}\dot{y}_r + \frac{2k_x l_{rt}^2}{I_i}x_r + \frac{l_{rt}^2}{I_i}F_{xr} \tag{4.18}$$

where I_i and I_k are the inertias around the x- and z-axes respectively. The y_r block, surrounded by the dotted lines in Figure 4.5, models the y-axis inclination motion equation. The input variables are the force and x-axis speed, while the output variable is the y-axis movement. This block diagram can be understood with reference to the block diagram in Figure 4.2 from the previous section.

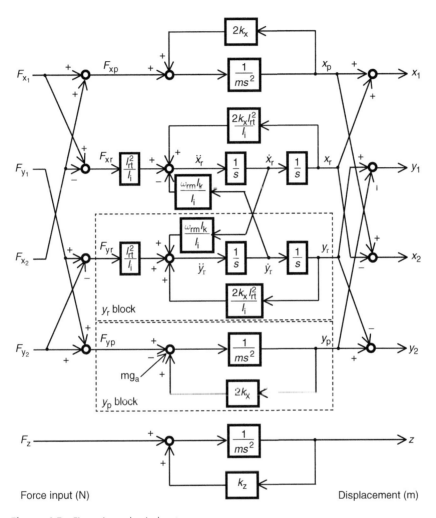

Figure 4.5 Five-axis mechanical system

Figure 4.5 shows a block diagram of both the x- and y-axis variables for translational and inclination movement. The radial forces applied to the magnetic bearing rotors are input variables. These variables are transformed into translational and inclination forces using (4.14). Therefore, from these forces, the displacements are obtained using (4.15–4.18). The displacements are transformed into rotor movement using (4.13). In addition to the 4-axis motion equation, a thrust bearing block diagram is added to the bottom of the block diagram. Note that, for simplicity, the thrust movement is taken to be independent of the other axes. The block diagram has force inputs and displacement outputs and it describes the constraints governing shaft motion.

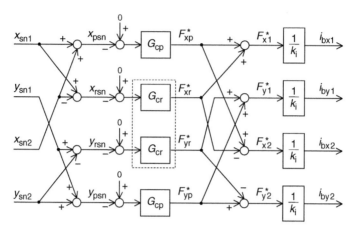

Figure 4.6 Controller configuration for translational and rotational axes

4.2.2 Controller structures

A simple controller structure has independent controls for each axis. Shaft displacements x_1, y_1, x_2 and y_2 are detected by displacement sensors and then amplified by controller G_c to generate radial force commands. These radial force commands produce current signals that generate radial forces F_{x_1}, F_{y_1}, F_{x_2} and F_{y_2} on the magnetic bearing rotor. The controllers for all axes are independently generating radial force commands.

If the gyroscopic effect, or any other interference in x- and y-axes, is considered, then controllers should be designed for the translational and inclination axes. Figure 4.6 shows a block diagram. From the detected shaft movement, displacements on the translational and inclination axes are calculated. These displacements x_{psn}, x_{rsn}, y_{rsn} and y_{psn} are then amplified by radial position controllers G_{cp} and G_{cr} to produce the radial force reference commands F_{xp}^*, F_{xr}^*, F_{yr}^* and F_{yp}^*. From these commands the required radial force commands for magnetic bearings on the x- and y-axes are calculated, so that currents can be supplied to produce the actual radial forces that follow the references. The inclination position block (surrounded by the dotted line) can be improved with additional cross-coupling blocks so that the two-axis interference is cancelled.

4.3 Thrust magnetic bearing and requirement of five-axis suspension

4.3.1 Thrust magnetic bearing

Figure 4.7 shows the cross section of a thrust magnetic bearing. The thrust magnetic bearing generates magnetic force in the axial direction. A ferromagnetic plate is attached to the shaft. On both sides of the plate, cylindrical electromagnets

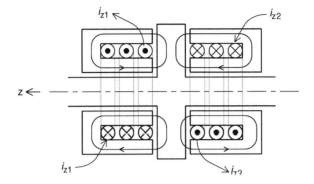

Figure 4.7 Thrust magnetic bearing

are located so that attractive magnetic forces are generated. The currents i_{z1} and i_{z2}, produce magnetic forces in opposite directions. These current values are adjusted so that the required magnetic force is generated from the difference of these attractive forces.

The magnetic bearing currents are obtained from the sum of bias and force currents so that

$$i_{z1} = I_b + i_z \qquad (4.19)$$

$$i_{z2} = I_b - i_z \qquad (4.20)$$

And the axial magnetic F_z force is written as

$$F_z = k_z I_b i_z \qquad (4.21)$$

Therefore the axial force is proportional to the current i_z. Hence an axial positioning system can be constructed as follows: using the axial position of the shaft, an axial force command is generated by a controller which produces a current command i_z^* to control the winding currents i_{z1} and i_{z2} that are supplied by inverters.

4.3.2 Requirement of five-axis suspension

Table 4.1 compares inverter and wiring requirements between magnetic bearing and bearingless motor systems with five-axis magnetic suspension. In five-axis active suspension using magnetic bearings, five displacement sensors are required for detection of the shaft positions x_{sn1}, y_{sn1}, x_{sn2}, y_{sn2} and z. Two radial magnetic bearings and one thrust magnetic bearing are needed. Since magnetic force regulation in one axis requires two single-phase inverters because of the push–pull operation of the magnetic forces, a total of 10 single-phase inverters are required. Therefore 20 wires are needed between the magnetic bearings and

Table 4.1 Comparison of five-axis active suspension

Magnetic bearing with a motor drive		
Displacement sensors	5	x_{sn1}, y_{sn1}, x_{sn2}, y_{sn2}, z
Magnetic bearing inverters	10	Two single-phase inverters for one axis
Magnetic bearing power wires	20	
Motor inverter	1	
Motor inverter wires	3	
Bearingless motor		
Displacement sensors	5	x_{sn1}, y_{sn1}, x_{sn2}, y_{sn2}, z
Magnetic bearing inverters	2+2	Two 3-phase inverters and two single-phase inverters
Magnetic bearing power wires	10	
Motor inverter	1	
Motor inverter wires	3	

inverters. In addition to the magnetic bearings, the motor needs one 3-phase inverter and three wires.

In a bearingless motor with five-axis active suspension, the requirements are shown in the bottom half of Table 4.1. For the magnetic suspension, two 3-phase inverters and two single-phase inverters are needed. Two radial-axis positions can be regulated by one 3-phase inverter so four radial-axis positions require two 3-phase inverters with six wires. In addition, two single-phase inverters with four wires are required for a thrust magnetic bearing. Therefore in total 10 wires are required. In addition to the suspension, a 3-phase inverter with three wires is required for the motor drive. The motor windings of two tandem bearingless motor units can be connected in series. One can see that there is a more simple power electronic and wire requirement for a bearingless drive.

The above comparison is based on a bearingless motor with 4-pole and 2-pole windings. In some winding configurations, the requirements may be greater.

References

[1] G. Schweitzer, H. Bleuler and A. Traxler, "Active Magnetic Bearings", Hochschulverlag AG an der ETH Zurich, 1994, ISBN 3 7281 2132 0.
[2] Y. Okada *et al.*, "Basic and Application of Magnetic Bearings", JSME & Yokendo Ltd, 1995, ISBN 4-8425-9521-3 (*in Japanese*).

5

Power electronic circuits for magnetic bearings

Akira Chiba

In magnetic suspension controllers, current or voltage commands are generated via the sensing of the displacement of a suspended object. Power electronic circuits apply voltage to the winding terminals to produce current in the windings. The rate of change of current should be fast enough to follow the current command. However, the amplitude and frequency ranges are limited by the voltage and current ratings of the power electronic circuits and the bearing winding inductances. If the current response is not fast enough, then the feedback loop becomes unstable, which results in shaft touch down.

In this chapter, the structures and principles of power electronic circuits are explained (focusing on switched-mode amplifiers) and the principles of pulse width modulation (PWM) are introduced. For a PWM scheme, the linear and nonlinear current regulations are shown. The limitation of amplitude and frequency range of current supply is derived, with a brief introduction of IGBT and MOSFET power devices as well as gate drive circuits.

5.1 Structure and principles of power electronic circuits

Table 5.1 shows typical voltage and current regulators for magnetic bearings. Linear analogue amplifiers have push–pull transistors at the output stage, as shown in Figure 5.1(a). This circuit is often used in audio amplifiers. Basically, push–pull circuits amplify the current capability. In addition to the current capability enhancement, a high-gain linear amplifier can be integrated, such as a power operational amplifier. Voltage or current followers can be constructed with external resistor circuits. The linear amplifiers have the advantage of precise current and voltage regulation as well as low noise and they have a current rating of less than 10 A. Operation at the rated current is available

Table 5.1 Voltage and current regulation

	Advantage	Disadvantage
Linear analogue amplifier	Precise regulation Low noise	High cost Low efficiency Large dimension
Switched-mode PWM amplifier	High efficiency Compact Low cost	Noise generation
Hybrid amplifier	Precise regulation Better efficiency	Difficult to set operating point

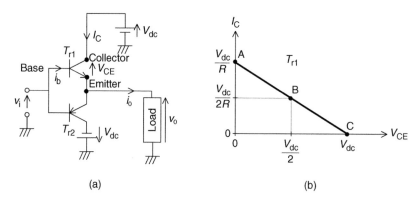

(a)   (b)

Figure 5.1 Push–pull transistors and operating points: (a) push–pull transistors; (b) operating points

only with effective cooling. Cooling of the output stage transistors is done with large heat sinks. Therefore the amplifier dimensions are large, resulting in high cost. The efficiency is low because of high losses in the push–pull transistors.

Switched-mode amplifiers enhance efficiency. Since the losses in the power devices are reduced the heat sinks are much smaller and, as a result, switched-mode amplifiers are compact in dimension so that the cost is low. Switched-mode operation of power devices is widely used in industry, e.g., for general-purpose inverters in ac drives and computer power supplies. Domestic appliances, most air conditioners, high specification refrigerators and washing machines use switched-mode power inverters to improve efficiency. Hence switched-mode amplifiers are dominant in magnetic bearing drivers.

Hybrid amplifiers take advantage of linear and switched-mode amplifiers. At low current the push–pull transistors operate as a linear amplifier but at high current they operate in switched mode. To take advantage of a hybrid amplifier, it is quite important to modify the winding structure in magnetic bearings.

5.1.1 Linear transistor amplifiers

Figure 5.1(a) shows the circuit configuration of the push–pull amplifier. Positive and negative dc voltages are supplied. The output voltage v_o follows the input voltage v_i, and if v_i is sinusoidal then v_o is also sinusoidal with the same amplitude. The current drive capability is increased by the collector current gain h_{FE}, which is normally in the range of 20–100, e.g., a base current i_b of 0.1 A is required for an output current i_o of 5 A if $h_{FE} = 50$.

Figure 5.1(b) shows the possible operating points of T_{r1} on the voltage and current plane. The possible operating points are on the line ABC. At point A, T_{r1} is on so that the collector–emitter voltage V_{CE} is almost zero with the maximum collector current I_C, i.e., V_{dc}/R, where R is a load resistance. The device power consumption is almost zero. The operating point moves along the load line as V_{CE} is increased. At point B, I_C and V_{CE} are half of the maximum current and voltage so that the output voltage $v_o = V_{dc}/2$. The power consumption of T_{r1} is a product of V_{CE} and I_C so that $P_{tr1} = V_{dc}^2/4R$. The operating point travels towards point C if v_o further decreases. At point C, v_o is minimum, and i_o is zero, i.e., T_{r1} is off. The power consumption in T_{r1} is zero. One can draw a similar load line with negative voltage for the transistor T_{r2}.

In linear amplifiers, the operating points travel along the load line and considerable power is consumed by the transistors because they are neither fully on nor fully off. Only at operating points A and C the power consumption is zero. The principle of switched-mode amplifiers is that the operating points are always set to A or C to minimize losses. Therefore the output voltage is not continuous but has two discrete values, i.e., $+V_{dc}$ or $-V_{dc}$. The required voltage value is somewhere between $+V_{dc}$ and $-V_{dc}$ so that the time lengths of operating points (i.e., the "on" and "off" times) are adjusted so that the average voltage corresponds to the reference voltage. The details are explained in the section "PWM Operation".

Figure 5.2 shows the circuit configuration of a linear amplifier with two stages. An oscillator is connected to the push–pull amplifier to drive the transistors and apply a voltage to the winding terminals. The oscillator can be replaced with operational amplifiers or D/A converters. It may be thought that an oscillator may be connected directly to the winding terminals to apply the voltage; however, voltage reduction occurs because of the output impedance. This is explained below.

General-purpose oscillators, D/A converters and operational amplifiers usually have an output impedance of several hundred ohms with an output voltage range of -12 V to $+12$ V. Let us consider a sinusoidal waveform. If the output terminals of the oscillator are opened then a sinusoidal waveform, having amplitude of 12 V, can be obtained. However, if an electrical circuit with low input impedance is connected to the oscillator terminals, the output voltage amplitude is decreased. The oscillator output voltage v_i is written as

$$v_i = \frac{Z_i}{Z_i + Z_o} v_{osc} \tag{5.1}$$

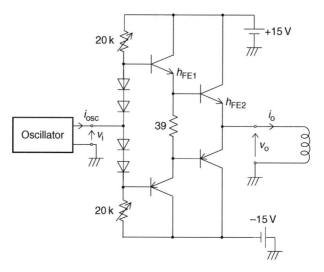

Figure 5.2 Two stage push–pull amplifier

where v_{osc} is the open-circuited voltage amplitude, Z_o is the output impedance of the oscillator and Z_i is the input impedance of an electrical circuit connected to the oscillator output. For example, the amplitude is reduced to 6 V if the input impedance is the same as the output impedance of the oscillator. Hence if the winding has an impedance of only a few ohms then the output voltage amplitude is reduced significantly. Therefore the reduction of output impedance is important. In most cases, the input impedance of the connected circuit is designed to be high enough to make the output impedance negligible so that v_i is almost equal to v_{osc}.

The two-stage amplifier circuit reduces the output impedance. In one transistor, the collector current is amplified by h_{FE} times the base current. Therefore the required input current i_{osc} is written as

$$i_{osc} = \frac{i_o}{h_{FE1} h_{FE2}} \tag{5.2}$$

where h_{FE1} and h_{FE2} are the current gains in transistors 1 and 2. For example, if $i_o = 1$ A, $h_{FE1} = 100$ and $h_{FE2} = 60$ then $i_{osc} = 1.6$ mA. Hence the input impedance of the 2-stage amplifier is 12 V/1.6 mA $= 7.5$ kΩ, which is sufficiently high compared with the output impedance of the oscillator.

The adjustable 20 kΩ resistors set the bias dc current for less waveform distortion and the series diodes compensate the voltage drop between base and emitter terminals. The first stage transistors are low current types; however, the second stage transistors have a high enough current rating for the load current. For example, a maximum current rating of 5 A is usually required for a 1 A load current and heat sinks are required for the second stage transistors.

5.1.2 Linear operational amplifiers

Transistor amplifiers with push–pull circuits can be integrated into one package with the differential amplifiers. These semiconductor devices are sold in the market as power operational amplifiers. General-purpose operational amplifiers can drive output currents between 10 and 30 mA. However, the rated current for a power operational amplifier is usually between 1 and 10 A. Normally an effective cooling method is required to operate at the rated current. In most cases, de rating is necessary to match the heat sinks and airflow.

A voltage follower circuit is easily fabricated from a power operational amplifier. The negative input is connected to the output and the positive input is connected to the input voltage v_i. The output voltage v_o follows the input voltage v_i. However, the voltage gain is unity and current drive capability is increased.

Figure 5.3 shows a linear current driver using a power operational amplifier. The output current i_o is following the input voltage v_i. If the gain is set to unity then 1 A of current is provided at an input voltage of 1 V. The resistance R_6 has a low value to avoid losses because the main current flows through it. The other resistances R_1, R_2, R_3 and R_5 have values in the range of several kΩ. The relationship between the input voltage and output current is written as

$$i_o = \frac{-R_2}{R_1 R_6} v_i \qquad (5.3)$$

In the case of resistance values given in the figure, the current gain is unity. In designing the current driver, the value of R_6 should be determined by considering the load resistance, supply voltage, heat generation in power operational amplifier and R_6. It should be noted that the resistors should have low temperature drift and wide frequency range. Noise and high frequency oscillation should be avoided with practical wiring and smoothing capacitors.

The diodes provide a regenerative current path for the inductive load. The output voltage is limited to the power supply voltages. Chemical capacitors are necessary in parallel with the power supply to maintain the power supply voltage in case of sudden increase of output current. In addition to the chemical

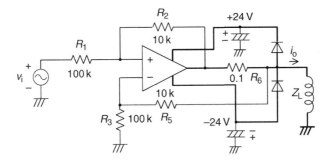

Figure 5.3   Linear current driver

capacitors, ceramic capacitors with a high frequency range are also necessary, as with most electrical circuits.

Example 5.1   Figure 5.4(a) shows an output stage of a transistor circuit with a load resistance of $10\,\Omega$ and supply voltage of $\pm 100\,\text{V}$.

1. Suppose that the output voltage is sinusoidal with maximum amplitude. Draw the waveforms for the load voltage v_L and current i_L, as well as voltage v_{CE1} and current i_{C1}. Also draw the instantaneous power consumption in T_{r1}. Derive average power consumption of T_{r1}.
2. Suppose that the transistors are operating in switched mode, and the output voltage has a square waveform. Draw the waveforms and derive the average power consumption of T_{r1} assuming that the voltage drop for the on state of the transistors is $3\,\text{V}$.
3. Efficiency is defined as the ratio of load power consumption to the power provided by the DC source power. Derive the efficiency for the cases of linear and switched-mode operations.
4. How much power is consumed in linear operation compared to switched-mode operation?

Answer

Figure 5.4(b) shows the waveforms for linear operation. The load voltage and current are sinusoidal and T_{r1} supplies current in the positive half period. The transistor voltage v_{CE1} is given as

$$v_{CE1} = 100 - v_L = 100(1 - \sin \omega t)$$

Instantaneous power consumption is the product of v_{CE1} and i_{c1}, i.e.,

$$100(1 - \sin \omega t) \times 10 \sin(\omega t)$$

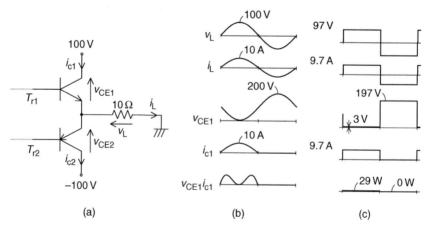

(a)   (b)   (c)

Figure 5.4 Linear and switched-mode operations: (a) output stage; (b) linear operation; (c) switched-mode operation

as shown in the figure. Therefore average power consumption is

$$\frac{\omega}{2\pi} \int\limits_{0}^{\pi/\omega} 1000(1 - \sin \omega t) \sin \omega t \, dt = 68 \, \text{W}$$

In switched-mode operation, the waveforms are shown in Figure 5.4(c). A transistor on-state voltage drop of 3 V is assumed. Therefore the output voltage amplitude is |97 V or 97 V and the peak value of the load current is +9.7 A or −9.7 A. The power consumption in T_{r1} over a half cycle is the product of 3 V and 9.7 A, i.e., 29 W. Hence the average power consumption is 15 W.

In linear operation, the load power consumption is $100/\sqrt{2} \times 10/\sqrt{2} = 500$ W. The supplied power from the power source is the sum of load power and transistor power losses $= 500 + 68 + 68 = 636$ W, giving an efficiency of $500/636 = 78.6\%$.

In switched-mode operation, the load power consumption is $97 \times 9.7 = 941$ W. The supplied power is a sum of load power and transistor loss $= 941 + 15 + 15 = 971$ W, giving an efficiency of $941/971 = 97\%$.

The power consumption ratio is $68/15 = 4.5$. Note that switched-mode operation results in significantly less power consumption in the semiconductor devices. Small heat sinks with less airflow are therefore required so that small and lightweight voltage and current drivers can be used. It should be noticed that if the voltage is increased, but the current is decreased to maintain the same power, then the efficiency is higher. Increasing the number of coil turns by using thinner wires will increase the coil impedance allowing for high voltage operation. Also, if the voltage drop during the on state is decreased, more efficient drivers are obtained. Recent power device developments have led to the voltage drop during the on-state in IGBTs to decrease down to about 1.6 V. MOSFETs provide low device losses with low on-state resistance in low voltage circuits. For further power consumption calculations, switching losses should be included with the on-state power consumption.

5.1.3 Switched-mode amplifiers

In this section, three types of switched-mode amplifiers are introduced. The first is a single-phase H-bridge inverter; the second is a simplified single-phase inverter for unidirectional current. This inverter is widely used in magnetic bearings. The third is a chopper circuit for low voltage and current drivers.

Figure 5.5 shows the two main circuits that are used in a single-phase inverter. In Figure 5.5(a), four power-switching devices (e.g., IGBTs) are connected in an H-bridge configuration. The power-switching devices can be other self-turn-off devices such as MOSFETs, transistors, SITs, and so on. These power-switching devices are encapsulated with an anti-parallel diode. The magnetic

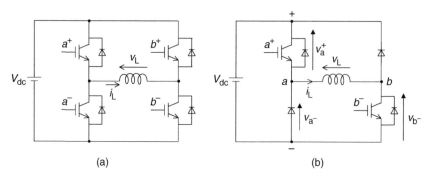

(a)　　　　　　　　　　　　　　　(b)

Figure 5.5 Single-phase inverters: (a) single-phase inverter; (b) single-phase inverter with unidirectional current

bearing winding is represented by an inductor and the terminal voltage and current are v_L and i_L, respectively. The circuit operation has several modes depending on the current direction and on–off states of the switching devices.

Figure 5.6 shows the operation modes. In modes 1 and 2, the gate voltage is applied to IGBT a^+ and b^-. If the winding current is positive then the current flows from positive dc terminal, through IGBT a^+, the winding and the IGBT b^- and returns to the negative dc terminal as shown in mode 1. If the winding current is negative, current does not go through IGBTs a^+ and b^-; instead, the current goes through the anti-parallel diodes of a^+ and b^- so that the stored magnetic energy in the winding flows back into the dc power source. In modes 1 and 2, positive voltage is applied to the winding terminals and the current path is automatically changed depending on the winding current direction.

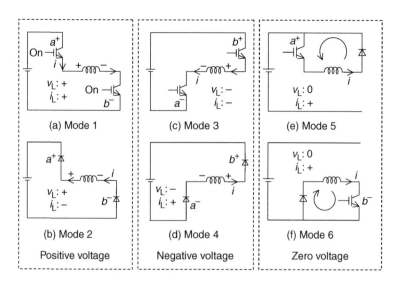

Figure 5.6 Operation modes

In modes 3 and 4, negative voltage is applied to the winding terminals. The gate voltages are applied to IGBTs a^- and b^+. Depending on the current direction of the winding, the current path is as in mode 3 or 4.

One may notice that the single-phase inverter is designed to provide positive and negative voltage as well as bidirectional current flow. In magnetic bearing controls that utilize bias current, only unidirectional current is required. Suppose only positive current is required, then modes 2 and 3 are not necessary. Hence the single-phase inverter can be simplified as shown in Figure 5.5(b).

Diodes can replace IGBTs a and b^1. This simplification reduces the cost, so this circuit is widely used in magnetic bearings. For one radial magnetic bearing with two-axis active positioning, four electromagnets are constructed, hence four inverters with eight wirings are required and, in total, eight switching devices are necessary (down from 16 when using full H-bridge inverters). This circuit configuration is also used widely in switched reluctance motor drives. However, these power modules are not widely available on the market in 3-phase inverter form.

Figure 5.6(e,f) shows additional modes, i.e., the zero-voltage modes. In these modes the winding terminal voltage is almost zero. They are useful for reducing the switching frequency. In mode 5, gate voltage is applied to only the a^+ switching device. The winding current circulates through the diode and a^+ device providing a short-circuit path. In mode 6, switching device b^- is on, providing a short-circuit path with another diode. In the zero-voltage modes, the winding current is almost constant. These modes of operation are often described as freewheeling.

Figure 5.7 shows an operating sequence, including the zero-voltage mode. In mode 1, gate voltages a^+ and b^- are high so that the IGBTs are on. The load

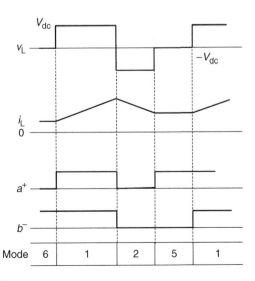

Figure 5.7 Operation sequence

current i_L increases as positive voltage is applied. In mode 2, the diodes are on so that the winding current decreases due to negative voltage being applied. In mode 5, the current value is kept constant and zero voltage is applied to the terminals.

5.1.4 Low cost chopper circuit

In some applications, cost reduction is more important than efficiency and performance of the power electronic circuits. This is especially true in small voltage and current applications where the minimum number of switching power devices should be used for cost reduction. A drop in efficiency is allowed because the voltage and current are not high. In these applications, a chopper circuit (which is basically a dc-to-dc converter) with variable voltage is employed.

Figure 5.8 shows the main circuit of a chopper circuit. Only one switching power device is required for one winding. There are two operating modes. In mode 1, the switching device is turned on and positive voltage is applied to the winding. In mode 2, the switching device is turned off, and a zero voltage loop occurs through the diode and winding (freewheeling again).

Figure 5.9 shows the operating sequence for the chopper circuit. In mode 1, the winding voltage v_L is positive and the current i_L increases. The on-state voltage v_{gs} of the switching device is high and the winding current i_{sw} is flowing through the switching device. In mode 2, the current i_L is almost constant since the winding is short-circuited.

The main problem associated with the chopper circuit is easily understood from the waveforms. A fast increase in winding current is possible in mode 1; however, during mode 2 (zero voltage) the current decay is quite slow. The current decay leads to energy dissipation via winding loss and diode voltage drop. Unsymmetrical current response characteristics result in a nonlinear transfer function of the current regulator. The nonlinearity may cause a stability problem with the negative feedback loops in a magnetic suspension system. To enhance the current response, one solution is to connect a resistor in series with the winding. In zero voltage mode, the current decreases as a function of $\exp(-R_t/L)$, where R is the sum of a winding resistance and a connected resistor. Therefore an increase in R results in a fast current decay. Moreover, the

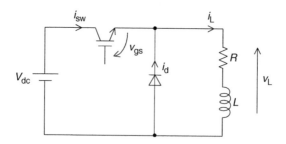

Figure 5.8 *Chopper*

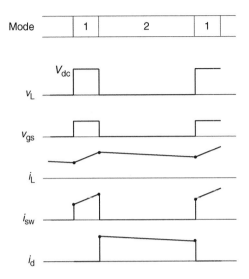

Figure 5.9 Chopper operating waveforms

rate of change of current increase is reduced because of the voltage drop at the resistor terminals. Therefore an external resistor compensates for the unsymmetrical current response to some extent. The disadvantage is the additional power consumption in the resistor.

The other way to enhance the response of current decay is to connect the external resistor in series with the freewheeling diode. The loss in the resistor is decreased because resistor current is only present during the zero-voltage mode.

With considerable trade-off in the design a low cost power electronic circuit is possible for low voltage and current applications.

5.2 PWM operation

In this section, the operation of a single-phase inverter with unidirectional current is analysed and discussed. First, the principles of PWM are explained. Then an analysis of voltage and current is carried out and the derived voltage and current relationships are applied to a magnetic bearing example. A calculation method for the winding resistance is shown in order to obtain the time constant. The example is also analysed by computer simulation for better understanding.

5.2.1 Principles

Figure 5.10 shows the principles of PWM operation in a single-phase inverter with unidirectional current. The triangular waveform v_c has a frequency of f_c and the maximum and minimum values are $+1$ and -1, respectively. The voltage

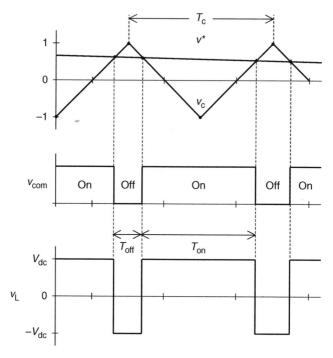

Figure 5.10 Output voltage of single-phase inverter with unidirectional current

command is v^* and a comparator provides a digital signal v_{com} from the comparison of v_c and v^* as shown in the figure. Using the comparator output signal the switching devices are operated. If v_{com} is high then the switching devices are turned on so that the dc bus voltage V_{dc} is applied to the winding. On the other hand, $-V_{dc}$ is applied when v_{com} is low.

Let us define the carrier period and the on and off periods as T_c, T_{on} and T_{off} respectively. Therefore the average output voltage in winding terminals is

$$\bar{v}_L = \frac{T_{on}}{T_c} V_{dc} - \frac{T_{off}}{T_c} V_{dc} \tag{5.4}$$

where

$$T_{off} = T_c - T_{on} \tag{5.5}$$

In addition, the relationship between the voltage command and the on-to-period ratio (the duty cycle) is written as

$$\frac{T_{on}}{T_c} = \frac{1}{2} v^* + \frac{1}{2} \tag{5.6}$$

Substituting these equations into (5.4) yields

$$\bar{v}_L = v^* V_{dc} \tag{5.7}$$

This equation indicates that the voltage across the winding terminals is proportional to the voltage command. The winding voltage can be controlled by the voltage command when the dc bus voltage V_{dc} is kept constant. The voltage can be both positive and negative. Note that a dead time, which is normally required in a typical inverter, is not needed. Therefore better linearity is obtained.

5.2.2 Analysis of RL circuit

A magnetic bearing winding can be represented by a series R-L circuit. When the dc voltage V_{dc} is applied as a step function, the voltage and current equations can be obtained using the Laplace transform:

$$\frac{V_{dc}}{s} = L[sI(s) - I_0] + RI(s) \tag{5.8}$$

where I_0 is the initial current. Solving the equation for $I(s)$ and transforming into the time domain provides the current:

$$i(t) = \frac{V_{dc}}{R}\left(1 - e^{-\left(\frac{R}{L}\right)t}\right) + I_0 e^{-\left(\frac{R}{L}\right)t} \tag{5.9}$$

The first term gives the current component due to the applied voltage. The second term is obtained from the initial current.

For the negative voltage period, substituting $-V_{dc}$ into the above equation gives

$$i(t) = -\frac{V_{dc}}{R}\left(1 - e^{-\left(\frac{R}{L}\right)t}\right) + I_1 e^{-\left(\frac{R}{L}\right)t} \tag{5.10}$$

where I_1 is the initial current of the negative voltage period.

Example 5.2 For the radial magnetic bearing shown in Figure 2.14 derive and calculate the following:

1. The winding resistance.
2. The time constant.
3. Suppose that the PWM carrier frequency is 10 kHz with $T_{on} = 55\,\mu s$ and $V_{dc} = 24\,V$. Find the current values at the switching points in the period. Also find the average current value.
4. Draw the waveforms for the triangular carrier signal, voltage command, comparator output, terminal voltage and winding current by a digital simulation.

Answer

1. The conductor cross-sectional area S is given as a product of the wire cross-sectional area and the number of parallel strands, i.e., $N_{st} = 2$, so that

$$S = \pi \left(\frac{D_w}{2}\right)^2 \times N_{st}$$

$$= \pi \left(\frac{0.8 \times 10^{-3}}{2}\right)^2 \times 2 = 1 \times 10^{-6}\,\text{m}^2$$

(5.11)

where D_w is the diameter of a wire.

Let us assume that the length of the straight coil part is the sum of the iron core stack length l and an additional length l_{ad} of 4.5 mm at the both ends. The coil end is assumed to be a semicircle of diameter D_{end}. The wire length is the sum of the straight and circular lengths. The straight length L_{st} is

$$L_{st} = 2N(l + 2l_{ad})$$

$$= 2 \times 50 \times (50 + 2 \times 4.5) \times 10^{-3} = 5.9\,\text{m}$$

(5.12)

Figure 5.11 shows an enlarged figure of iron core cross section. The conductors fill the shaded area in the stator slots. To simplify the calculation, let us assume that the coil-end diameter D_{end} is equal to the arc between the slot coil centres so that

$$D_{end} = 2\pi \left(\frac{D}{2} + g + \frac{l_s}{2}\right) \frac{35}{360}$$

$$= 2\pi \left(\frac{50 \times 10^{-3}}{2} + 0.4 \times 10^{-3} + \frac{15 \times 10^{-3}}{2}\right) \frac{35}{360}$$

(5.13)

$$= 20.1 \times 10^{-3}\,\text{m}$$

Therefore the wire length of the circular part of the coil is

$$L_{end} = \pi D_{end} N = \pi \times 20.1 \times 10^{-3} \times 50 = 3.16\,\text{m}$$

(5.14)

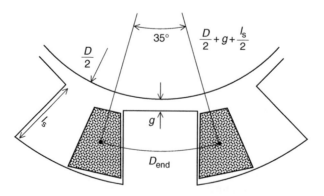

Figure 5.11 Slot centre

Hence the total length of wire in a phase coil is

$$L_c = L_{st} + L_{end} = 5.9 + 3.16 = 9.06 \, \text{m} \tag{5.15}$$

Two coils wound around neighbouring magnetic poles are connected in series so that the total wire length is $2L_c$. The resistivity of copper σ is $2.1 \times 10^{-8} \, \Omega\text{m}$ at a temperature of $75 \, °\text{C}$. Therefore the resistance is

$$R = \sigma \times \frac{2I_c}{S}$$
$$- 2.1 \times 10^{-8} \times \frac{2 \times 9.06}{1 \times 10^{-6}} - 0.38 \, \Omega \tag{5.16}$$

2. The time constant is given as a ratio of inductance and resistance

$$T = \frac{L}{R} = \frac{8.6 \times 10^{-3}}{0.38} = 0.022 \, \text{sec} \tag{5.17}$$

3. From (5.9), the current I_1 can be obtained after $55 \, \mu\text{s}$ of positive voltage application using

$$I_1 = i_1(55 \, \mu\text{s}) = 0.1533 + 0.9975 I_0 \tag{5.18}$$

And also the current value at the end of the negative voltage mode, from (5.10), is

$$I_0 = i_2(45 \, \mu\text{s}) = -0.1254 + 0.9980 I_1 \tag{5.19}$$

Substituting I_1 given in (5.18) into (5.19) and solving for I_0 gives

$$I_0 = 6.238 \, \text{A} \tag{5.20}$$

Substituting I_0 into (5.18) yields

$$I_1 = 6.376 \, \text{A} \tag{5.21}$$

Hence the current values at the switching instances are derived. The instantaneous current values can be obtained from (5.9) and (5.10) by substituting I_0 and I_1.

The average current value can be derived from the integration of the instantaneous current waveforms; however, there is a simple way to obtain the average current from the average voltage. The average voltage is calculated from (5.4) so that

$$\bar{v}_L = \frac{T_{on}}{T_c} V_{dc} - \frac{T_{off}}{T_c} V_{dc}$$
$$= 0.55 \times 24 - 0.45 \times 24 = 2.4 \, \text{V} \tag{5.22}$$

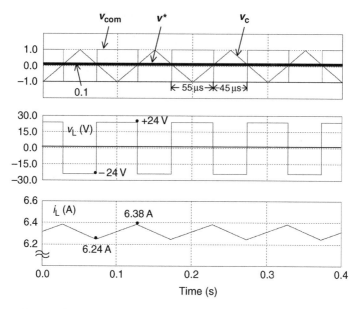

Figure 5.12 Waveforms in single-phase inverter with unidirectional current

The average current is obtained by dividing the average voltage by the resistance

$$\bar{I}_L = \frac{\bar{v}_L}{R} = \frac{2.4}{0.38} = 6.316\,\text{A} \tag{5.23}$$

4. The top graph of Figure 5.12 shows the waveforms for carrier signal v_c, voltage command v^* and comparator output v_{com}. In the middle graph the winding terminal voltage v_L characteristic is illustrated while the bottom graph shows the winding current waveform. The power electronic simulation is carried out using the PSIM program, which is a fast, efficient and cost-competitive digital simulation program for modelling power electronic circuits, including switches and controllers.

5.3 Current feedback

In the previous chapter, inverter and chopper operations were introduced and discussed. The single-phase inverter with unidirectional current direction is often used in magnetic bearing applications so this chapter focuses on the operation of this inverter configuration.

A high frequency response is required in a magnetic bearing system. Therefore current feedback is necessary for fast response. Moreover the current feedback compensates for the problems associated with the following characteristics:

1. A transfer delay of $1/(R+sL)$ is caused by the winding resistance and inductance.
2. The switching speed of the power devices used in inverters is limited. This switching delay causes a nonlinear transfer function between the voltage command and output voltage of the inverter.
3. The output voltage of the inverter drifts if the dc supply voltage varies.
4. The voltage drop across a switching device when turned on is not zero. The on-state voltage drop results in a nonlinearity in the voltage transfer function.
5. A temperature drift of winding resistance causes a time variant system.

In order to suppress the influence of these problems, current feedback is usually employed. The current feedback strategies are as follows:

1. A linear current controller.
2. A nonlinear current controller such as a hysteresis function or a delta modulation controller, etc.

In this section, the characteristics of a linear current controller are discussed first, followed by a description of a nonlinear controller.

5.3.1 Linear current controller

Figure 5.13 shows a block diagram of a linear current controller system. The current i_L is detected then compared with the current command i_L^*. The current error i_e is amplified by a controller block G_c and, from the current controller, a voltage command v^* is generated. A limiter block keeps the voltage command within the PWM triangular carrier waveforms. In the G_{inv} block, the on-and-off logic for the main power devices is determined and the voltage is applied to the winding. The winding transfer function is simply represented by $1/(R+sL)$ and the current i_L is the output. The transfer function of this feedback loop is then

$$\frac{i_L}{i_L^*} = \frac{1}{1+\dfrac{R+sL}{G_c G_{inv}}} \tag{5.24}$$

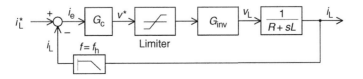

Figure 5.13 Current feedback

This equation indicates that the transfer function is close to unity if $G_c G_{inv}$ is high. However, a gain limitation exists because of the limited switching frequency.

Let us look at the current regulating characteristics from computer simulation. In the simulation, the main circuit current is detected and a high-cut filter provides current feedback signal, which is compared with the current command. A proportional and integral (PI) controller, having proportional gain K_{PI} and integrator time constant T_{PI}, amplifies the current error so that the output is limited to between -1 and 1. This is the amplitude of a PWM triangular waveform. The limiter output, i.e., feedback voltage v^*, is compared with a PWM carrier triangular waveform at $10\,kHz$. The comparator output logic is fed to the main power-switching devices.

Figure 5.14(a–g) shows the current waveforms. The waveforms are plotted from $0.5\,ms$ in order to see the steady-state conditions. To observe the transient response, the current command is decreased and increased at 1 and 2 ms, respectively. In this simulation the zero-voltage loop is not used. The waveforms show the current command and winding current. The current waveform includes a triangular component, which is often called the current ripple. In Figure 5.14(a), the current command has a step function change from 4 to 3.9 A at 1 ms. With the effect of feedback, the current follows the command after a 0.5 ms transient period. A similarly good response is also seen after the step at 2 ms. Hence it is found that the current response is good with a small step change of the current command.

Figure 5.14(b–g) shows the current response when the current command has a large step change. The current command is decreased from 4 to 3 A at 1 ms, then increased back to 4 A at 2 ms. Just after the step command, the feedback voltage is saturated so that the winding terminal voltage is fixed to the minimum value for a period. The current has a ramp function decrease. In Figure 5.14(b) it can be seen that it takes about 0.4 ms to follow the current command. The current slew rate is determined by the applied voltage and also the winding inductance. To enhance the current slew rate, an increase in inverter dc voltage is necessary. In addition, after reaching the current command at a time of 1.3 ms, the current then overshoots the target command. To improve the current response the gain of the proportional controller should be increased.

Figure 5.14(c) shows the waveforms with $K_{PI} = 20$, which is 10 times that of the original controller. It is seen that the current response is much faster in the time range of 1.4–1.7 ms compared with the previous figure. However, voltage command v^* includes high frequency oscillation. This oscillation causes abnormal switching which increases the switching frequency so that it may cause excessive heat generation in power devices. The high frequency oscillation should be removed with a high-cut filter.

Figure 5.14(d) shows the effect of the high-cut filter with a cut-off frequency of $f_h = 10\,kHz$. The voltage command v^* does not include a high frequency component; however, a $10\,kHz$ sinusoidal ripple is added. The current response is now excellent, at around 1.4–1.7 ms in this simulation.

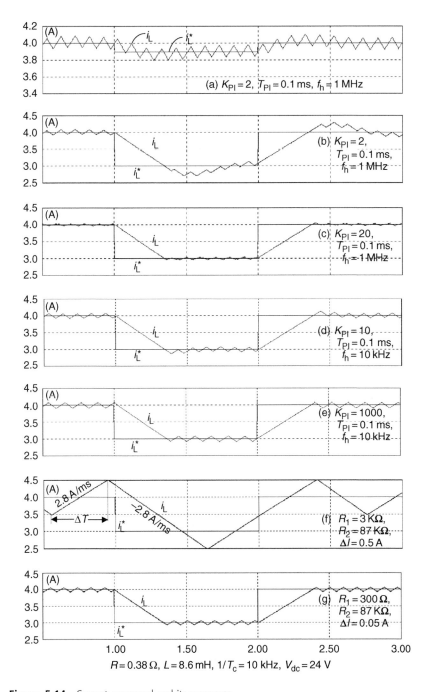

Figure 5.14 Current command and its responses

In Figure 5.14(e) the proportional gain is increased significantly. Now there is no abnormal switching of the triangular function and the current response is excellent. The feedback voltage is mostly saturated. Even though the current follows to the command in a time range of 1.5–2 ms, the feedback voltage is saturated. This operation is very close to the nonlinear current control method.

In this section, the linear current feedback method was introduced. The current response is good for large step response with an increased proportional gain and a high-cut filter. If the proportional gain is significantly increased then the operation is close to the nonlinear feedback method.

5.3.2 Non-linear current controller

Figure 5.15(a–c) shows the principles of a hysteresis current controller. In Figure 5.15(a), the detected current i_L is compared with the current command i_L^* so that the error i_e is obtained. From the current error, a hysteresis block generates an on-off command for the main power-switching devices.

Figure 5.15(b) shows the function of a hysteresis current controller. The horizontal axis is the current error as an analogue function. The vertical axis is the output digital signal, which takes a value of 0 or 1. Let us define the digital signal as S. This is 1 if $i_e > \Delta i$ and 0 if $i_e < -\Delta i$. If the absolute value of i_e is less than Δi then S is left unchanged from the previous value.

Suppose that i_e is at the operating point a. If i_e is decreased gradually, the operating point travels from a to b and then to c. If i_e is decreased further then S becomes 0. Next, let us suppose that the operating point is at d. If i_e is increased, the operating point travels from d to e and then to f. If i_e is increased further, then S becomes 1.

Figure 5.16 shows the waveforms of a hysteresis current controller. The upper figure shows the current command i_L^* as a solid curve while $i_L^* + \Delta i$ and $i_L^* - \Delta i$ are dotted curves. The middle and the lower graphs show the current error i_e and on-off logic S. Suppose that the original operating point is a; as the current value is less than $i_L^* - \Delta i$, S is 1. If a positive voltage is applied to the main circuit then current increases rapidly and the operating point moves from a to b and then to c. At the operating point c, S becomes 0. Since there is a delay in the power devices, a negative voltage is applied to the winding terminals at d so that the current decreases when moving between points e, f and a.

Note that the current waveform is composed of straight lines connecting at the switching points (i.e., a linear approximation of the circuit). The change of current can be written as

$$i_L = \frac{V_{dc}}{L} t \tag{5.25}$$

where V_{dc} is the peak value of applied voltage, L is the winding inductance and t is time. This equation is valid only when the switching frequency is high with respect to the time constant of the RL circuit.

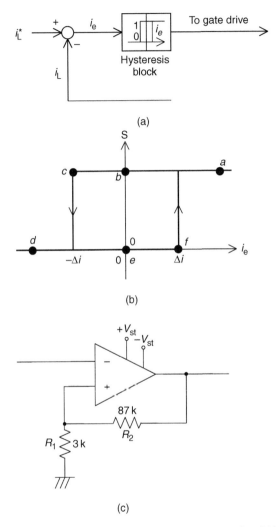

Figure 5.15 Hysteresis current regulator: (a) hysteresis current controller; (b) hysteresis function; (c) hysteresis comparator with operational amplifier

Example 5.3 A hysteresis current controller is regulating an *RL* series circuit current with hysteresis widths of 0.5 or 0.05 A, and a current command of 4 A at $t = 0$. The current command is decreased to 3 A at 1 ms then increased to 4 A at 2 ms. Sketch the current waveform. The dc voltage of the inverter is 24 V and load inductance is 8.6 mH. Also derive the switching frequency in steady state. Construct a hysteresis function using an operational amplifier and resistors. Construct the digital simulation blocks and show the current waveforms.

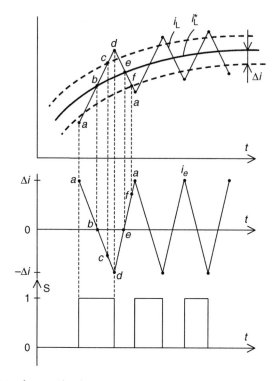

Figure 5.16 Waveforms with a hysteresis current controller

Answer

From (5.25), the rate of change of current is given as $V_{dc}/L = 2.8\,\text{A/ms}$. Figure 5.14(f) shows the current waveform for the case of $\Delta i = 0.5\,\text{A}$. Although it is not drawn, the current is zero at $t = 0$. A line is drawn having that slew rate. The line has an intersection with $i_L^* + \Delta i$, at which point the current rate is changed to $-2.8\,\text{A/ms}$. The second line has an intersection with $i_L^* - \Delta i$ at $t = 0.6\,\text{ms}$ as shown. Then the current is increased up to 4.5 A at $t = 0.9\,\text{ms}$ where it is switched again and decreases down to 2.5 A at $t = 1.7\,\text{ms}$. During this period the current command has a step-function change.

In the figure, ΔT is defined as a term of half cycle. Substituting $i_L = 2\Delta i$ and $t = \Delta T$ into (5.25) then

$$\Delta T = 2\Delta i \frac{L}{V_{dc}} \qquad (5.26)$$

One cycle termination T_s is $2\Delta T$ and the switching frequency f_s is an inverse function of T_s so that

$$f_s = \frac{1}{4\Delta i}\frac{V_{dc}}{L} \qquad (5.27)$$

Substituting values, $f_s = 1.4\,\text{kHz}$.

Figure 5.15(c) shows a hysteresis comparator circuit formed from an operational amplifier and resistors. The resistors R_1 and R_2 determine the hysteresis width. Let us suppose that a saturation voltage of the operational amplifier output is V_{st}. Hence the hysteresis voltage Δv is

$$\Delta v = \frac{R_1}{R_1 + R_2} V_{st} \tag{5.28}$$

Suppose that the current detection gain is K_{ig} so that the current function is converted into a voltage function using $v = K_{ig} i_L$. Therefore the hysteresis current width is obtained from

$$\Delta i = K_{ig} \frac{R_1}{R_1 + R_2} V_{st} \tag{5.29}$$

Substituting $K_{ig} = 1$, $R_1 = 3\,\text{k}\Omega$, $R_2 = 87\,\text{k}\Omega$, $V_{st} = 15\,\text{V}$ then Δi is 0.5 A.

Figure 5.14(g) shows the current response with $R_1 = 300\,\Omega$. The current hysteresis is 0.05 A. Hence we get a fast response and small error.

5.4 Current driver operating area

In the previous section, current control strategies were investigated in order to improve the system response. However, it was shown that the slew rate of the current was limited by the dc supply voltage of the inverter. This fact indicates that the operating frequency range of a current driver is limited. At a low frequency the output current is limited by the rated value, which is determined by power devices and dc power source. At high frequency the maximum output voltage limits the current response. If a current, with frequency ω and amplitude I_p, is required then the inverter should have an output voltage of $\omega L I_p$, where L is the inductance of a winding. Since the inverter output voltage is limited, the possible current amplitude decreases and is inversely proportional to ω.

In a magnetic bearing system, the current driver operating area should be wide enough for the requirement. If the current command, generated by the radial positioning loop, is outside the operating area then the current driver is saturated and the negative feedback is opened, producing unstable radial positioning. This is because the magnetic suspension system is inherently unstable; so it is quite important to check the operating area of a current driver well in advance.

5.4.1 Operating area

Figure 5.17(a) shows the definition of the operating area of a current driver. The horizontal axis has a logarithmic frequency scale. The vertical axis is the current amplitude and also has a logarithmic scale. At low frequency the operating area

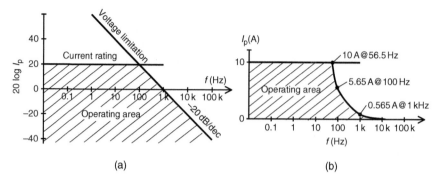

Figure 5.17   Current drive operating area: (a) current drive operating in log scale; (b) operating area for ±24 V inverter with 8.6 mH load

is limited by the current rating. However, the voltage limitation line, which has an incline of −20 dB/dec, provides the high frequency current limit. Hence the operating area is the shaded area.

Figure 5.18(a) shows the voltage waveform at the maximum output. The voltage waveform is square in shape with peak amplitude equal to the inverter dc voltage. The fundamental component, which has an amplitude of $4/\pi$ of the voltage waveform, is also drawn. Let us assume that the winding resistance and power device voltage-drops can be neglected with respect to winding inductance voltage. Therefore the amplitude of the sinusoidal current I_p is

$$I_p = \frac{4}{\pi} \frac{V_{dc}}{\omega L} \tag{5.30}$$

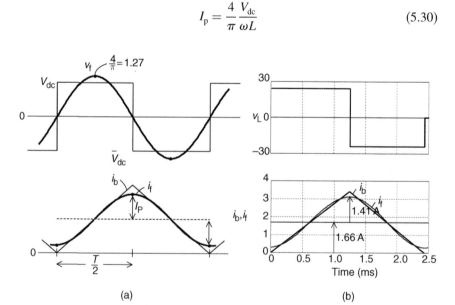

Figure 5.18   The maximum voltage and current: (a) ideal maximum voltage and current waveforms; (b) the maximum voltage and current in simulation

This equation is also written in terms of the current frequency f as

$$I_p = \frac{2}{\pi^2} \frac{V_{dc}}{L} \frac{1}{f} \tag{5.31}$$

In the lower graph the current waveform i_b, which is triangular in shape, is drawn. The sum of the dc and fundamental component i_f (having the peak amplitude of I_p) is also illustrated. The waveform shows the maximum current and (5.31) provides the voltage limitation line shown previously in Figure 5.17(a).

Example 5.4 Show the operating area of a 10 A current driver with a dc voltage of 24 V and a winding inductance of 8.6 mH. Draw the figure with a linear vertical axis. Also show the digital simulation result with a series winding resistance of 0.38 Ω.

Answer
Substituting $V_{dc} = 24$ V and $L = 8.6$ mH into (5.31) yields $I_p = 565$ A-s per revolution. Therefore I_p is 5.65 A, 1.41 A, 0.565 A and 0.0565 A at frequencies of 100 Hz, 400 Hz, 1 kHz and 10 kHz, respectively. The operating area is drawn as shown in Figure 5.17(b) in a linear scale. The linear scale provides a straightforward image of the operation area. After 56.5 Hz, the current amplitude is quite limited.

Figure 5.18(b) shows the digital simulation result. A square wave voltage command of 400 Hz with maximum amplitude is fed to the single-phase inverter (which has unidirectional current). Some amount of energy will be consumed by the winding resistance, so that the negative half cycle of the voltage waveform is slightly lower in magnitude than the positive half cycle. In the negative half cycle, the diodes provide a current recovery path to return stored magnetic energy back to the dc voltage source. The diodes are turned off when the current reaches zero. The current waveform is almost triangular in shape; however, it is smaller in the negative half cycle. The Fourier analysis of the current waveform shows dc and fundamental components of 1.66 and 1.41 A respectively. The curve of i_f shows the sum of the dc and fundamental components. The amplitude of the fundamental component agrees with the calculated value from (5.31).

5.4.2 Acceleration test

It is quite important to investigate the influence of insufficient current from the current driver. In magnetic bearing applications the required amplitude and frequency of the current command increases as the rotational speed increases up to the critical speed. If the current command is outside the operating area then the radial positioning becomes unstable. Figure 5.19 shows a simulation of the rotational acceleration, including unbalance and sensor target eccentricity.

The current command and current are drawn in the bottom graph. The current follows the command up to point A; however, from point A to B, a discrepancy

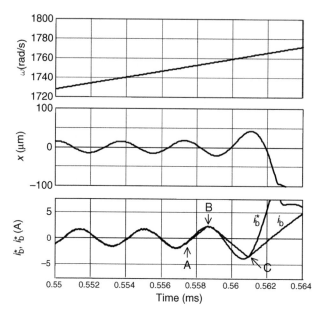

Figure 5.19 Acceleration test with fatal touch down

is seen. This is because enough voltage is not available to force the current to follow the command. Whilst they converge again at *B*, after *B* a significant error is seen. At point *C* the voltage is insufficient such that a fatal error is seen after point *C*, where the shaft radial position cannot be regulated and the shaft touches down to an emergency bearing. The required current peak is about 2 A at a frequency of 280 Hz. This limit corresponds to the case in Example 5.4.

To avoid this problem, the following solutions can be implemented:

1. An increase in the dc voltage of the inverter.
2. A decrease in the number of series turns in the magnetic bearing winding to reduce the inductance. To compensate a reduction of series turns, parallel turns may be used with an increase in current rating.
3. The eccentric errors of the shaft, rotor and sensor target ring should be reduced and better mechanical balancing should be achieved.
4. The controller should be improved. The frequency characteristics of the radial position controller should be shaped so that current requirement is less. Feed-forward signal injection may be considered.

5.5 Power devices and gate drive circuits

In this section, the characteristics of various power devices and gate driver circuits are described and waveform measurements are given. Insulated gate bipolar transistors and metal oxide silicon field effect transistors are discussed because these devices are popular. An example of a gate drive circuit is shown and the principles and operation are explained. In addition, the electrical failure mechanisms associated with the circuit are discussed and explained. The fundamental functions of a differential probe and isolators are introduced.

5.5.1 Power devices

Insulated gate bipolar transistors and metal oxide silicon field effect transistor are the most popular semiconductor devices used in inverters. Power transistors, which require considerable base current, have been used for over a decade. The IGBT is an improved power transistor with voltage controlled MOS gate characteristics. There are three terminals: gate, collector and emitter. Turn-on and turn-off is achieved by applying a specific voltage between the gate and emitter terminals. For example, if the gate terminal voltage is 15 V higher than the emitter terminal then the device conducts current from collector to emitter. If the gate–emitter voltage is zero or negative then the device is off. Note that the on-state and off-state are controlled by the voltage and, unlike in the power transistor, a high gate current is not required. Gate current injection is required only for a short period during turn-on to charge the gate capacitance. During turn-off, the charged energy is extracted. The required power for a gate drive circuit is much less than the equivalent base drive circuit for a power transistor and fast switching is obtained. However, IGBTs do have the good output characteristics of power transistors. The voltage and current ratings are available up to more than 1000 V and 1000 A. The on-state forward voltage drop is 1.6 to 3 V. Hence IGBTs are preferred in applications of more than 100 V.

Metal Oxide Silicon Field Effect Transistors have a longer history compared to IGBTs. The MOSFET has a fast switching speed and low on-state resistance. The three terminals are gate, drain and source. The drain and source correspond to the collector and emitter of the IGBT. The on-state voltage drop is a product of the drain–source resistance R_{on} and the drain current, so a low voltage drop is achieved with low R_{on} and low current. Hence MOSFETs are preferred in low voltage systems, e.g., 12, 24 and 48 V applications.

5.5.2 Gate drive circuit

Figure 5.20 shows the structure of a main circuit with gate drive interface circuits. A digital processor provides the inverse logic $\bar{S}_{on}$ and a logic power supply of 5 V. When $\bar{S}_{on} = 0$, the main switches are on. In the gate drive circuits a photo coupler is driven by the current loop. The inverse logic with current loop provides a fail-safe logic for the off-state in power devices in a case of accidental wire disconnection.

The gate drive circuits need isolated dc power supplies V_{g1} and V_{g2}. Let us examine the negative voltage terminal potential of V_{g1}. The potential is equal to the source voltage of S_{w1}. If the winding voltage is positive then the negative

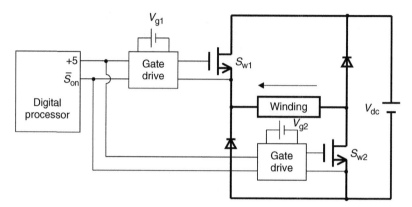

Figure 5.20 Power device drive

terminal potential is equal to V_{dc}. If the winding voltage is negative then the negative terminal potential is zero. Hence the potential voltage has changed in accordance with winding voltage. Therefore V_{g1} should be isolated using a transformer. In recent dc/dc converters the input dc power is converted to high frequency ac and applied to a primary winding in an integrated transformer to provide isolation. The output from the secondary winding of the transformer is rectified to obtain a dc voltage source. The use of a high frequency reduces the transformer size.

The isolation circuit also has frequency-dependent characteristics. In power devices, the switching speed is as fast as $0.5\,\mu s$ where the slew rate is $600\,V/\mu s$ for the case of $V_{dc} = 300\,V$. V_{g1} is supposed to maintain a constant voltage during this fast voltage transition so that excellent regulation at high frequency is required.

Figure 5.21 shows an example of a gate drive circuit and Figure 5.22 shows the operational waveforms. Twisted-pair wires are connected to the left side of the photo coupler, which has a high slew rate in common mode rejection ratio (CMRR) mode, for example $1000\,V/\mu s$ to avoid malfunction of the output stage. When $\bar{S}_{on} = 0$, current flows from the $+5\,V$ supply and through the $1\,k\Omega$ resistance and the photo diode. For the photo diode, there is an integrated circuit (IC) with a buffer that gives $+15\,V$ at the photo coupler output terminal. A complimentary transistor buffer amplifies the current; v_{tr} has an amplitude of $+15$ and $0\,V$ with respect to the negative terminal of V_{g1}. This signal is the inverse of $\bar{S}_{on}$. The gate resistor R_g is to restrict the turn-on and turn-off speed of main switching devices. If the switching speed is too fast then turn-off delays in the diodes (integrated into the power devices) can cause inrush current. The gate resistor and gate capacitance provide an RC series circuit configuration so that the gate–source voltage waveform is delayed as shown in Figure 5.22.

In power devices with low voltage and current ratings, the gate–source capacitance is rather low. In this case, the gate–source voltage can be directly driven by the photo coupler without an isolated dc power supply to reduce the cost.

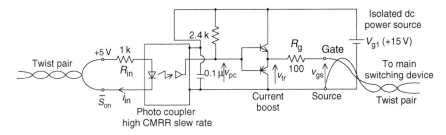

Figure 5.21 Gate drive circuit

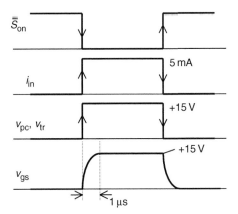

Figure 5.22 Waveforms

The gate drive circuit may be replaced by capacitors using recently developed charge pump gate driver ICs.

5.5.3 Waveform checking and isolation

Single-phase inverters with a unidirectional current are not common for inverter circuits so that these power modules are not manufactured. Therefore these circuits have to be fabricated from components. The circuits may be tested by checking the waveforms. Incorrect waveform checking using an oscilloscope may cause a fatal electric breakdown in the inverter circuits. To avoid the problem let us see how the possible failure occurs when checking the waveform.

Figure 5.23(a,b) shows two possible two fault connections for an oscilloscope when attempting to see the voltage waveforms of the switching power devices. When we want to see if the power devices are working correctly, we look at the drain–source voltage of the two switching devices. Unfortunately, once the oscilloscope probes are connected as shown in Figure 5.23(a), the switching device S_{w1} has a fatal inrush current because the source is connected to the bottom rail of the dc source via the oscilloscope. The inrush current occurs because the probe lead grounds are connected together by the oscilloscope channels and the

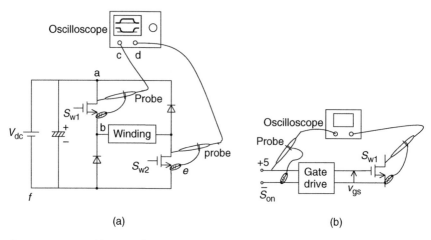

Figure 5.23 Possible typical failure: (a) failure caused by waveform checking; (b) failure caused by waveform checking

internal impedance of V_{dc} is small so that the inrush current is extremely high. This will burn out the switching device and the probe leads.

Figure 5.23(b) shows another possible connection of an oscilloscope. The on and off delay of the main switching device is being checked. Channel 1 is connected to the input terminals of gate drive circuit and channel 2 is connected to the main switching device. In this case, a failure may be avoided if isolation is provided. However, there are possible problems:

• If dc power supply of the digital processor is not isolated, V_{dc} is applied to $\bar{S}_{on}$, causing the digital processor circuit to break down.
• Even though the power supply is isolated, the output device of $\bar{S}_{on}$ may break down because the output voltage is driven by the main circuit voltage.

In order to avoid the problems associated with oscilloscope ground connections, differential probes or isolated amplifiers should be used. Differential probes allow the measured voltage to float and the ability to simply measure the potential difference between the probe terminals. Oscilloscope distributors supply this type of probes. An isolated amplifier is an alternative measurement method. A four-channel isolated amplifier provides a wide frequency range with less temperature drift of the offset voltage. Probes can be connected to the different voltage potentials and the output of the amplifier is connected to oscilloscope. For the cases in Figure 5.23 the inrush current and electric failure can be avoided by use of either of these two methods.

6

Primitive model and control strategy of bearingless motors

Akira Chiba

In this chapter, a primitive bearingless machine is introduced. This machine can generate radial forces so that it can be used as a magnetic bearing; it is used here to illustrate the principles of magnetic suspension in bearingless motors. A typical bearingless stator with 4-pole and 2-pole windings and a cylindrical rotor made of silicon steel are described.

6.1 Principles of radial force generation

Figure 6.1 shows the cross section of a primitive bearingless motor under different conditions. In Figure 6.1(a), there is a symmetrical 4-pole flux distribution. The solid curves illustrate the flux paths circulating around the four conductors $4a$, these conductors are located in the stator slots. The 4-pole flux wave Ψ_{4a} produces airgap poles in the order N, S, N and S in the airgap sections 1, 2, 3 and 4 respectively. Since the flux distribution is symmetrical, the flux density magnitudes in airgap sections 1, 2, 3 and 4 are of the same value at the same point in the pole section. There are attractive magnetic forces between the rotor poles and stator iron. The amplitudes of these attractive radial forces are the same, but the directions are equally distributed so that the sum of radial force acting on the rotor is zero.

Figure 6.1(b) shows the principle of radial force generation. Two conductors $2a$ are located in the stator slots. With the current direction as shown in the figure, a 2-pole flux wave Ψ_{2a} is generated. In airgap section 1, the flux density is increased because the direction of the 4-pole and 2-pole fluxes is the same. However, in airgap section 3, the flux density is decreased because the direction of these fluxes is opposite. The magnetic forces in the airgap sections 1 and 3

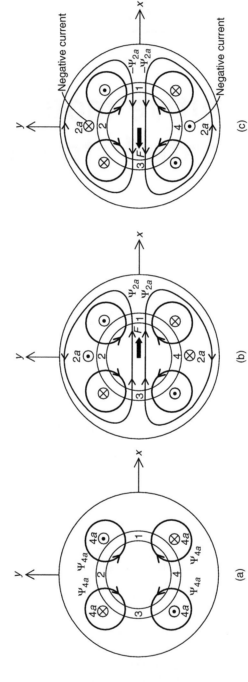

Figure 6.1 Principles of radial force generation: (a) 4-pole symmetrical flux; (b) x-direction radial force; (c) negative x-direction force

are no longer equal, i.e., the force in airgap 1 is larger than in airgap 3. Hence a radial force F results in the x-axis direction. It follows that the amplitude of the radial force increases as the current value in conductors $2a$ increases.

Figure 6.1(c) shows how a negative radial force in the x-axis direction is generated. The current in conductors $2a$ is reversed so that the flux density in airgap section 1 now decreases while that in airgap section 3 increases. Hence the magnetic force in airgap section 3 is larger than that in airgap section 1, producing a radial force in the negative x-axis direction.

Figure 6.2 shows radial force generation in the y-axis direction. Two conductors $2b$, which have an MMF centred on the y-axis, are added to the stator. A similar flux density imbalance occurs but this time between airgap sections 4 and 2, hence producing a force on the y axis. The polarity of the current will dictate the direction of the force.

These are the principles of radial force generation in x- and y-axis directions. The force values are almost proportional to the current in conductors $2a$ and $2b$ (assuming constant 4-pole current). The vector sum of these two perpendicular radial forces can produce a radial force in any desired direction and with any amplitude.

Figure 6.3 shows the flux plots for a primitive bearingless motor. In Figure 6.3(a), a symmetrical 4-pole flux distribution is seen with an excitation of the 4-pole winding. The 4-pole conductors are placed in stator slots as shown. Only the u-phase conductors are shown to represent current density, although the other windings are simultaneously excited using a symmetrical 3-phase current set. In Figure 6.3(b) the 2-pole 3-phase winding is also excited so that the flux distribution is unbalanced, which results in a radial force in the x-axis direction.

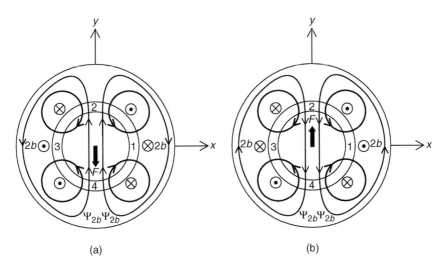

(a)   (b)

Figure 6.2 y-Direction radial force: (a) negative y-direction radial force; (b) y-direction radial force

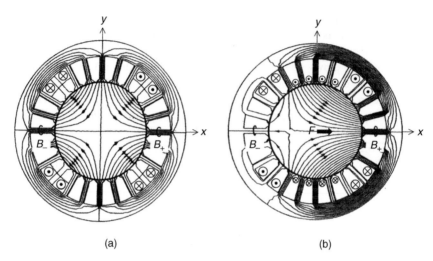

(a)   (b)

Figure 6.3 Radial force generation in a primitive bearingless motor: (a) symmetrical 4-pole flux distribution; (b) radial force generation

Figure 6.4 shows another example of a surface-mounted permanent magnet (SPM) bearingless motor. The permanent magnets are magnetized to generate a 4-pole flux distribution. In Figure 6.4(a) a symmetrical 4-pole flux distribution is seen. In Figure 6.4(b) the 2-pole winding is excited. With the superimposed 2-pole flux wave, the flux distribution is not symmetrical; the flux density in permanent magnet A is high while in C it is low. Hence a radial force is generated in the x-axis direction.

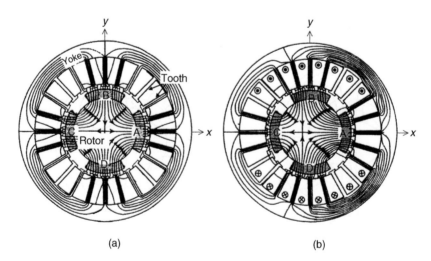

(a)   (b)

Figure 6.4 Radial force generation in a 4-pole permanent magnet machine: (a) symmetrical flux distribution; (b) radial force generation

6.2 Two-pole bearingless motor

In the previous section the 4-pole winding is the main motor winding so that the 2-pole winding is used as the suspension winding in order to generate radial forces. It is possible to change the role of the windings. In this section a bearingless motor with a 2-pole motor winding is shown. This 2-pole configuration is useful in induction type bearingless motors.

Figure 6.5 shows the radial force generation. In Figure 6.5(a) the 2-pole conductors $2a$ are excited as a motor winding so that a strong 2-pole flux distribution is produced. The 4-pole conductors $4a$ produce a weak 4-pole flux wave. The flux density in airgap section 1 is increased, but decreased in airgap section 3. Hence radial force is generated in the x-axis direction. Negative-direction radial force can be generated by reversing the current in conductors $4a$.

Figure 6.5(b) shows the radial force generation in the y-axis direction. The conductors $4b$, which are electrically perpendicular (which is 45 mechanical deg) to conductors $4a$, are excited. A 4-pole flux wave is generated. This flux wave is superimposed on the strong motor flux Ψ_{2a} so that the flux density in airgap sections $1'$ and $2'$ is increased whereas it is decreased in airgap sections $3'$ and $4'$. Therefore radial force is generated in the y-axis direction; a negative current in conductors $4b$ will generate negative y-axis radial force.

The flux distribution in Figure 6.5(a) is similar to the one in the previous section (Figure 6.4(a)). The radial force is generated by the flux density variation in the magnetic poles. However, in Figure 6.5(b), the flux distribution is slightly different. The airgap sections $1'$, $2'$, $3'$ and $4'$ are not exactly positioned at the centre of the 2-pole flux wave, so if a salient-pole rotor is employed, the generated radial force is reduced in Figure 6.5(b) compared with that in Figure 6.5(a). Hence a 2-pole motor drive is suitable for a machine with a cylindrical rotor,

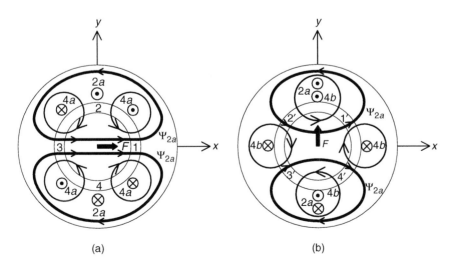

(a)   (b)

Figure 6.5   2-Pole bearingless motor: (a) x-direction radial force; (b) y-direction radial force

like an induction and SPM machines. In the case of a salient-pole synchronous machine, the unsymmetrical characteristics can be compensated in a decoupling controller, which results in unsymmetrical current injection. The 2-pole drive is also suitable for large airgap structures.

6.3 MMF and permeance

In this section, the MMF of a 2-phase winding is expressed as a sinusoidal function with an approximation to the fundamental component. The permeance variations caused by airgap length variation due to eccentric rotor displacement are derived to provide a basis for inductance-variation and radial-force expressions for use in further sections. The following assumptions are made:

(a) The MMF spatial distribution is approximated to the fundamental component.
(b) Magnetic saturation is neglected.
(c) Airgap permeance distribution is smooth. Stator slot harmonics are neglected. The rotor and stator surfaces are cylindrical and smooth.
(d) The magnetic reluctance of the iron core is negligible. The iron permeability is infinite.
(e) Eccentric rotor displacement is small with respect to the airgap length between the rotor surface and stator inner surface. This displacement is also small compared to the rotor radius.

Figure 6.6 shows the winding arrangement for a primitive bearingless motor. Two sets of 2-phase windings are wound on the stator. The 4-pole windings are 4a and 4b and the 2-pole windings are 2a and 2b. The positive current directions of each winding are shown in the figure. Note that current directions in the 2a and 2b windings are arranged such that the MMF directions are aligned on the x- and y-axis directions respectively. The 4a winding is arranged so that the MMF direction in airgap 1 is also aligned to the x-axis and the 4b winding is arranged to be perpendicular to the 4a winding in electrical terms with a phase-lead angle of 90 electrical deg (45 mechanical deg) in the counter-clockwise direction. If co-sinusoidal and sinusoidal currents, with a frequency of $2f$, are supplied to the 4a and 4b windings then a magnetic field, revolving in a counter-clockwise direction with speed f, is generated. For the case of the 2a and 2b windings, with co-sinusoidal and sinusoidal currents with frequency f, the rotational direction is also counter-clockwise with a rotational speed of f. The flux lines show the case for a point in time when there are positive currents in 2a and 4a and zero in 2b and 4b, which results in force in the x-axis.

The MMF space distributions for the 4a, 4b, 2a and 2b windings with unity current can be written as

$$A_{4a} = N_4 \cos(2\phi_s) \tag{6.1}$$

$$A_{4b} = N_4 \sin(2\phi_s) \tag{6.2}$$

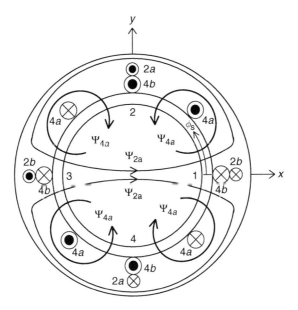

Figure 6.6   4-Pole and 2-pole winding arrangements

$$A_{2a} = N_2 \cos(\phi_s) \tag{6.3}$$

$$A_{2b} = N_2 \sin(\phi_s) \tag{6.4}$$

where N_4 and N_2 are the amplitudes of the fundamental component of MMF distribution. The angular coordinate ϕ_s is a counter-clockwise rotational angular position starting from the x-axis. Note that the positive direction of MMF is defined in the radial direction from the rotor to the stator. For example, a flux Ψ_{4a} generated by a positive dc current in the $4a$ winding is positive because it flows from the rotor to the stator teeth in airgap section 1, where $\phi_s = 0$.

Figure 6.7(a) shows a rotor having an eccentric displacement from the stator centre. The perpendicular x- and y-axes are fixed to the stator centre. The rotor centre is displaced in the positive direction along the x- and y-axes. Figure 6.7(b) shows an enlarged figure illustrating the coordinate system. Let us assume that the airgap length between the rotor and stator is g_0 when the rotor is centred in the stator bore. If the rotor displacements are x and y then the airgap length g between the rotor and the stator is

$$g = g_0 - x \cos(\phi_s) - y \sin(\phi_s) \tag{6.5}$$

The inverse of the airgap length can be written with an assumption that the displacements are small compared to the nominal airgap length g_0 so that

$$\frac{1}{g} = \frac{1}{g_0} \left(1 + \frac{x}{g_0} \cos \phi_s + \frac{y}{g_0} \sin \phi_s \right)$$

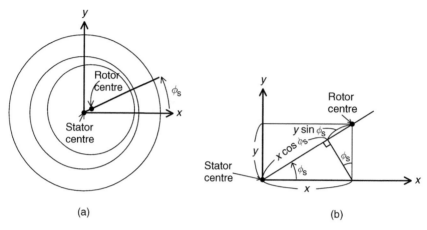

Figure 6.7 Airgap length variation with rotor eccentric displacement: (a) coordinate definition; (b) stator and rotor centres

The permeance P_0 at an angular position ϕ_s is

$$P_0(\phi_s) = \frac{\mu_0 Rl}{g_0}\left(1 + \frac{x}{g_0}\cos\phi_s + \frac{y}{g_0}\sin\phi_s\right) \qquad (6.6)$$

where R and l are the rotor radius and axial length.

Example 6.1 Figure 6.8 shows an alternative representation of the winding and coordinate arrangement. Note that the polarity of winding $4a$ is reversed. The definition of the stationary angular position ϕ_s is defined in the clockwise

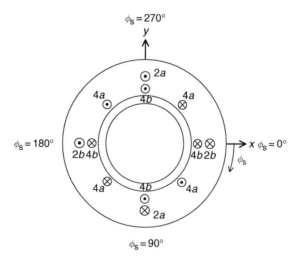

Figure 6.8 Alternative winding and coordinate arrangement

direction and the rotor also has clockwise rotation. As shown in Figure 6.6, the x- and y-axes are defined and $2a$ and $2b$ are arranged to generate MMF on these axes. Windings $4a$ and $2a$ are advanced by 90 electrical deg in the clockwise direction with respect to $4b$ and $2b$. This arrangement of the windings and coordinates will be used in the examples throughout this chapter. Find the equations corresponding to equations (6.1–6.4). Also show the airgap length distribution and permeance.

Answer
The MMFs for unity current are

$$\left.\begin{array}{l} A_{4a} = N_4 \cos(2\phi_s) \\ A_{4b} = -N_4 \sin(2\phi_s) \\ A_{2a} = N_2 \cos(\phi_s) \\ A_{2b} = -N_2 \sin(\phi_s) \end{array}\right\} \tag{6.7}$$

The airgap length is written as

$$g = g_0 - x\cos(\phi_s) + y\sin(\phi_s) \tag{6.8}$$

Note that the sign of y is positive because the ϕ_s direction is reversed. Therefore the permeance distribution is

$$P_0 = \frac{\mu_0 R l}{g_0}\left(1 + \frac{x}{g_0}\cos\phi_s - \frac{y}{g_0}\sin\phi_s\right) \tag{6.9}$$

6.4 Magnetic potential and flux distribution

In this section, the airgap flux distribution is derived using the MMF and permeance distribution. In the calculation process it is quite important to assume that the magnetic potential of the rotor is not zero because flux distribution is unsymmetrical when a rotor has radial displacement.

Figure 6.9 shows a simplified magnetic circuit and the electrical equivalent circuit. In Figure 6.9(a), a cylindrical iron rotor and a surrounding iron stator are shown. On the stator, conductors $2a$ are located in slots. These conductors are connected in series to form a winding $2a$. With unity current in the winding, a sinusoidal space MMF distribution is generated.

In Figure 6.9(b), an electrical equivalent circuit is drawn. The conductances represent the airgap permeances given by (6.6). The conductance values are not equal; rather the values are dependent on the rotor radial displacement and angular position ϕ_s. In series with the airgap permeances are dc voltage sources. These voltage sources represent the MMF of winding $2a$. The voltage values are half of A_{2a} as given in (6.3). The voltage sources are not included at $\phi_s = 90$ and 270 because A_{2a} is zero. The values of the voltage sources are also dependent

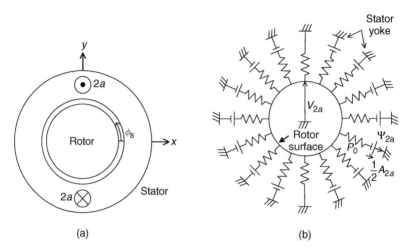

Figure 6.9 Magnetic potential and equivalent circuit: (a) magnetic circuit; (b) equivalent electrical circuit

on ϕ_s. The ground symbols connected to the voltage sources indicate that the stator yoke magnetic potential is assumed to be zero and that the stator tooth and yoke reluctances are also assumed to be zero.

The ring circuit connecting all conductances represents the rotor surface permeance, which is assumed to be infinite. The voltage V_{2a} is the magnetic potential of the rotor. In total, 16 branches of conductances and voltage sources are drawn in the figure to illustrate the calculation concept; however, the mathematical calculation can be processed using a large number of branches since it is really a distributed network. The current values in the branches correspond to the airgap flux. Hence the flux Ψ_{2a} in one branch is written as

$$\Psi_{2a} = P_0 \left(\frac{1}{2} A_{2a} + V_{2a} \right) \qquad (6.10)$$

An integral of the flux around the rotor surface should be zero according to Gauss' law, so the following equation is obtained

$$\int_0^{2\pi} \Psi_{2a} \mathrm{d}\phi_s = 0 \qquad (6.11)$$

Substituting (6.10) into (6.11) and solving for the rotor magnetic potential results in

$$V_{2a} = -\frac{\frac{1}{2} \int_0^{2\pi} P_0 A_{2a} \mathrm{d}\phi_s}{\int_0^{2\pi} P_0 \mathrm{d}\phi_s} \qquad (6.12)$$

Substituting (6.3) and (6.6) into this equation and solving the integral yields a simplified result for the rotor magnetic potential:

$$V_{2a} = -\frac{N_2}{4g_0}x \qquad (6.13)$$

From this equation, we can observe the following:

(a) The magnetic potential is zero if the rotor is positioned at the stator centre.
(b) When the rotor has x-axis displacement, the rotor magnetic potential is proportional to the displacement.
(c) The airgap flux distribution is no longer symmetrical as shown in (6.10).

Note that the rotor magnetic potential calculation is quite important in the analysis of bearingless motors. If zero magnetic potential is assumed as in conventional motor analysis then an unsymmetrical flux distribution is not correctly defined and the radial force calculation is inaccurate.

In a similar manner, a magnetic potential also exists when exciting the winding 2b with unity current:

$$V_{2b} = -\frac{N_2}{4g_0}y \qquad (6.14)$$

where the rotor magnetic potential is proportional to the y-axis rotor displacement.

It is possible to derive the magnetic potentials that excite the 4-pole windings 4a and 4b. In these cases the resulting magnetic potential is zero and does not depend on the rotor radial displacement. This can be simply explained by way of example: 2a in (6.12) is substituted by 4a, which is a periodic function of $2\phi_s$. The permeance given in (6.6) is a periodic function of ϕ_s; therefore the integration of this product is zero.

The airgap flux distributions produced by windings 2a, 2b, 4a and 4b when excited by a unity current are

$$\Psi_{2a} = P_0\left(\frac{1}{2}A_{2a} - \frac{N_2x}{4g_0}\right) \qquad (6.15)$$

$$\Psi_{2b} = P_0\left(\frac{1}{2}A_{2b} - \frac{N_2y}{4g_0}\right) \qquad (6.16)$$

$$\Psi_{4a} = P_0\left(\frac{1}{2}A_{4a}\right) \qquad (6.17)$$

$$\Psi_{4b} = P_0\left(\frac{1}{2}A_{4b}\right) \qquad (6.18)$$

The rotor magnetic potential is not zero when the 2-pole winding is excited with a radial displacement.

Example 6.2 Derive rotor magnetic potentials using the alternative coordinates shown in the previous example.

Answer
The definition of rotor magnetic potential in (6.12) is not dependent on the winding arrangement and angular position. Thus, the denominator in (6.12) is

$$
\int_0^{2\pi} \frac{\mu_0 Rl}{g_0} \left(1 + \frac{x}{g_0}\cos\phi_s - \frac{y}{g_0}\sin\phi_s\right) d\phi_s = 2\pi\frac{\mu_0 Rl}{g_0}
$$

and the numerator is

$$
\frac{1}{2}\int_0^{2\pi} \frac{\mu_0 Rl}{g_0} \left(1 + \frac{x}{g_0}\cos\phi_s - \frac{y}{g_0}\sin\phi_s\right) N_2\cos\phi_s d\phi_s = \frac{\pi}{2}\frac{\mu_0 Rl}{g_0^2}N_2 x
$$

Therefore we get exactly the same expression as (6.13) for the rotor magnetic potential. In the calculation of the y-axis rotor potential, the same expression as (6.14) is obtained.

6.5 Inductance matrix

In this section, the inductance matrix is derived using the airgap flux distributions, as found in the previous section. Some elements of the inductance matrix are shown to be a function of the radial rotor displacement.

Let us suppose that the flux linkages of windings 4a, 4b, 2a and 2b are λ_{4a}, λ_{4b}, λ_{2a} and λ_{2b} respectively. Let us also suppose that instantaneous currents of windings 4a, 4b, 2a and 2b are i_{4a}, i_{4b}, i_{2a} and i_{2b} respectively. The flux linkage and current relationships are expressed in a matrix form as

$$
\begin{bmatrix} \lambda_{4a} \\ \lambda_{4b} \\ \lambda_{2a} \\ \lambda_{2b} \end{bmatrix} = \begin{bmatrix} L_{4a} & M_{4a4b} & M_{4a2a} & M_{4a2b} \\ M_{4a4b} & L_{4b} & M_{4b2a} & M_{4b2b} \\ M_{4a2a} & M_{4b2a} & L_{2a} & M_{2a2b} \\ M_{4a2b} & M_{4b2b} & M_{2a2b} & L_{2b} \end{bmatrix} \begin{bmatrix} i_{4a} \\ i_{4b} \\ i_{2a} \\ i_{2b} \end{bmatrix} \tag{6.19}
$$

The inductances defined in the above matrix form can be derived by integration of the product of airgap flux and winding distribution such that

$$
L_{4a} = \frac{1}{2}\int_0^{2\pi} \Psi_{4a}A_{4a}d\phi_s \tag{6.20}
$$

$$
L_{4b} = \frac{1}{2}\int_0^{2\pi} \Psi_{4b}A_{4b}d\phi_s \tag{6.21}
$$

$$M_{4a4b} = \frac{1}{2} \int_0^{2\pi} \Psi_{4b} A_{4a} d\phi_s \tag{6.22}$$

$$L_{2a} = \frac{1}{2} \int_0^{2\pi} \Psi_{2a} A_{2a} d\phi_s \tag{6.23}$$

$$L_{2b} = \frac{1}{2} \int_0^{2\pi} \Psi_{2b} A_{2b} d\phi_s \tag{6.24}$$

$$M_{2a2b} = \frac{1}{2} \int_0^{2\pi} \Psi_{2a} A_{2b} d\phi_s \tag{6.25}$$

$$M_{4a2a} = \frac{1}{2} \int_0^{2\pi} \Psi_{2a} A_{4a} d\phi_s \tag{6.26}$$

$$M_{4b2a} = \frac{1}{2} \int_0^{2\pi} \Psi_{2a} A_{4b} d\phi_s \tag{6.27}$$

$$M_{4a2b} = \frac{1}{2} \int_0^{2\pi} \Psi_{2b} A_{4a} d\phi_s \tag{6.28}$$

$$M_{4b2b} = \frac{1}{2} \int_0^{2\pi} \Psi_{2b} A_{4b} d\phi_s \tag{6.29}$$

Substituting (6.1–6.4), (6.15–6.18) and (6.6) into the above equations and solving the integrations result in a simple mathematical form. Denoting them in a matrix form

$$\begin{bmatrix} L_{4a} & M_{4a4b} \\ M_{4a4b} & L_{4b} \end{bmatrix} = \frac{\pi \mu_0 R l N_4^2}{4g_0} \begin{bmatrix} 1 & 0 \\ 0 & 1 \end{bmatrix} \tag{6.30}$$

$$\begin{bmatrix} L_{2a} & M_{2a2b} \\ M_{2a2b} & L_{2b} \end{bmatrix} = \frac{\pi \mu_0 R l N_2^2}{4g_0} \begin{bmatrix} 1 & 0 \\ 0 & 1 \end{bmatrix} \tag{6.31}$$

$$\begin{bmatrix} M_{4a2a} & M_{4a2b} \\ M_{4b2a} & M_{4b2b} \end{bmatrix} = \frac{\pi \mu_0 R l N_2 N_4}{8g_0^2} \begin{bmatrix} x & -y \\ y & x \end{bmatrix} \tag{6.32}$$

From the above-calculated results, one can see the following:

(a) The self-inductances L_{4a}, L_{4b}, L_{2a} and L_{2b} are a product of airgap permeability, axial length, rotor radius and winding turns, and the inverse of the airgap length.
(b) The mutual inductance M_{4a4b} between the 4-pole windings is zero because windings 4a and 4b are perpendicular to each other. The same is true for M_{2a2b}.

(c) The mutual inductances $M_{4a2a}, M_{4b2a}, M_{4a2b}$ and M_{4b2b} represent the coupling between the 4-pole and 2-pole windings. These mutual inductances are proportional to the rotor radial displacements x and y.

(d) These mutual inductances are zero when the rotor is positioned at the centre. This fact is easily understood because the 4-pole and 2-pole windings are symmetrically wound and are not coupled when the rotor is centred. If these inductances are zero there is no induced voltage in the 2-pole winding when a 4-pole revolving magnetic field is generated. On the other hand, no induced voltage appears at the 4-pole terminals when the 2-pole winding current generates a 2-pole revolving magnetic field. Therefore the voltage requirement for the suspension winding is low.

(e) Radial force is associated with the radial-displacement-dependent inductance terms (6.32) because they represent an imbalance in stored magnetic energy in the airgap. The derivatives (with respect to the radial displacement) of (6.32) are constant, producing constant radial force gains.

Figure 6.10 shows an example of the measured inductances with respect to the radial rotor position. In Figure 6.10(a), the inductances are shown when the rotor is displaced in the x-axis direction; M_{4a2a} and M_{4b2b} are proportional to the rotor displacement x and the mutual inductances M_{4b2a} and M_{4a2b} are almost zero. This corresponds to (6.32). In Figure 6.10(b) it is the mutual inductances M_{4b2a} and M_{4a2b} that are almost proportional to the displacement y. The airgap length of the measured test machine is 0.4 mm. Good linearity is seen up to about half of the airgap length. It is interesting to see this linearity because the equations are

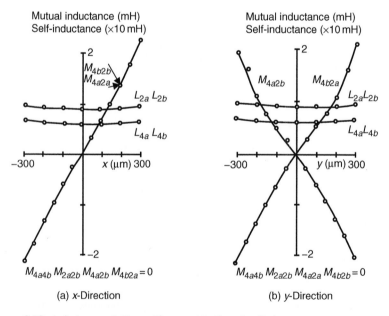

Figure 6.10 Inductance variations with respect to the rotor displacements

derived with an assumption that the rotor radial displacement is small compared to the airgap length.

The self-inductances are constant according to (6.30) and (6.31); however, these inductances have a slight minimum value at the centre position. The approximation used to obtain (6.6) neglects the second and higher order terms of radial displacements and there is a small variation with eccentricity in the constant permeance term in (6.6), which is neglected.

Example 6.3   Derive the mutual inductance matrix between the 4-pole and 2-pole windings for the alternative winding and coordinate system definitions shown in Example 6.1.

Answer
According to the definition in (6.26) M_{4a2a} can be derived for the alternative winding and coordinate system definition as

$$M_{4a2a} = \frac{1}{2} \int_0^{2\pi} \Psi_{2a} A_{4a} d\phi_s$$

$$= \frac{1}{2} \int_0^{2\pi} \left\{ \frac{\mu_0 Rl}{g_0} \left(1 + \frac{x}{g_0} \cos\phi_s - \frac{y}{g_0} \sin\phi_s \right) \right.$$

$$\left. \times \left(\frac{1}{2} N_2 \cos\phi_s - \frac{N_2 x}{4g_0} \right)(-N_4)\cos 2\phi_s \right\} d\phi_s \qquad (6.33)$$

$$= \frac{\pi \mu_0 Rl N_2 N_4}{8g_0^2} x$$

A similar calculation can be done for the other three elements so that the mutual inductances are derived as

$$\begin{bmatrix} M_{4a2a} & M_{4a2b} \\ M_{4b2a} & M_{4b2b} \end{bmatrix} = \frac{\pi \mu_0 Rl N_2 N_4}{8g_0^2} \begin{bmatrix} -x & y \\ y & x \end{bmatrix} \qquad (6.34)$$

6.6 Radial force and current

In this section the radial forces are derived from partial derivatives of the stored magnetic energy where the stored magnetic energy is calculated from the inductance functions.

In the previous section the inductance functions are derived mathematically. Based on these functions, the relationships between flux linkages and winding currents can be written as

$$\begin{bmatrix} \lambda_{4a} \\ \lambda_{4b} \\ \lambda_{2a} \\ \lambda_{2b} \end{bmatrix} = \begin{bmatrix} L_4 & 0 & M'x & -M'y \\ 0 & L_4 & M'y & M'x \\ M'x & M'y & L_2 & 0 \\ -M'y & M'x & 0 & L_2 \end{bmatrix} \begin{bmatrix} i_{4a} \\ i_{4b} \\ i_{2a} \\ i_{2b} \end{bmatrix} \qquad (6.35)$$

where L_4 and L_2 are constants, i.e., the self-inductances. M' is the derivative of mutual inductance with respect to the rotor radial displacement. Let us define a 4×4 inductance matrix as $[L]$ and define the current vector as $[i]$. Then the stored magnetic energy W_m is given by

$$W_m = \frac{1}{2}[i]^t[L][i] \tag{6.36}$$

Expansion of the matrix results in

$$\begin{aligned}
W_m = \frac{1}{2}L_4 i_{4a}^2 + \frac{1}{2}L_4 i_{4b}^2 + \frac{1}{2}L_2 i_{2a}^2 + \frac{1}{2}L_2 i_{2b}^2 \\
+ M' x\, i_{4a} i_{2a} - M' y\, i_{4a} i_{2b} + M' y\, i_{4b} i_{2a} + M' x\, i_{4b} i_{2b}
\end{aligned} \tag{6.37}$$

The first four terms represent the stored magnetic energy associated with the self-inductances. The last four terms are the energy terms of the mutual inductances between the 4-pole and 2-pole windings. The radial forces can be derived from the partial derivatives of the stored magnetic energy, assuming a magnetically linear system where

$$\begin{bmatrix} F_x \\ F_y \end{bmatrix} = \begin{bmatrix} \dfrac{\partial W_m}{\partial x} \\[2mm] \dfrac{\partial W_m}{\partial y} \end{bmatrix} \tag{6.38}$$

Substituting (6.37) into (6.38) yields a simple mathematical expression because the first line in (6.37) disappears since the first 4 terms are not functions of radial displacement:

$$\begin{bmatrix} F_x \\ F_y \end{bmatrix} = M' \begin{bmatrix} i_{4a} i_{2a} & +i_{4b} i_{2b} \\ -i_{4a} i_{2b} & +i_{4b} i_{2a} \end{bmatrix} \tag{6.39}$$

The radial forces are expressed as a sum of the current products between the 4-pole and 2-pole windings. The radial forces are proportional to the derivative of mutual inductance M'. This equation can be also written in matrix forms:

$$\begin{bmatrix} F_x \\ F_y \end{bmatrix} = M' \begin{bmatrix} i_{4a} & i_{4b} \\ i_{4b} & -i_{4a} \end{bmatrix} \begin{bmatrix} i_{2a} \\ i_{2b} \end{bmatrix} \tag{6.40}$$

$$\begin{bmatrix} F_x \\ F_y \end{bmatrix} = M' \begin{bmatrix} i_{2a} & i_{2b} \\ -i_{2b} & i_{2a} \end{bmatrix} \begin{bmatrix} i_{4a} \\ i_{4b} \end{bmatrix} \tag{6.41}$$

In (6.40) the radial forces are expressed as a product of the 2×2 matrix of the 4-pole winding currents and the 2-pole winding current vector. This expression is quite useful when i_{4a} and i_{4b} are assigned as the motor currents and i_{2a} and i_{2b} as the suspension winding currents. On the other hand, in (6.41) the 4-pole current vector expresses radial forces so that this equation is useful when the 4-pole winding is assigned as the suspension winding.

Let us assume that the 4-pole winding is assigned as the motor winding so that the 2-phase sinusoidal currents are

$$i_{4a} = I_4 \cos 2\omega t, \ i_{4b} = I_4 \sin 2\omega t \tag{6.42}$$

and (6.40) is rewritten as

$$\begin{bmatrix} F_x \\ F_y \end{bmatrix} = M' I_4 \begin{bmatrix} \cos 2\omega t & \sin 2\omega t \\ \sin 2\omega t & -\cos 2\omega t \end{bmatrix} \begin{bmatrix} i_{2a} \\ i_{2b} \end{bmatrix} \tag{6.43}$$

From this equation, the following points can be made:

(a) Relationships between the radial forces and suspension winding currents are modulated by the revolving motor magnetic field.
(b) The radial forces are proportional to the motor excitation current I_4.
(c) The radial forces are proportional to the radial force winding currents i_{2a} and i_{2b}.

In comparison to the magnetic bearing, a bias current is not needed in the suspension windings of a bearingless motor. The bias current is replaced by an excitation current I_4 in the motor windings. Hence the current requirement in the radial force winding and the driving inverter is proportional to the required radial force. This leads to a lower current requirement in the suspension windings.

In addition, the radial forces are naturally proportional to the suspension winding currents, even without linearization around the operation point (as usually employed in the magnetic bearing using a bias current). It can be said that bearingless motors have inherently linear characteristics due to differential suspension winding operation.

Let us examine a case when sinusoidal currents are used in the 2-pole windings to generate a revolving magnetic field. In this case the 2-pole windings are assigned as the motor windings and the 4-pole windings are used as the suspension windings. For counter-clockwise rotor rotation the currents are

$$i_{2a} = I_2 \cos \omega t, \ i_{2b} = I_2 \sin \omega t \tag{6.44}$$

so that (6.41) is written as

$$\begin{bmatrix} F_x \\ F_y \end{bmatrix} = M' I_2 \begin{bmatrix} \cos \omega t & \sin \omega t \\ -\sin \omega t & \cos \omega t \end{bmatrix} \begin{bmatrix} i_{4a} \\ i_{4b} \end{bmatrix} \tag{6.45}$$

From this equation the following points are apparent:

(a) The radial forces are proportional to the 4-pole winding currents i_{4a} and i_{4b}. Hence it is seen that the radial forces are proportional to the radial force currents in both the 4-pole and 2-pole motor cases.
(b) The radial forces are modulated by sinusoidal and co-sinusoidal functions. The frequency of the 2×2 matrix is half that in (6.43) because the modulation frequency is determined by the revolving speed of the magnetic field. In (6.45) the magnetic field has one mechanical rotation for one current cycle since the motor windings are 2-pole. However, two current cycles are required for one mechanical rotation when the motor windings are 4-pole, as in the case of (6.43).

Example 6.4 Derive the magnetic energy and radial force equations for the alternative winding and coordinate system definitions. Also find expressions for the current for clockwise rotation of the magnetic field.

Answer
In the 4×4 matrix shown in (6.35), negative signs are seen in elements (1,4) and (4,1). In the alternative definition, these signs are in elements (1,3) and (3,1) only. The stored magnetic energy can be calculated. The result is composed of self-inductance and mutual inductance functions. The self-inductance energy is obtained from the first line in (6.37) and the mutual inductance energy is

$$M'(-x\,i_{4a}i_{2a} + y\,i_{4a}i_{2b} + y\,i_{4b}i_{2a} + x\,i_{4b}i_{2b}) \tag{6.46}$$

so that the radial forces can be written in matrix forms such that

$$\begin{bmatrix} F_x \\ F_y \end{bmatrix} = M' \begin{bmatrix} -i_{4a} & i_{4b} \\ i_{4b} & i_{4a} \end{bmatrix} \begin{bmatrix} i_{2a} \\ i_{2b} \end{bmatrix} \tag{6.47}$$

$$\begin{bmatrix} F_x \\ F_y \end{bmatrix} = M' \begin{bmatrix} -i_{2a} & i_{2b} \\ i_{2b} & i_{2a} \end{bmatrix} \begin{bmatrix} i_{4a} \\ i_{4b} \end{bmatrix} \tag{6.48}$$

Suppose the 4-pole winding is used as the motor winding. In order to generate a revolving magnetic field in the clockwise direction, the current should be as given in (6.42). Note that a clockwise-revolving magnetic field is generated because the polarity winding $4a$ is reversed. Hence the radial forces are written as

$$\begin{bmatrix} F_x \\ F_y \end{bmatrix} = M'I_4 \begin{bmatrix} -\cos 2\omega t & \sin 2\omega t \\ \sin 2\omega t & \cos 2\omega t \end{bmatrix} \begin{bmatrix} i_{2a} \\ i_{2b} \end{bmatrix} \tag{6.49}$$

Note the negative sign in element (1,1).
If the 2-pole winding is used as the motor winding and the revolving magnetic field is in the clockwise direction then

$$i_{2a} = I_2 \cos \omega t, \quad i_{2b} = -I_2 \sin \omega t$$

Hence the radial forces are

$$\begin{bmatrix} F_x \\ F_y \end{bmatrix} = M'I_2 \begin{bmatrix} -\cos \omega t & -\sin \omega t \\ -\sin \omega t & \cos \omega t \end{bmatrix} \begin{bmatrix} i_{4a} \\ i_{4b} \end{bmatrix} \tag{6.50}$$

6.7 The dc excitation of the primitive bearingless motor

In this section a dc current is used to excite the motor winding in the primitive bearingless motor. Since dc current generates only a static magnetic field, there is no rotational torque production. Hence it acts as a radial magnetic bearing. A basic controller configuration is introduced. It is shown that the system has good characteristics and can operate as a 3-phase inverter-driven magnetic bearing.

Let us consider that one of the 4-pole winding currents i_{4a} is set to dc current I_4, while i_{4b} is zero. This current value provides the condition at $\omega t = 0$ in (6.42) so that the relationships between the radial forces and suspension winding currents are given by (6.43); substituting $\omega t = 0$ gives

$$\begin{bmatrix} F_x \\ F_y \end{bmatrix} = M' I_4 \begin{bmatrix} 1 & 0 \\ 0 & -1 \end{bmatrix} \begin{bmatrix} i_{2a} \\ i_{2b} \end{bmatrix} \tag{6.51}$$

This equation can be simplified to

$$\begin{bmatrix} F_x \\ F_y \end{bmatrix} = M' I_4 \begin{bmatrix} i_{2a} \\ -i_{2b} \end{bmatrix} \tag{6.52}$$

This equation shows the simple relationships between radial force and winding current for a condition shown in the Figures 6.1 and 6.2. Note that the radial force direction in Figure 6.2(a) is negative corresponding to the negative sign in the above equation.

Using the relationships in (6.52) we can design a controller block diagram. In a radial magnetic suspension system the radial force commands of the two perpendicular axes are given by the position controllers, e.g. the PID controllers. Current commands are generated from the controller so that the actual radial forces precisely follow the force commands. Suppose that the radial force commands are generated on the x- and y-axes as F_x^* and F_y^*. The current commands i_{2a}^* and i_{2b}^* should be generated using the above equation so that

$$\begin{bmatrix} i_{2a}^* \\ i_{2b}^* \end{bmatrix} = \frac{1}{M' I_4} \begin{bmatrix} F_x^* \\ -F_y^* \end{bmatrix} \tag{6.53}$$

The derivative of mutual inductance M' is a constant determined by the machine dimensions and number of winding turns. The excitation dc current I_4 can be kept constant. Thus a simple structure of a radial magnetic suspension system can be designed as illustrated in Figure 6.11.

Figure 6.11 shows a primitive bearingless motor with 4-pole and 2-pole windings. Two shaft displacement sensors are installed to detect the shaft movement in two perpendicular axes. The detected radial positions x and y are compared with these references x^* and y^*, which are zero in most cases, so that radial

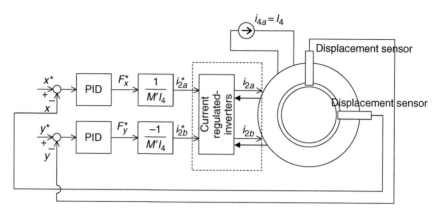

Figure 6.11 DC-excited primitive bearingless motor

position errors are amplified by the radial position controllers as shown by the PID blocks. Radial force commands F_x^* and F_y^* are generated, and from these radial force commands the suspension winding current commands i_{2a}^* and i_{2b}^* are generated by division by the constant $M'I_4$. Note that a minus sign is seen in the y-axis block. The current commands are fed to an inverter to regulate instantaneous current values, which are forced to follow the commands so that radial forces are generated in the bearingless motor. The radial forces therefore track the radial force commands.

In the motor winding, a dc current I_4 is supplied to the a-phase winding. The supplied dc current generates a static 4-pole flux distribution. This static field acts as the bias flux as seen in the conventional radial magnetic bearing.

Although 2-phase windings are described for simplicity, 3-phase windings with a 3-phase inverter are usually used so that a 3-phase radial magnetic bearing can be constructed. The 2-phase coordinates can be easily transformed into 3-phase coordinates, as often done in standard electrical motors. This concept is briefly introduced below; however, the details will be shown in the next chapter. The 3-phase and 2-phase current relationships can be related to each other using the matrix

$$\begin{bmatrix} i_{2u}^* \\ i_{2v}^* \\ i_{2w}^* \end{bmatrix} = [C_{32}]^t \begin{bmatrix} i_{2a}^* \\ i_{2b}^* \end{bmatrix} \tag{6.54}$$

where $[C_{32}]^t$ is the connection matrix given by

$$[C_{32}]^t = \frac{\sqrt{2}}{\sqrt{3}} \begin{bmatrix} 1 & 0 \\ -\frac{1}{2} & \frac{\sqrt{3}}{2} \\ -\frac{1}{2} & -\frac{\sqrt{3}}{2} \end{bmatrix} \tag{6.55}$$

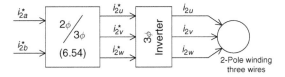

Figure 6.12   3-Phase radial magnetic bearing

Note that the 3-phase windings have only two degrees of freedom because the sum of the three currents should be zero. Hence the 3-phase currents are directly derived from the 2-phase instantaneous current values. The blocks shown in Figure 6.12 with 3-phase windings can replace the dotted area in Figure 6.11.

In Figure 6.12 the 2-phase current commands are transformed into 3-phase current commands by equations (6.54) and (6.55). The 3-phase current commands are fed to the 3-phase inverter, which regulates the instantaneous 3-phase currents of the 3-phase suspension windings.

The advantages of having this dc-excited bearingless motor operating as a radial magnetic bearing are summarized as follows:

(a) A low-cost 3-phase inverter power module, including power devices, main circuit connections, gate drive circuits and protection circuits, can be applied to a radial magnetic bearing application.
(b) Only a 3-wire connection is required for 2-axis active positioning.
(c) Software algorithms developed for motor drives can be easily applied.

6.8 AC excitation and revolving magnetic field

In this section, 2-phase ac currents are supplied to the primitive bearingless motor winding set to generate a revolving magnetic field. The rotor is made up of circular laminations of silicon steel so that torque is not generated even though the magnetic field is rotating. Hence this ac-excited bearingless motor is not practical but it does aid analysis and understanding.

Let us suppose that 2-phase currents excite the 4-pole windings as previously shown. Substituting the command values for radial forces and the currents into (6.43) produces current commands. In doing so, it is important to check if the inverse matrix exists for the 2×2 square matrix. From a simple calculation it can be seen that the determinant of the matrix is -1. Hence an inverse matrix does exist and it is written as

$$\begin{bmatrix} \cos \phi & \sin \phi \\ \sin \phi & -\cos \phi \end{bmatrix}^{-1} = \begin{bmatrix} \cos \phi & \sin \phi \\ \sin \phi & -\cos \phi \end{bmatrix} \tag{6.56}$$

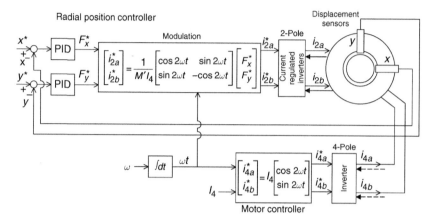

Figure 6.13   *Primitive bearingless drive with 4-pole drive*

where ϕ is equal to $2\omega t$. This matrix is characterized in such a way that the inverse matrix is exactly the same as the original matrix. This is a kind of unitary matrix. Substituting the commands into (6.43) and solving for the radial force current commands yields

$$\begin{bmatrix} i_{2a}^* \\ i_{2b}^* \end{bmatrix} = \frac{1}{M'I_4} \begin{bmatrix} \cos 2\omega t & \sin 2\omega t \\ \sin 2\omega t & -\cos 2\omega t \end{bmatrix} \begin{bmatrix} F_x^* \\ F_y^* \end{bmatrix} \tag{6.57}$$

This equation shows how the radial force current commands are calculated from the radial force commands. Figure 6.13 shows the block diagram of the primitive bearingless motor excited with 2-phase ac currents in the 4-pole windings. The radial force commands are fed from the PID controllers to the modulation block, which implements (6.57), so that the suspension winding currents follow the force commands.

The angular speed ω of the current is fed to an integration block with a reset. The integrated output is expressed as ωt. The angular speed can be assumed to be almost a constant in the steady-state conditions and together with the amplitude I_4, the 4-pole current commands i_{4a}^* and i_{4b}^* are generated. These commands are fed to an inverter that regulates the instantaneous currents. The 2-phase ac current produces a 4-pole revolving magnetic field; note that the 2-phase winding case is drawn in the figure for simplicity although a 3-phase winding and 3-phase inverter can be used as shown in the previous section.

If the 2-pole winding is employed as the motor winding then the motor currents are regulated to follow the currents as given in (6.44). In the modulation block, the suspension winding current is obtained from (6.45); substituting the command values for the radial forces and the 4-pole winding current gives

$$\begin{bmatrix} i_{4a}^* \\ i_{4b}^* \end{bmatrix} = \frac{1}{M'I_2} \begin{bmatrix} \cos \omega t & -\sin \omega t \\ \sin \omega t & \cos \omega t \end{bmatrix} \begin{bmatrix} F_x^* \\ F_y^* \end{bmatrix} \tag{6.58}$$

Note that the angular frequency of the motor currents and modulation is ω for the 2-pole motor case and that the negative sign is not in the same position.

Example 6.5 Derive the equations for determining the currents of the suspension windings from the radial force commands for both the 4-pole and 2-pole motor cases using the alternative coordinate system definition.

Answer
From the Example 6.4, an inverse matrix relationship for the 4-pole motor can be derived, where

$$\begin{bmatrix} i^*_{2a} \\ i^*_{2b} \end{bmatrix} = \frac{1}{M'I_4} \begin{bmatrix} -\cos 2\omega t & \sin 2\omega t \\ \sin 2\omega t & \cos 2\omega t \end{bmatrix} \begin{bmatrix} F^*_x \\ F^*_y \end{bmatrix} \tag{6.59}$$

And for the 2-pole motor case:

$$\begin{bmatrix} i^*_{4a} \\ i^*_{4b} \end{bmatrix} = \frac{1}{M'I_2} \begin{bmatrix} -\cos \omega t & -\sin \omega t \\ -\sin \omega t & \cos \omega t \end{bmatrix} \begin{bmatrix} F^*_x \\ F^*_y \end{bmatrix} \tag{6.60}$$

6.9 Inductance measurements

In the previous sections, the windings were assumed to be 2-phase. This assumption was made to simplify the analysis. However, in practice, 3-phase windings are usually used because the total number of wires is only 3, compared to 4 for 2-phase windings. Therefore it is quite important to examine the 3-phase winding system. In this section, an example of the 3-phase winding system is introduced. In addition, the inductances of the 3-phase windings are transformed into the 2-phase system and it is shown that considerable simplification is possible by use of 2-phase coordinates.

Figure 6.14 shows a cross-sectional view of a stator core with winding conductors. There are 24 slots in the stator. The stator slots contain, alternately, one 4-pole coil-side and one 2-pole coil-side. Coil sides $4u$, $4v$ and $4w$ are for the 4-pole windings of phase u, v and w respectively. Coil-sides $2u$, $2v$ and $2w$ are for the 2-pole windings of phase u, v and w, respectively. The coils of $2u$ are arranged so that a positive current generates an MMF in the x-axis direction only. This arrangement corresponds to winding $2a$ as shown in Figure 6.6. The coils of $2v$ and $2w$ are arranged so that the vector sum of their MMFs is in the y-axis direction when there is positive current in the coils of $2v$ and an equal but opposite current in the coils of $2w$. This arrangement corresponds to winding $2b$. The coils of $4u$ are arranged to correspond to winding $4a$ and the vector sum

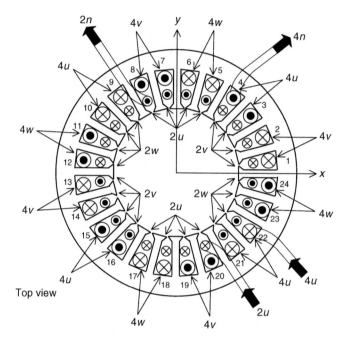

Figure 6.14 The 2-pole and 4-pole winding sets with 3-phase in 24 slot stator

of the MMFs due to $4v$ and $4w$ align with the $4b$ winding MMF (again with positive and negative currents in $4v$ and $4w$). The correspondence between the 2-phase and 3-phase windings is important for 3-phase to 2-phase transformation using the matrix in (6.55).

The coil connections are shown for the u-phase windings only. Coil side $4u$ enters at slot 22 and exits at slot 4. The entry-end is connected to the $4u$ terminal and the exit-end is connected to the star point $4n$. For $2u$, the winding enters at slot 20 from terminal $2u$, exits from slot 8 and is connected to the 2-pole star point $2n$.

Figures 6.15 and 6.16 show the open coil maps. The cylindrical stator core is opened between slots 1 and 24 and linearized. This figure is familiar to coil manufacturers. The first coil of $4u$ starts from terminal $4u$ then goes down slot 22 and returns up slot 3. There are several turns of the coil round these two slots before the phase is connected to coils that span slots 21 and 16, 10 and 15 and finally 9 and 4, with the phase winding finishing at slot 4. Coil manufactures wind the copper wires as shown in the figure and then insert coils into stator slots.

Figure 6.17 illustrates a stator core with both 4-pole and 2-pole windings. The short-pitched (inner) coil ends are for the 4-pole windings, and the long-pitched (outer) windings are for the 2-pole windings. In total there are 6 wires for the winding terminals.

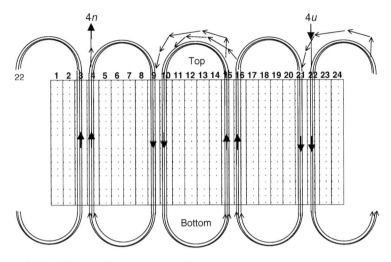

Figure 6.15 4-Pole winding arrangement of 4u

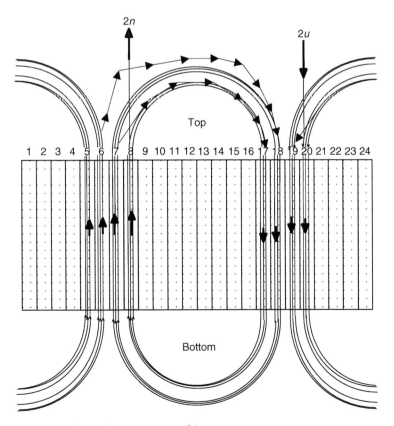

Figure 6.16 2-Pole winding arrangement of 2u

Figure 6.17 A stator core and windings with both 4-pole and 2-pole sets

Since there are six terminals for each of the 3-phase windings in the bearing-less motor, the relationships between the flux linkages λ_{4u}, λ_{4v}, λ_{4w}, λ_{2u}, λ_{2v}, λ_{2w} and currents i_{4u}, i_{4v}, i_{4w}, i_{2u}, i_{2v}, i_{2w} can be expressed by a 6×6 matrix:

$$
\begin{bmatrix} \lambda_{4u} \\ \lambda_{4v} \\ \lambda_{4w} \\ \lambda_{2u} \\ \lambda_{2v} \\ \lambda_{2w} \end{bmatrix} =
\begin{bmatrix}
L_{4u} & M_{4u4v} & M_{4u4w} & M_{4u2u} & M_{4u2v} & M_{4u2w} \\
* & L_{4v} & M_{4v4w} & M_{4v2u} & M_{4v2v} & M_{4v2w} \\
* & * & L_{4w} & M_{4w2u} & M_{4w2v} & M_{4w2w} \\
* & * & * & L_{2u} & M_{2u2v} & M_{2u2w} \\
* & * & * & * & L_{2v} & M_{2v2w} \\
* & * & * & * & * & L_{2w}
\end{bmatrix}
\begin{bmatrix} i_{4u} \\ i_{4v} \\ i_{4w} \\ i_{2u} \\ i_{2v} \\ i_{2w} \end{bmatrix}
\tag{6.61}
$$

The asterisks* indicate matrix elements that are not written because the matrix is symmetrical. Some of the inductances will be functions of the radial rotor displacement.

Figure 6.18 shows a circuit diagram that can be used to measure the inductances. All the inductances cannot be measured at once. In the figure, $4u$ is connected to a single-phase ac power supply. The self-inductance L_{4u} is obtained from

$$
L_{4u} = \frac{\sqrt{\left(\dfrac{V_{4u}}{I_{4u}}\right)^2 - R_{4u}^2}}{\omega}
$$

where V_{4u} and I_{4u} are the rms values of voltage and current at terminal $4u$, R_{4u} is the winding resistance and ω is the angular frequency of the power source.

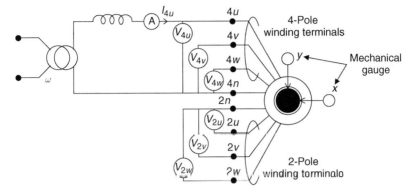

Figure 6.18 Inductance measurements

The mutual inductances are calculated, for example, from $M_{4u4v} = V_{4v}/(\omega I_{4u})$. The rotor radial positions can be fixed to any desired position within the airgap length between the stator and rotor. The radial displacements are measured by mechanical dial gauges or eddy current position sensors. It is important to move the rotor along one radial axis while keeping the displacement of the other axis zero. Initially the x-axis displacement is zero; V_{2u} can be monitored using an oscilloscope. The induced voltage V_{2u} is proportional to the x-axis displacement so that if the rotor is moved in a positive x-axis radial direction then V_{2u} increases and it is in phase with respect to V_{4u}. If there is a negative displacement along the x-axis, the polarity of V_{2u} is the inverse of V_{4u}. In order to keep the y-axis displacement zero, the difference between V_{2v} and V_{2w} is monitored using an oscilloscope. This voltage difference is proportional only to the y-axis displacement.

Once the radial position is fixed, all the inductances are measured by rotating the winding excitation. First, $4u$ is excited to measure six inductances, next the winding $4v$ is excited to measure five inductances and finally $4w$ is excited to measure four inductances. Then, windings $2u$, $2v$ and $2w$ are excited independently in order to measure all the inductances in the above equation. Once this is done the rotor is moved to the next radial position. When the rotor is displaced, one will see vibration caused by strong radial attractive magnetic force at high current; this vibration should be minimized to reduce measurement errors. However, the current level should be close to the normal motor excitation level to obtain the inductance values at that operating point. To do this, the rotor should be tightly fixed at each measuring point before exciting a winding. One way to fix the rotor is to fabricate a shaft with a large diameter end-disc with a tapped hole, which fixes the rotor shaft to a stationary position with a bolt.

Figure 6.19 shows the inductance variations with respect to a normalized radial displacement x_n, which is equal to x divided by the nominal airgap length g_0. In the figure, the rotor is displaced by about $g_0/2$. In Figure 6.19(a), the self-inductances and mutual inductances of the 4-pole and 2-pole windings are shown. These inductances are almost constant as expected. If the rotor is moved further,

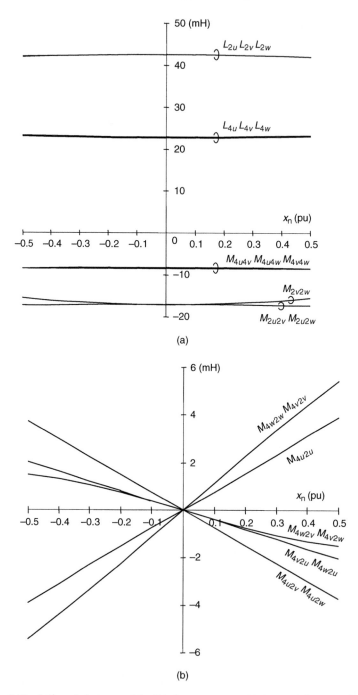

Figure 6.19   3-Phase inductances: (a) self-inductances, mutual inductances between the motor windings and mutual inductances between the suspension windings; (b) mutual inductances between motor and suspension windings

these inductances start to vary as a quadric curve, having the minimum at $x_n = 0$. In Figure 6.19(b), the mutual inductances between the 4-pole and 2-pole windings are shown. Some mutual inductances, for example M_{4u2v} and M_{4u2w}, are the same value because of the symmetrical winding arrangement. In total nine mutual inductances are shown. It can be seen that the slopes of these inductances with respect to the rotor radial displacement are not constant. In a simple theoretical inductance calculation, with a fundamental MMF approximation, only two slopes are obtained; however, the gradient is seen to vary in the figure. This is because of the effects of space harmonics, especially the third harmonic component, in the MMF space distributions. The third harmonic influence is eliminated when transformed into the 2-phase coordinates.

Let us transform these inductance functions into the 2-phase coordinates. Equation (6.61) can be written in a matrix form such that

$$[\lambda_6] = [L_6][i_6] \tag{6.62}$$

The 3-phase to 2-phase transform matrix was shown in (6.54) and (6.55). With this transform matrix, the current and flux linkage relationships between the 3-phase and 2-phase representation can be written as

$$\begin{bmatrix} i_{4u} \\ i_{4v} \\ i_{4w} \end{bmatrix} = [C_{32}]^t \begin{bmatrix} i_{4a} \\ i_{4b} \end{bmatrix} \tag{6.63}$$

$$\begin{bmatrix} \lambda_{4u} \\ \lambda_{4v} \\ \lambda_{4w} \end{bmatrix} = [C_{32}]^t \begin{bmatrix} \lambda_{4a} \\ \lambda_{4b} \end{bmatrix} \tag{6.64}$$

$$\begin{bmatrix} \lambda_{2u} \\ \lambda_{2v} \\ \lambda_{2w} \end{bmatrix} = [C_{32}]^t \begin{bmatrix} \lambda_{2a} \\ \lambda_{2b} \end{bmatrix} \tag{6.65}$$

Let us define a 6×4 transform matrix as follows:

$$[C_6]^t = \begin{bmatrix} [C_{32}]^t & [0] \\ [0] & [C_{32}]^t \end{bmatrix} \tag{6.66}$$

where [0] indicates a 3×2 zero matrix. In 2-phase coordinates the relationship between the flux linkages and currents are expressed by a 4×4 inductance matrix as

$$[\lambda_4] = [L_4][i_4] \tag{6.67}$$

where

$$[\lambda_4] = \begin{bmatrix} \lambda_{4a} \\ \lambda_{4b} \\ \lambda_{2a} \\ \lambda_{2b} \end{bmatrix} \tag{6.68}$$

$$[i_4] = \begin{bmatrix} i_{4a} \\ i_{4b} \\ i_{2a} \\ i_{2b} \end{bmatrix} \tag{6.69}$$

These flux linkage and current vectors correspond to vectors in 3-phase systems and are linked by simple matrices

$$[i_6] = [C_6]^t[i_4] \tag{6.70}$$

$$[\lambda_6] = [C_6]^t[\lambda_4] \tag{6.71}$$

The original $[C_6]$ provides an inverse matrix of the transpose matrix $[C_6]^t$. Hence from (6.71), $[\lambda_4] = [C_6][\lambda_6]$ and substituting this equation into the left side of (6.67) yields

$$[C_6][\lambda_6] = [L_4][i_4] \tag{6.72}$$

Substituting (6.62) into the above equation and eliminating $[\lambda_6]$,

$$[C_6][L_6][i_6] = [L_4][i_4] \tag{6.73}$$

Substituting (6.70) an inductance matrix in 2-phase coordinates is derived:

$$[L_4] = [C_6][L_6][C_6]^t \tag{6.74}$$

This equation provides the transformation from 3-phase inductances to 2-phase inductances. Therefore (6.74) can be written as

$$
\begin{bmatrix} L_{4a} & M_{4a4b} & M_{4a2a} & M_{4a2b} \\ * & L_{4b} & M_{4b2a} & M_{4b2b} \\ * & * & L_{2a} & M_{2a2b} \\ * & * & * & L_{2b} \end{bmatrix}
= \sqrt{\frac{2}{3}}
\begin{bmatrix} 1 & \frac{-1}{2} & \frac{-1}{2} & 0 & 0 & 0 \\ 0 & \frac{\sqrt{3}}{2} & \frac{-\sqrt{3}}{2} & 0 & 0 & 0 \\ 0 & 0 & 0 & 1 & \frac{-1}{2} & \frac{-1}{2} \\ 0 & 0 & 0 & 0 & \frac{\sqrt{3}}{2} & \frac{-\sqrt{3}}{2} \end{bmatrix}
$$

$$
\times
\begin{bmatrix} L_{4u} & M_{4u4v} & M_{4u4w} & M_{4u2u} & M_{4u2v} & M_{4u2w} \\ * & L_{4v} & M_{4v4w} & M_{4v2u} & M_{4v2v} & M_{4v2w} \\ * & * & L_{4w} & M_{4w2u} & M_{4w2v} & M_{4w2w} \\ * & * & * & L_{2u} & M_{2u2v} & M_{2u2w} \\ * & * & * & * & L_{2v} & M_{2v2w} \\ * & * & * & * & * & L_{2w} \end{bmatrix}
$$

$$
\times \sqrt{\frac{2}{3}}
\begin{bmatrix} 1 & 0 & 0 & 0 \\ \frac{-1}{2} & \frac{\sqrt{3}}{2} & 0 & 0 \\ \frac{-1}{2} & \frac{-\sqrt{3}}{2} & 0 & 0 \\ 0 & 0 & 1 & 0 \\ 0 & 0 & \frac{-1}{2} & \frac{\sqrt{3}}{2} \\ 0 & 0 & \frac{-1}{2} & \frac{-\sqrt{3}}{2} \end{bmatrix}
\tag{6.75}
$$

Further calculation of the above matrix equation generates the set of equations below.

$$L_{4a} = \frac{1}{6}(4L_{4u} + L_{4v} + L_{4w} - 4M_{4u4v} - 4M_{4u4w} + 2M_{4v4w})$$

$$M_{4a4b} = \frac{1}{2\sqrt{3}}(-L_{4v} + L_{4w} + 2M_{4u4v} - 2M_{4u4w})$$

$$L_{4b} = \frac{1}{2}(L_{4v} + L_{4w} - 2M_{4v4w})$$

$$L_{2u} = \frac{1}{6}(4L_{2u} + L_{2v} + L_{2w} - 4M_{2u2v} - 4M_{2u2w} + 2M_{2v2w})$$

$$M_{2a2b} = \frac{1}{2\sqrt{3}}(-L_{2v} + L_{2w} + 2M_{2u2v} - 2M_{2u2w})$$

$$L_{2b} = \frac{1}{2}(L_{2v} + L_{2w} - 2M_{2v2w})$$

$$M_{4a2a} = \frac{1}{6}(4M_{4u2u} - 2M_{4v2u} - 2M_{4w2u} - 2M_{4u2v} - 2M_{4u2w} + M_{4v2v} + M_{4v2w}$$
$$+ M_{4w2v} + M_{4w2w})$$

$$M_{4a2b} = \frac{1}{2\sqrt{3}}(2M_{4u2v} - 2M_{4u2w} - M_{4v2v} - M_{4w2v} + M_{4v2w} + M_{4w2w})$$

$$M_{4b2a} = \frac{1}{2\sqrt{3}}(2M_{4v2u} - 2M_{4w2u} - M_{4v2v} - M_{4v2w} + M_{4w2v} + M_{4w2w})$$

$$M_{4b2b} = \frac{1}{2}(M_{4v2v} + M_{4w2w} - M_{4v2w} - M_{4w2v})$$

The 2-phase inductances above can be calculated by substituting the values shown in Figure 6.19; these are shown in Figure 6.20 for the mutual inductances between the 4-pole and 2-pole windings for both the x and y displacements. In Figure 6.20(a), the inductance values M_{4a2a}, M_{4b2a}, M_{4a2b} and M_{4b2b} are shown, but note that M_{4a2b} and M_{4b2a} are zero while M_{4a2a} and M_{4b2b} are proportional to the radial displacement and have the same slope as $M' = 40$ H/m. Quite simple inductance functions are found as a result of the 3-phase to 2-phase transformation. This is the main reason why the analysis was done using 2-phase winding theory in the previous sections of this chapter. In Figure 6.20(b), the mutual inductances are drawn with respect to the y displacement. M_{4a2a} and M_{4b2b} are zero, while M_{4b2a} and M_{4a2b} are almost the same as the absolute value, i.e., a slope of 40 H/m.

The validity of equation (6.35) can be confirmed. Measuring the inductances is the first step in obtaining the machine parameters, and these measurements can be done with simple equipment. The radial force constants can also be obtained, which are necessary to estimate the loop gains of the radial position regulation loops. It is also useful to check to see if windings are correct. If one turn is

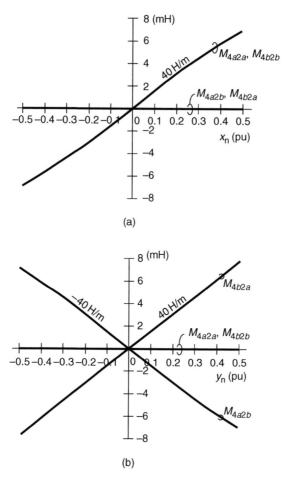

Figure 6.20   Mutual inductance variations in 2-phase coordinates: (a) mutual inductance variation in x-axis; (b) mutual inductance variation in y-axis

missing in a certain phase, the inductance values will have slight differences between phases. As shown in (6.30–6.32), the inductance values are proportional to the square of the number of turns so that it is possible to estimate the increase or decrease of inductance when there is one turn too many or too few.

7

Analysis in rotational coordinates and magnetic suspension strategy for bearingless drives with 2-pole and 4-pole windings

Akira Chiba

In the previous chapter a primitive bearingless motor was analysed. The primitive bearingless motor has a cylindrical rotor without permanent magnets or squirrel cage conductors so that rotational torque is not generated. The analysis of the primitive bearingless motor is similar to the bearingless motor though it only acts as a radial magnetic bearing; but it is helpful in developing the theory for the bearingless motor.

In this chapter vector control theory for permanent magnet motors is introduced; this theory is also called field-oriented control theory. It derives torque and current relationships using the instantaneous rotational position of the revolving magnetic field [1]. In short, rotational torque is generated by an interaction of the revolving magnetic field and the component of MMF that is perpendicular to the field on the d–q axes. The principle is also applicable to field-excited synchronous machines and induction motors. In addition to the basic principles of operation, rotational speed and torque regulation strategies are developed. To aid understanding, vector control theory is explained here by first studying the dc motor as an example. After the torque analysis, additional windings are introduced and the radial forces are derived for a dc motor with a 4-pole and 2-pole winding layout. Variables in the stationary coordinates, i.e., the dc motor model, are transformed into rotational coordinates to obtain the relationships for the ac motor. The derived equations are applied to a cylindrical permanent magnet motor and a salient-pole synchronous motor.

7.1 Vector control theory of electrical motors

Vector control theory provides a control strategy to regulate the instantaneous rotational torque in electrical machines. In a dc motor the generated shaft torque is proportional to the armature current so that precise dc current regulation produces torque regulation. However, in ac motors both the amplitude and phase angle of the currents (which are now ac) need to be regulated to control the torque. Hence it is important to study the relationships between torque and current.

Figure 7.1 shows the cross section of a 4-pole dc motor. It has a cylindrical slotted rotor (referred to as the armature in a dc machine) and a stator, which has four magnetic poles and a surrounding yoke. The airgap length between the rotor iron and stator pole faces is small but the gap between the stator yoke and the rotor is large so that the stator poles provide magnetic saliency. The 4-pole magnetic field Ψ_m may be generated by permanent magnets; however, for the analysis, virtual conductors 'm' are defined and currents in these virtual conductors generate Ψ_m so that the stator poles are excited with the polarities shown by N and S.

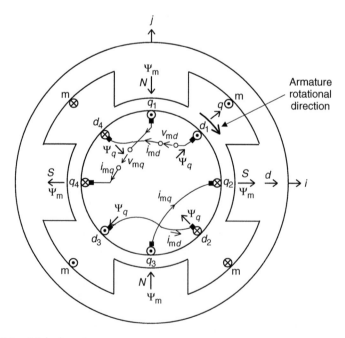

Figure 7.1   4-Pole dc motor

On the armature there are two sets of designated conductors. q_1, q_2, q_3 and q_4 are the designated quadrature-axis conductors; the actual conductors rotate but they are connected through brushes and a commutator to the terminals so that the relative angular positions of q_1, q_2, q_3 and q_4, with respect to the stator, do not change. This is due to the switching action of the commutator. These conductors are connected in series and are called the q-axis coils because their MMFs sit on the q-axis of the stator poles. Four brushes on the d-axis terminate the armature q-axis coils. Because the conductors are under the stator pole faces, when a positive current flows in q_1 a force is generated in the right-hand direction (Fleming's left-hand rule). Similar forces are generated at the other three conductors so that torque is generated in the clockwise direction. In addition to the q-axis coils another set of conductors are present on the d-axis. These generate an electrically perpendicular MMF with respect to the q-axis conductors. They are also connected in series to form d-axis coils but positive current in these coils does not contribute to torque generation because they are not under the pole faces; however, the amplitude of the 4-pole magnetic field Ψ_m is increased.

Suppose that the voltage and current applied to the series-connected q-axis coils are v_{mq} and i_{mq} respectively, as shown in Figure 7.1. Let us also suppose that the d- and q-axis inductances are L_d and L_q so that

$$v_{mq} = R_s i_{mq} + PL_q i_{mq} + \omega \lambda_m + \omega L_d i_{md} \tag{7.1}$$

where R_s is the coil resistance, P is the derivative operator with respect to time ($= d/dt$), ω is the electrical angular frequency, λ_m is the flux linkage caused by the stator magnetic field and i_{md} is the d-axis coil current. The first term of the equation is the coil resistance voltage drop and the second term is the back-emf due to the coil inductance. The third and fourth terms are described as speed-induced voltages (and can be obtained from the "flux-cutting" rule).

For the d-axis coil the voltage v_{md} is described by

$$v_{md} = R_s i_{md} + PL_d i_{md} - \omega L_q i_{mq} \tag{7.2}$$

Note that the sign of speed-induced voltage is negative due to the definitions of positive voltage and current direction in the figure. The sign of the speed-induced voltage is described in a later example.

The voltage and current relationships can be written in a matrix form

$$\begin{bmatrix} v_{md} \\ v_{mq} \end{bmatrix} = \begin{bmatrix} R_s & 0 \\ 0 & R_s \end{bmatrix} \begin{bmatrix} i_{md} \\ i_{mq} \end{bmatrix} + \begin{bmatrix} PL_d & -\omega L_q \\ \omega L_d & PL_q \end{bmatrix} \begin{bmatrix} i_{md} \\ i_{mq} \end{bmatrix} + \begin{bmatrix} 0 \\ \omega \lambda_m \end{bmatrix} \tag{7.3}$$

Instantaneous power p_{ins} can be defined as the total instantaneous input power of the system at the terminals so that

$$p_{ins} = \begin{bmatrix} i_{md} & i_{mq} \end{bmatrix} \begin{bmatrix} v_{md} \\ v_{mq} \end{bmatrix} \tag{7.4}$$

Substituting (7.3) into (7.4) yields

$$p_{ins} = R_s(i_{md}^2 + i_{mq}^2) + (L_d i_{md} P i_{md} + L_q i_{mq} P i_{mq})$$
$$+ \omega(L_d - L_q)i_{md}i_{mq} + \omega\lambda_m i_{mq} \qquad (7.5)$$

The first term is the copper loss in the coils and the second term is the stored magnetic energy in the inductances; these terms do not contribute to torque production. The third term produces reluctance torque, which is generated by a difference between the d- and q-axis inductances due to the magnetic saliency. The fourth term is the excitation torque due to the permanent magnets.

The relationship between the shaft power p_s and torque T_r is

$$P_s = \omega_m T_r \qquad (7.6)$$

where the shaft angular speed ω_m is defined as

$$\omega_m = \frac{\omega}{p_p} \qquad (7.7)$$

and p_p is the number of pole-pairs. Hence the torque is

$$T_r = p_p \left[(L_d - L_q)i_{md}i_{mq} + \lambda_m i_{mq} \right] \qquad (7.8)$$

The first and second terms are the reluctance and excitation torque respectively.

In vector control theory the torque is expressed as a vector product. Suppose that current vector $[i]$ and flux linkage vector $[\lambda]$ are defined by d- and q-axis components where

$$[i] = (i_{md}, i_{mq})$$
$$[\lambda] = (\lambda_d, \lambda_q)$$

A vector (cross) product of these vectors can be calculated assuming $\lambda_d = \lambda_m + L_d i_{md}$ and $\lambda_q = L_q i_{mq}$ so that

$$|[\lambda] \times [i]| = (\lambda_d, \lambda_q) \times (i_{md}, i_{mq})$$
$$= \lambda_d i_{mq} - \lambda_q i_{md}$$
$$= (L_d - L_q)i_{md}i_{mq} + \lambda_m i_{mq}$$

Note that this equation corresponds to the derived torque for one pole-pair. Hence we can see that the torque per pole-pair is the vector product of the flux linkage and current. The absolute value of the vector product for torque can be written as

$$T_r = p_p |[\lambda]|\,|[i]| \sin\theta$$

where θ is an angle between $[\lambda]$ and $[i]$ so that only the current component which is perpendicular to the flux linkage vector contributes to torque production. In the torque controller, the main axis can be oriented on the flux linkage vector, which defines the magnetic field direction, and the quadrature axis is perpendicular to the main axis so that the current component on the quadrature axis is regulated to obtain the required torque. Hence vector control theory is also called field-oriented control theory.

There are several possible strategies for regulating the instantaneous vectors. The voltage and torque equations give an interesting insight, and note that these equations are also valid for permanent magnet synchronous motors as shown in the next section. The regulating strategies for the torque can be summarized as follows

(a) $i_{md} = 0$ strategy: let the d-axis current be zero. Hence the first term in the brackets in the torque equation is zero so that $T_r = p_p \lambda_m i_{mq}$. The torque is proportional to the q-axis current when the permanent magnet flux linkage is constant. This simple strategy is often employed in non-salient-pole synchronous motors especially with surface-mounted permanent magnet (SPM) rotors.

(b) Maximizing torque/current ratio strategy: the d- and q-axis currents are determined so that the torque-to-line current ratio is maximized. The current is limited by the motor and inverter current ratings. For the maximum current limit value, torque can be maximized by making i_{md} a function of i_{mq} so that both d- and q-axis currents are generated from a torque command. In addition, during low speed operation the copper loss is dominant, hence this strategy results in minimum heat generation.

(c) Power factor $= 1$ strategy: for the middle-speed range, maximizing the power factor minimizes the product of voltage and current. The power factor of the fundamental voltage and current components (i.e., the cosine of the phase angle as apposed to distortion factor) can be made unity by regulating the d-axis current as a function of i_{mq}.

(d) Field weakening strategy: for the high-speed range, the speed-induced voltage due to $\omega \lambda_m$ may be higher than the rated inverter voltage. The motor terminal voltage can be reduced if the d- and q-axis inductances are relatively high and the d-axis current is negative. This has the effect of reducing the total speed-induced voltage. It is advantageous if L_d is less than L_q (as is the case with interior permanent magnet (IPM) motor, where q-axis saliency is present) so that only low i_{md} is required to weaken the field. Also the subsequent reluctance torque will contribute to the torque production.

The effectiveness of these strategies is dependent on machine parameters L_d, L_q and λ_m. In addition, these strategies can be chosen to match the rotational speed and torque operating point.

Example 7.1   In Figure 7.2 two conductors are rotating in a magnetic field of flux density B. The rotational velocity of the conductors is ω_{lc}; (a) shows the top view and (b) shows the side view. Answer the following questions.

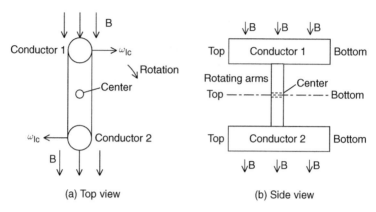

Figure 7.2 Primitive rotating conductors

(a) What is the speed-induced voltage polarity in conductors 1 and 2?

(b) If the bottom terminals of conductors 1 and 2 are connected together, what is the voltage polarity of the circuit?

(c) After conductors 1 and 2 are connected, the voltage polarity and current direction are defined for motoring operation, i.e., positive current flows down the second conductor and up the first, while the voltage at the top of the second conductor is positive with respect to the first. Show the polarity of the speed-induced voltage.

(d) For the case when the voltage polarity is reversed with respect to the motoring operation, show the polarity of speed-induced voltage.

(e) Suppose that the conductors are composed of several series-connected wires so that a coil is formed and flux is generated by current in another coil on the d-axis having a self-inductance L_d. The electrical rotational angular speed is ω. The voltage polarity and current direction of the q-axis coil are as in Figure 7.1. Draw a side view for the conductors q_1 and q_4. Show the speed-induced voltage due to the d-axis coil and also show the speed-induced voltage due to the d-axis permanent magnets.

(f) Draw a side view for the d-axis conductors d_1 and d_4 of the d-axis coil in Figure 7.1 and show the speed-induced voltage.

Answer

The polarity of the speed-induced voltage is defined using Fleming's right-hand rule as shown in Figure 7.3. The thumb, forefinger and middle finger indicate the directions of speed v_{lc}, flux density B and induced voltage e.

(a) The voltage potential of the bottom terminal 1B of conductor 1 is positive with respect to the top terminal 1T as shown in Figure 7.4(a) while the voltage potential of the top terminal 2T of conductor 2 is positive with respect to the bottom 2B, again as shown in Figure 7.4(a).

(b) The top of the conductor 2 is positive with respect to the top of conductor 1 as shown in Figure 7.4(a).

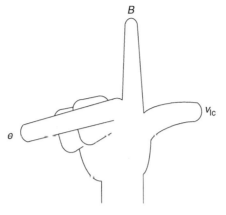

Figure 7.3   Fleming's right-hand rule

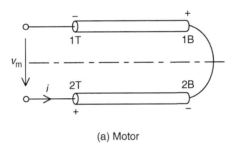

(a) Motor

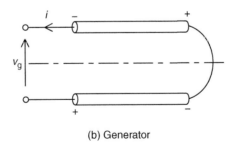

(b) Generator

Figure 7.4   Side views

(c) Positive – The induced voltage is opposite to the terminal voltage as shown
    in Figure 7.4(a).
(d) Negative – The induced voltage has the same polarity as the terminal
    voltage as shown in Figure 7.4(b). Note that the polarity of the speed-
    induced voltage is dependent on the voltage definition. Even in the case of
    a generator, the current direction can be defined in the opposite direction,
    while voltage polarity is the same as that in a motor.

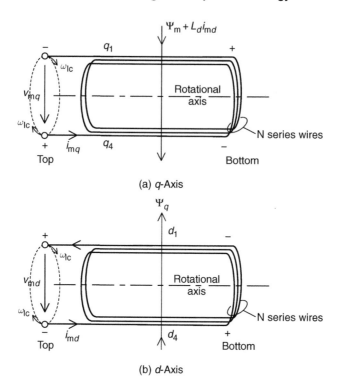

Figure 7.5 Side views

(e) $v_{mq} = \omega L_d i_{md} + \omega \lambda_m$. Refer to Figure 7.5(a).

(f) $v_{md} = -\omega L_q i_{mq}$. Refer to Figure 7.5(b). The voltage polarity is in the opposite direction to that in Figure 7.5(a). The voltage polarity is based on Figure 7.1.

7.2 Coordinate transformation and torque regulation

In the previous section, torque was expressed as a function of the d- and q-axis currents using a dc motor model. In vector control theory, using rotating reference axes and a 3-phase to 2-phase coordinate transformation, it can be shown that the synchronous and induction motors are equivalent to the dc motor (indeed, in a dc motor, while the armature draws dc current, the current in the actual armature coils is ac with a square-wave shape due to the switching action of the commutator). In this section the relationships between the dc motor and the synchronous motor are shown. The theory of the induction motor is rather more complicated and advanced readers are recommended to refer to Reference [1].

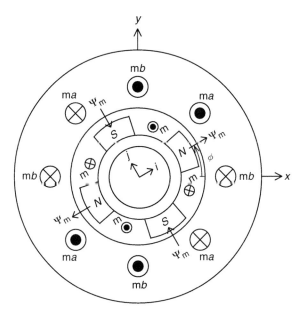

Figure 7.6 2-Phase ac machine

Figure 7.6 shows the cross section of a 2-phase synchronous motor. The rotor has four salient poles and field-winding conductors m are arranged so that 4-pole magnetization is produced. The i and j axes are aligned on the magnetic poles. These axes are fixed to the rotor so that they rotate with rotor rotation. The rotor angular position is defined by the angle ϕ. Note that the direction of rotation is opposite to that in Figure 7.1. This is in order to have corresponding movement with respect to the defined axes. The relative movement between armature and field is the same although in Figure 7.6 the field rotates and in Figure 7.1 the armature rotates. Located in the stator slots are conductors ma and mb which form a 4-pole 2-phase winding. The current directions of conductors ma and mb are defined so that the MMF directions correspond to the d_1-d_4 and q_1-q_4 conductors in Figure 7.1 at the point where $\phi = 0$. At this angular position it can be seen that the 2-phase motor in Figure 7.6 is equivalent to the dc motor in Figure 7.1. However, ma and mb are fixed to the stator and are not fed through brushes and a commutator. Hence the relative position of the conductors varies with the rotor angular position. On the other hand, the conductors in Figure 7.1 are stationary because the conductor currents are fed through brushes and a commutator.

To generate an equivalent MMF distribution, the conductor currents i_{ma} and i_{mb} can be varied with the rotor angular position using a rotational coordinate transformation matrix, as shown:

$$\begin{bmatrix} i_{ma} \\ i_{mb} \end{bmatrix} = \begin{bmatrix} \cos 2\phi & -\sin 2\phi \\ \sin 2\phi & \cos 2\phi \end{bmatrix} \begin{bmatrix} i_{md} \\ i_{mq} \end{bmatrix} \tag{7.9}$$

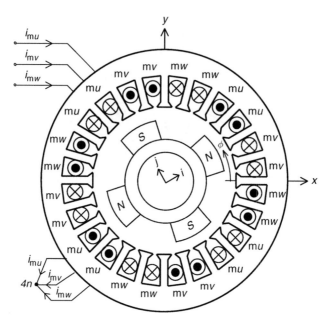

Figure 7.7   3-Phase ac machine

The equation matrix shows that the MMF distribution in Figure 7.1 is produced, and that the currents i_{ma} and i_{mb} are regulated by the d- and q-axis currents and also the rotor angular position ϕ.

Figure 7.7 shows an example of a winding arrangement in a practical 3-phase motor. On the stator are 24 slots that accommodate the motor conductors, which are indicated by letters and connected so that a 3-phase winding comprising of mu, mv and mw is formed. The current directions are balanced and defined so as to correspond with an equivalent 2-phase ac machine. The currents i_{mu}, i_{mv} and i_{mw} can be related to the equivalent current distribution shown in Figure 7.6 using the transformation

$$\begin{bmatrix} i_{mu} \\ i_{mv} \\ i_{mw} \end{bmatrix} = \frac{\sqrt{2}}{\sqrt{3}} \begin{bmatrix} 1 & 0 \\ \frac{-1}{2} & \frac{\sqrt{3}}{2} \\ \frac{-1}{2} & \frac{-\sqrt{3}}{2} \end{bmatrix} \begin{bmatrix} i_{ma} \\ i_{mb} \end{bmatrix} \tag{7.10}$$

This transformation is the 2-phase to 3-phase transformation matrix. The 2-phase current set has two degrees of freedom whereas the 3-phase current set has three degrees of freedom. However, the sum of the 3-phase currents is zero in the above calculation, whatever the values of the 2-phase currents. Since the current sum at the neutral point $4n$ is zero, the 3-phase system is a three-wire, three-terminal system with a star connection.

Figure 7.8 shows a block diagram of a control to calculate the instantaneous 3-phase current commands from the d- and q-axis current commands i_{md}^* and i_{mq}^*. In a vector controller the currents are controlled using the dc machine model.

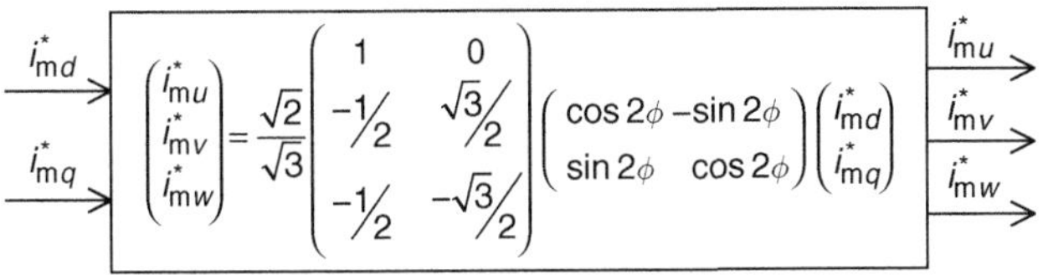

i^*_{md}
i^*_{mq}
$\begin{pmatrix} i^*_{mu} \\ i^*_{mv} \\ i^*_{mw} \end{pmatrix} = \frac{\sqrt{2}}{\sqrt{3}} \begin{pmatrix} 1 & 0 \\ -1/2 & \sqrt{3}/2 \\ -1/2 & -\sqrt{3}/2 \end{pmatrix} \begin{pmatrix} \cos 2\phi & -\sin 2\phi \\ \sin 2\phi & \cos 2\phi \end{pmatrix} \begin{pmatrix} i^*_{md} \\ i^*_{mq} \end{pmatrix}$
i^*_{mu}
i^*_{mv}
i^*_{mw}

Figure 7.8 3-Phase current block

Current regulation strategy
Motor
ω^*_m
+
−
ω_m
Speed regulator
T^*_r
i^*_{md}, i^*_{mq} Calculation
i^*_{md}
i^*_{mq}
3-Phase current
i^*_{mu}
i^*_{mv}
i^*_{mw}
Current-controlled inverter
Rotary encoder
ϕ Detection
Speed detection

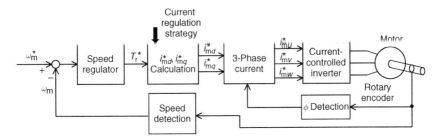

Figure 7.9 Speed regulation

The dc current commands are generated using the torque demand, the speed, etc. The 3-phase current commands i^*_{mu}, i^*_{mv} and i^*_{mw} are generated using (7.9) and (7.10).

Figure 7.9 shows an example of a speed-regulated motor drive system. On the rotor shaft is a rotary encoder, which is used to detect the instantaneous rotor angular position. The shaft speed ω_m is detected and compared with the speed command to generate a speed error. The speed error is amplified by a speed regulator to generate the torque command T^*_r. Using this command, the *d*- and *q*-axis current commands are calculated depending on the current regulation strategy. These current commands are then transformed into the 3-phase current commands. The motor currents are supplied and regulated to follow these current commands by an inverter. In steady-state conditions, the rotor runs at a constant speed with a constant load torque so that the torque command is equivalent to the load torque. The *d*- and *q*-axis currents are dc values whereas the 3-phase currents and their associated commands are 3-phase sinusoidal waveforms.

7.3 Vector control theory in bearingless motors

In the previous section, vector control theory was briefly explained using a dc machine model. This model was defined using conductors fixed to the rotor so that the dc machine model can be described in a rotational coordinate system model. In this section a rotational coordinate model of a 4-pole and 2-pole bearingless motor is introduced and the basic relationships between the flux linkages, currents and suspension forces are derived [2].

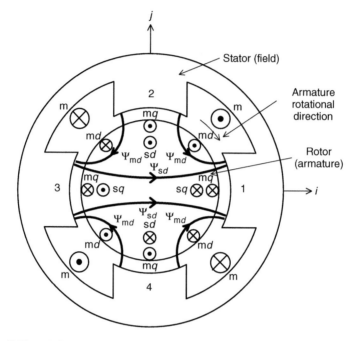

Figure 7.10 Winding arrangement in rotational coordinates

Figure 7.10 shows a bearingless motor with 4-pole motor windings md and mq and 2-pole suspension windings sd and sq, as well as the motor field winding m. The motor and suspension currents are rotor-mounted and fed through a commutator and brushes (which are not shown in the figure). The rotor is rotating in a counter-clockwise direction although the relative position of conductors is stationary with respect to the stator. Windings m, md and mq are motor windings corresponding to those in Figure 7.1. Conductors sd and sq are added as the suspension windings.

For the first step of the suspension force generation analysis, an inductance matrix has to be derived. The mutual inductances between the 4-pole and 2-pole windings play an important role in suspension force generation, and these mutual inductances should be represented as a function of the radial rotor displacement.

Figure 7.11 shows flux distributions when a rotor is displaced from the centre position. In Figure 7.11(a) the rotor is displaced in the i-axis direction. The 4-pole flux distribution generated by winding m is unbalanced so that more flux is concentrated under stator-pole 1. Hence some flux mutually links winding sd. Note that the flux linkage is in the same direction for both windings. Let us assume that the flux linkage is proportional to the rotor displacement; the flux linkage between the m and sd windings can then be expressed as $\lambda_m' i$, where λ_m' is a suspension force constant (Wb/m) and i is the i-axis rotor displacement. Similar flux linkage occurs when winding md is excited. In this case the mutual

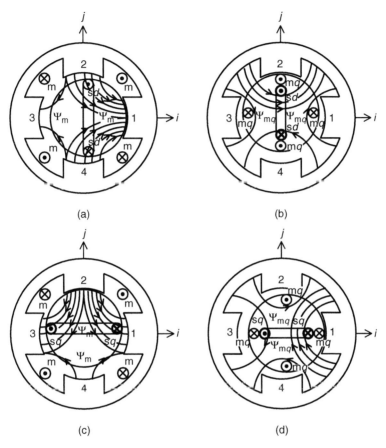

Figure 7.11 Mutual couplings when a rotor is displaced: (a) m-excited flux links to s*d*; (b) m*q*-excited flux links to s*d*; (c) m-excited flux links to s*q*; (d) m*q*-excited flux links to s*q*

inductance is expressed as $M'_d i$, where M'_d is a suspension force constant (H/m). A similar characteristic also occurs for the *j*-axis rotor displacement.

Figure 7.11(b) shows the flux distribution with winding m*q* excited. Suppose that the rotor was originally positioned at the stator centre producing a 4-pole symmetrical flux distribution. In that case some positive-direction flux flows around stator-pole 2 and links winding s*d*; however, negative flux around stator-pole 4 also links with winding s*d* so that the sum of the net flux linkage is zero. But when the rotor is displaced along the *j*-axis the positive flux is increased and the negative flux is decreased. Hence the flux linkage is no longer zero, so a mutual inductance exists between the m*q* and s*d* windings, which is expressed as $M'_q j$, i.e., the product of suspension force constant M'_q (H/m) and the *j*-axis displacement.

Figure 7.11(c,d) shows the variation of the rotor position in the same two directions, and these also lead to mutual inductances with winding s*q*.

Figure 7.11(c) illustrates the case when a rotor is displaced in j-axis direction. The mutual inductance between the m and sq windings is expressed as $-\lambda'_m j$, where j is the j-axis radial rotor displacement. The mutual inductance is negative because the flux linkage is reversed with respect to the polarity of the sq winding. Figure 7.11(d) illustrates the mutual inductance between windings mq and sq.

The mutual coupling of the suspension and md windings is the same as with the field winding since the md winding has the same MMF space distribution. The relationships between flux linkage and winding currents can be expressed in a matrix form

$$
\begin{bmatrix} \lambda_{md} \\ \lambda_{mq} \\ \lambda_{sd} \\ \lambda_{sq} \end{bmatrix} = \begin{bmatrix} L_d & 0 & M'_d i & -M'_d j \\ 0 & L_q & M'_q j & M'_q i \\ M'_d i & M'_q j & L_s & 0 \\ -M'_d j & M'_q i & 0 & L_s \end{bmatrix} \begin{bmatrix} i_{md} \\ i_{mq} \\ i_{sd} \\ i_{sq} \end{bmatrix} + \begin{bmatrix} \lambda_m \\ 0 \\ \lambda'_m i \\ -\lambda'_m j \end{bmatrix}
\tag{7.11}
$$

where λ_{md}, λ_{mq}, λ_{sd} and λ_{sq} are flux linkages of the motor and suspension d- and q-axis windings and λ'_m (Wb/m), M'_d (H/m) and M'_q (H/m) are the suspension force constants.

Using this matrix equation, the stored magnetic co-energy W'_m can be derived (neglecting magnetic saturation). If saturation does not exist then W'_m is equal to the stored magnetic energy W_m so that

$$
W'_m = W_m = \frac{1}{2} \begin{bmatrix} i_{md} & i_{mq} & i_{sd} & i_{sq} \end{bmatrix} \begin{bmatrix} \lambda_{md} \\ \lambda_{mq} \\ \lambda_{sd} \\ \lambda_{sq} \end{bmatrix}
\tag{7.12}
$$

Substituting (7.11) into (7.12) yields

$$
\begin{aligned}
W_m = \frac{1}{2} & \left(L_d i^2_{md} + L_q i^2_{mq} + L_s i^2_{sd} + L_s i^2_{sq} \right) \\
& + M'_d i i_{md} i_{sd} - M'_d j i_{md} i_{sq} + M'_q j i_{mq} i_{sd} + M'_q i i_{mq} i_{sq} \\
& + i_{md} \lambda_m + \lambda'_m i i_{sd} - \lambda'_m j i_{sq}
\end{aligned}
\tag{7.13}
$$

Although the equation is based on a non-holonomic system with only rotational movement, the derived suspension forces are valid because these forces are obtained from partial derivatives with respect to radial displacements. Hence the suspension forces are

$$
F_i = \frac{\partial W_m}{\partial i} = M'_d i_{md} i_{sd} + M'_q i_{mq} i_{sq} + \lambda'_m i_{sd}
\tag{7.14}
$$

$$
F_j = \frac{\partial W_m}{\partial j} = -M'_d i_{md} i_{sq} + M'_q i_{mq} i_{sd} - \lambda'_m i_{sq}
\tag{7.15}
$$

These equations can be written in a simple matrix form as

$$
\begin{bmatrix} F_i \\ F_j \end{bmatrix} = \begin{bmatrix} \lambda'_m + M'_d i_{md} & M'_q i_{mq} \\ M'_q i_{mq} & -\lambda'_m - M'_d i_{md} \end{bmatrix} \begin{bmatrix} i_{sd} \\ i_{sq} \end{bmatrix}
\tag{7.16}
$$

Note that this matrix equation is similar to the one derived in section 6.6 for the primitive bearingless motor with a cylindrical rotor. The i- and j-axis suspension forces are proportional to suspension currents and the coefficients $\lambda'_m + M'_d i_{md}$ and $M'_q i_{mq}$, where λ'_m is the derivative of the flux linkage due to the main field with respect to the radial rotor displacement and $M'_d i_{md}$ and $M'_q i_{mq}$ are derivatives of d- and q-axis winding flux linkages with respect to the radial displacement. Hence the suspension force characteristics of bearingless motors can be determined using the suspension force constants λ'_m, M'_d and M'_q.

For a cylindrical rotor, M'_d and M'_q are equal in value so that the matrix equation can be simplified. This is the case for the SPM motor. If thick magnets are used then $M'_d i_{md}$ and $M'_q i_{mq}$ can also be neglected (with respect to λ'_m) leading to further simplification. For an inset type of permanent magnet rotor (which has saliency on the magnet q-axis) with thick magnets, only $M'_d i_{md}$ can be neglected because of the low d-axis permeance.

In some cases, motors with interior permanent magnet rotors may have parameters λ'_m, M'_d and M'_q which vary depending on the d- and q-axis motor current values because of magnetic saturation in some parts of the rotor core (such as the rotor pole faces). In these cases, parameter identification is the key to successful magnetic suspension since they are functions of the current.

In induction motors, λ'_m is zero since there is no field excitation. In addition M'_d and M'_q are equal because the rotor is cylindrical. The d-axis current is the excitation current and the q-axis current is the torque-generating current. In synchronous reluctance motors also λ'_m is zero. Hence it can be seen that the suspension force equation governs the suspension force and current relationships in many types of bearingless motor which have 4-pole motor windings and 2-pole suspension windings.

7.4 Coordinate transformation from dc to ac bearingless machines

In the previous section the suspension force and current relationships were derived for a dc machine with 4-pole and 2-pole windings. In this section current regulating strategies with rotational and 2/3-phase transformations are shown and basic radial positioning schemes with indirect and direct vector controllers are introduced.

Figure 7.12 shows the structure of a 2-phase ac bearingless machine. This figure is similar to the 2-phase ac machine shown previously in section 7.2; however, the 2-pole windings sa and sb have been added. There are four series-connected windings with eight terminals as illustrated. sa and sb correspond to the sd and sq windings in Figure 7.10. The stator conductors are located in slots in the stator and there is an internal rotor. Two perpendicular axes,

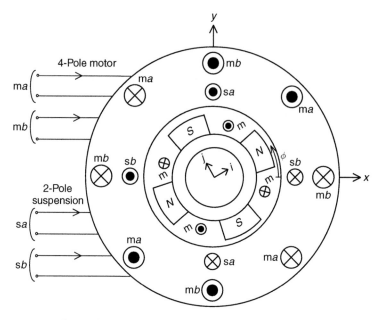

Figure 7.12   2-Phase ac bearingless machine

x and y, are arbitrarily fixed to the stator so that the radial position relationships between the i- and j-axes and the x- and y-axes are defined by

$$\begin{bmatrix} i \\ j \end{bmatrix} = \begin{bmatrix} \cos\phi & \sin\phi \\ -\sin\phi & \cos\phi \end{bmatrix} \begin{bmatrix} x \\ y \end{bmatrix} \tag{7.17}$$

Similar relationships are also defined for the suspension forces so that

$$\begin{bmatrix} F_i \\ F_j \end{bmatrix} = \begin{bmatrix} \cos\phi & \sin\phi \\ -\sin\phi & \cos\phi \end{bmatrix} \begin{bmatrix} F_x \\ F_y \end{bmatrix} \tag{7.18}$$

The 4-pole current relationships between i_{ma}, i_{mb}, i_{md} and i_{mq} were previously given in (7.9). The 2-pole current relationship can also be defined:

$$\begin{bmatrix} i_{sd} \\ i_{sq} \end{bmatrix} = \begin{bmatrix} \cos\phi & \sin\phi \\ -\sin\phi & \cos\phi \end{bmatrix} \begin{bmatrix} i_{sa} \\ i_{sb} \end{bmatrix} \tag{7.19}$$

Figure 7.13 shows an example of an actual 3-phase ac bearingless machine. The 4-pole conductors correspond to the 3-phase conductors as shown previously in section 7.2; however, there are additional 2-pole conductors in the stator slots to generate the suspension forces. The 2-pole conductors are su, sv and sw. The transformation between the 2-phase and 3-phase current representations was

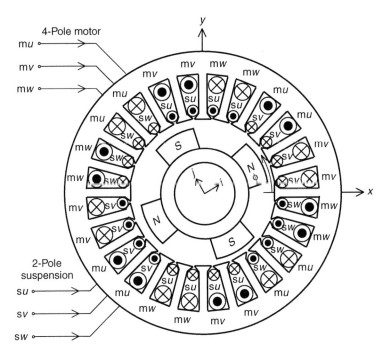

Figure 7.13 3-Phase ac bearingless machine

given in (7.10) for the 4-pole case. The same transformation exists for the 2-pole currents

$$\begin{bmatrix} i_{su} \\ i_{sv} \\ i_{sw} \end{bmatrix} = \frac{\sqrt{2}}{\sqrt{3}} \begin{bmatrix} 1 & 0 \\ \frac{-1}{2} & \frac{\sqrt{3}}{2} \\ \frac{-1}{2} & -\frac{\sqrt{3}}{2} \end{bmatrix} \begin{bmatrix} i_{sa} \\ i_{sb} \end{bmatrix} \tag{7.20}$$

There are six terminals for the two sets of 3-phase windings. However, two neutral points are formed – one is for mu, mv and mw and the other is for su, sv and sw (although these are not shown in the Figure 7.13).

Let us derive a relationship between the suspension forces and winding currents. Substitute (7.18) into the left-hand side of (7.16); an expression for $(i_{sd}, i_{sq})^t$ can be derived from (7.19) and (7.20) which can be substituted into the right-hand side of (7.16). Solving for suspension forces yields

$$\begin{bmatrix} F_x \\ F_y \end{bmatrix} = \begin{bmatrix} \cos\phi & -\sin\phi \\ \sin\phi & \cos\phi \end{bmatrix} \begin{bmatrix} \lambda'_m + M'_d i_{md} & M'_q i_{mq} \\ M'_q i_{mq} & -\lambda'_m - M'_d i_{md} \end{bmatrix}$$

$$\times \begin{bmatrix} \cos\phi & \sin\phi \\ -\sin\phi & \cos\phi \end{bmatrix} \frac{\sqrt{2}}{\sqrt{3}} \begin{bmatrix} 1 & \frac{-1}{2} & \frac{-1}{2} \\ 0 & \frac{\sqrt{3}}{2} & \frac{-\sqrt{3}}{2} \end{bmatrix} \begin{bmatrix} i_{su} \\ i_{sv} \\ i_{sw} \end{bmatrix} \tag{7.21}$$

Note that this matrix equation shows the relationship between the suspension forces in the stationary coordinates as a function of the suspension winding currents. It can be seen that the suspension forces are also functions of the rotor angular position and suspension force constants.

An insight into the current regulating scheme can be obtained from the above matrix equation. Let us suppose that the suspension force constants and rotor angular position are known so that the 3-phase current commands i_{su}^*, i_{sv}^* and i_{sw}^* are generated from the suspension force commands F_x^* and F_y^*.

$$
\begin{bmatrix} i_{su}^* \\ i_{sv}^* \\ i_{sw} \end{bmatrix} = \frac{\sqrt{2}}{\sqrt{3}} \begin{bmatrix} 1 & 0 \\ \frac{-1}{2} & \frac{\sqrt{3}}{2} \\ \frac{-1}{2} & \frac{-\sqrt{3}}{2} \end{bmatrix} \begin{bmatrix} \cos\phi & -\sin\phi \\ \sin\phi & \cos\phi \end{bmatrix}
$$

$$
\times \begin{bmatrix} \lambda_m' + M_d' i_{md} & M_q' i_{mq} \\ M_q' i_{mq} & -\lambda_m' - M_d' i_{md} \end{bmatrix}^{-1} \begin{bmatrix} \cos\phi & \sin\phi \\ -\sin\phi & \cos\phi \end{bmatrix} \begin{bmatrix} F_x^* \\ F_y^* \end{bmatrix} \quad (7.22)
$$

This matrix equation provides the basic operation of a decoupling controller. For the radial position regulating loops, suspension force commands are required in the stationary frame so that the winding current commands are controlled. Hence the actual suspension forces correspond to the force commands. The above equation provides a calculation method for the suspension winding currents so that the actual suspension forces follow these commands.

In order to execute the above calculation in a controller, information about the instantaneous d- and q-axis motor currents and the rotor angular position is needed. Figures 7.14 and 7.15 show some possible system block diagrams for the bearingless motor. Figure 7.14 shows an indirect system structure. The decoupling controller executes the above matrix equation. The d- and q-axis current commands generated in the motor controller are used in the decoupling controller,

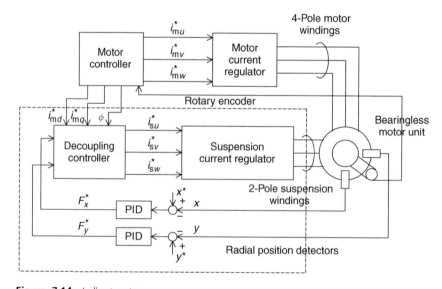

Figure 7.14   Indirect system

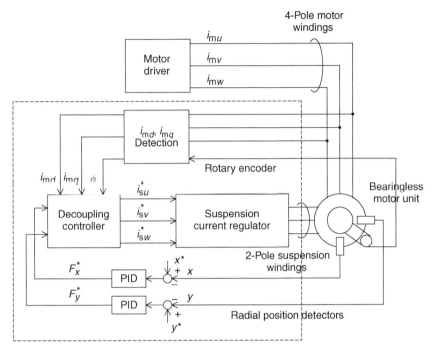

Figure 7.15 Direct system

assuming that the actual d- and q-axis motor currents follow these commands. This system structure is useful when the motor and decoupling controllers are built into one digital processor.

Figure 7.15 shows a direct system structure. In the direct system, the d- and q-axis motor currents are detected from the instantaneous line currents so that these currents are used in the decoupling controller. There is no signal link between motor and decoupling controllers. Hence this system is suitable in the case where the motor is driven by an independent inverter. This structure is good for applications where the motor is driven by a low-cost general-purpose inverter. These are now widely available.

7.5 System block diagrams of bearingless machines

In this section, system block diagrams are introduced which are based on the relationship between the suspension force and the associated winding currents as derived in the previous section. The block diagrams are drawn for a few basic synchronous motor structures, i.e., an SPM motor, a salient-pole permanent magnet motor and a synchronous reluctance motor.

In the cylindrical-rotor synchronous motor such as the SPM motor, there is no magnetic saliency so that the shaft torque is written as $T_r = p_p \lambda'_m i_{mq}$. Since d-axis current does not contribute to the torque production, the motor controller sets $i_{md} = 0$. In addition, armature reaction is not apparent since the flux linkage, which originates from winding inductances, is small compared to the permanent magnet flux linkage when thick permanent magnets are employed. Hence $\lambda'_m \gg M'_q i_{mq}$ and $M'_d i_{md} = 0$. Substituting these relations into (7.22) results in a simple matrix calculation

$$\begin{bmatrix} i^*_{su} \\ i^*_{sv} \\ i_{sw} \end{bmatrix} = \frac{1}{\lambda'_m} \frac{\sqrt{2}}{\sqrt{3}} \begin{bmatrix} 1 & 0 \\ \frac{-1}{2} & \frac{\sqrt{3}}{2} \\ \frac{-1}{2} & \frac{-\sqrt{3}}{2} \end{bmatrix} \begin{bmatrix} \cos 2\phi & \sin 2\phi \\ \sin 2\phi & -\cos 2\phi \end{bmatrix} \begin{bmatrix} F^*_x \\ F^*_y \end{bmatrix} \tag{7.23}$$

Figure 7.16 shows a block diagram for the SPM bearingless motor. The suspension force commands F^*_x and F^*_y are generated from the radial position regulators using the detected radial shaft position. The decoupling and 2-phase/3-phase blocks generate 2-pole suspension winding current commands using the above equation. A current regulator, i.e., a current-controlled inverter, regulates the 3-phase currents. In a motor speed controller, the rotational speed and angular position are detected. The detected speed is compared with its command and a torque command is generated via the speed regulator. Using this, the q-axis motor current command is generated. Note that the d-axis motor current command is zero. The 2-phase and 3-phase current commands are generated and a current regulator provides instantaneous current regulation of the motor winding currents. The radial position controller requires only λ'_m and not $M'_d i_{md}$ and $M'_q i_{mq}$, so the d- and q-axis motor current values are not required.

In the salient-pole synchronous machine, the structure of the system block diagram is rather complicated. The inverse matrix shown in (7.22) can be simplified to

$$\begin{bmatrix} \lambda'_m + M'_d i_{md} & M'_q i_{mq} \\ M'_q i_{mq} & -\lambda'_m - M'_d i_{md} \end{bmatrix}^{-1} = \frac{1}{K_f} \begin{bmatrix} \cos \theta_f & \sin \theta_f \\ \sin \theta_f & -\cos \theta_f \end{bmatrix} \tag{7.24}$$

where

$$K_f = \sqrt{(\lambda'_m + M'_d i_{md})^2 + (M'_q i_{mq})^2} \tag{7.25}$$

$$\theta_f = \tan^{-1} \left(\frac{M'_q i_{mq}}{\lambda'_m + M'_d i_{md}} \right) \tag{7.26}$$

and the current commands are defined by

$$\begin{bmatrix} i^*_{su} \\ i^*_{sv} \\ i_{sw} \end{bmatrix} = \frac{1}{K_f} \frac{\sqrt{2}}{\sqrt{3}} \begin{bmatrix} 1 & 0 \\ \frac{-1}{2} & \frac{\sqrt{3}}{2} \\ \frac{-1}{2} & \frac{-\sqrt{3}}{2} \end{bmatrix} \begin{bmatrix} \cos(2\phi + \theta_f) & \sin(2\phi + \theta_f) \\ \sin(2\phi + \theta_f) & -\cos(2\phi + \theta_f) \end{bmatrix} \begin{bmatrix} F^*_x \\ F^*_y \end{bmatrix} \tag{7.27}$$

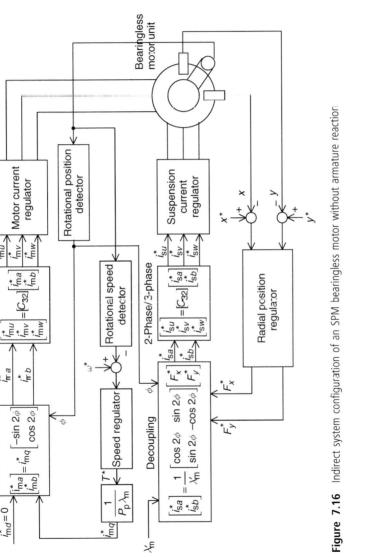

Figure 7.16 Indirect system configuration of an SPM bearingless motor without armature reaction.

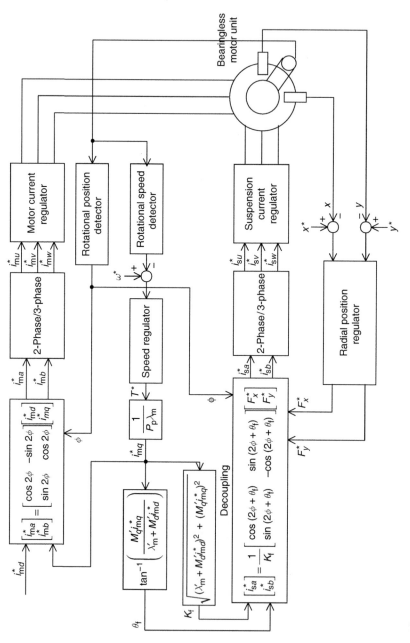

Figure 7.17 An indirect system configuration of a salient-pole bearingless motor with armature reaction compensation

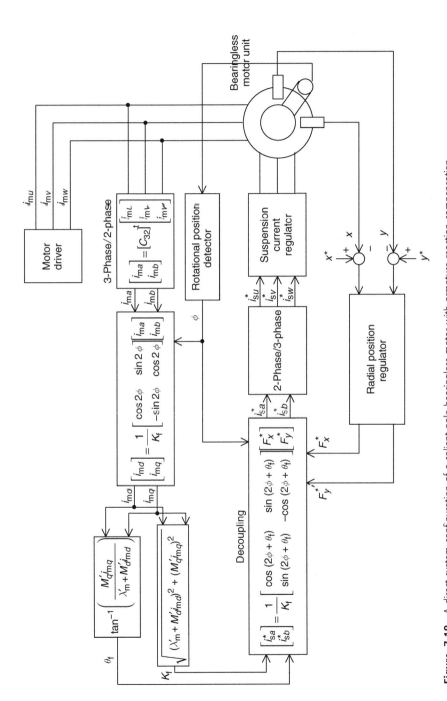

Figure 7.18 A direct system configuration of a salient-pole bearingless motor with armature reaction compensation

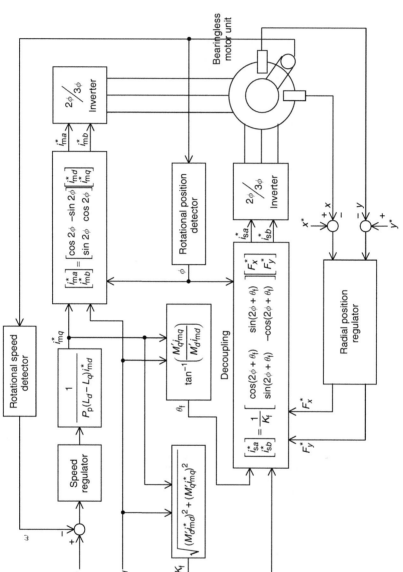

Figure 7.19 System block diagram of a synchronous reluctance bearingless motor

Figure 7.17 shows a block diagram for the salient-pole PM motor. The blocks corresponding to (7.25) and (7.26) are added and the function for the decoupling controller is modified to include θ_f.

Figure 7.18 shows a direct type of system configuration. The d- and q-axis motor currents are detected using the measured 3-phase line currents, then K_f and θ_f are generated in blocks corresponding to (7.25) and (7.26). Note that the motor-drive inverter is operating independently of the radial-position-regulating loops. Flux density detectors, e.g., Hall sensors or search coils, can also be integrated in motor airgap, and then, these sensors provide K_f, θ_f and ϕ. The detailed system requirements and machine parameter measurements, as well as detection methods, are described in a later section.

Figure 7.19 shows the block diagram for a synchronous reluctance type of bearingless motor. In synchronous reluctance motors there is no permanent magnet excitation so that λ_m in the torque equation is zero and λ'_m in the suspension force equation is also zero. The q-axis motor current command block and the K_f and θ_f blocks are modified. Detailed requirements for synchronous reluctance bearingless motors will be described in a later chapter.

References

[1] D.W. Novotny and T.A. Lipo, "Vector Control and Dynamic of AC Drive", Oxford University Press, Oxford, 1996, ISBN 0198564392.
[2] Akira Chiba, Tazumi Deido, Tadashi Fukao and M.A. Rahman, "An Analysis of Bearingless AC Motors", IEEE Transaction on Energy Conversion, Vol. 9, No. 1, March, 1994, pp. 61–68.

8

Field orientation, VA requirement and magnetic saturation

Akira Chiba

In this chapter some further fundamental aspects of the bearingless motor are discussed. The first point concerns the misalignment of the revolving magnetic field. A phase shift between the detected and actual field orientation produces interference in the two-axis radial force generation which causes operational discrepancies and hence stability problems in the radial position regulation loops. The phase shift may occur if the suspension force constants are not precise and accurate. These constants may vary depending on torque and suspension force outputs. The influence and measurement of the suspension force misalignment are therefore introduced.

In the second section an example is shown for the voltage and current (VA) requirement at the suspension winding terminals. For efficient operation and maximization of capacity, the required VA should be as low as possible. The requirement is dependent on the required suspension force and the radial sensor signal conditions. Since both motor and suspension voltages are proportional to the rotational speed, the suspension winding VA requirement is also expressed as a ratio with respect to the motor winding VA requirement.

In the final section the influence of magnetic saturation on the suspension force generation is discussed. A relationship between the suspension force, the current and the tooth flux density is shown for a primitive bearingless motor so that an operating point can be suggested.

8.1 Misalignment of field orientation

Figure 8.1(a,b) shows the principles of radial force interference between the x and y axes when the angular position of the revolving magnetic field has a phase lag angle of θ. Figure 8.1(a) shows the radial force generation without interference. The ma winding flux indicates the equivalent current directions for the revolving magnetic field Ψ_{ma}, which has a 4-pole distribution. For suspension, positive current in winding sa generates a 2-pole flux Ψ_{sa}. In air-gap section 1 the directions of Ψ_{ma} and Ψ_{sa} are the same whereas they are opposite in air-gap section 3 so that radial force is generated along the x-axis.

Let us suppose that the 4-pole excitation current is delayed by an electrical angle of 2θ so that the 4-pole revolving magnetic field is delayed by a mechanical angle θ with respect to the x–y coordinates. Winding ma', shown by the dotted line in Figure 8.1(a), indicates the position of the equivalent winding for a 4-pole magnetic field with the lag angle of θ. New x'- and y'-axes are drawn in Figure 8.1(b), which are rotated by θ and correspond to the rotation of the 4-pole magnetic field. Let us examine the case when a suspension current exists only in winding sa. This can be represented by two equivalent current components in the sa'- and sb'-windings, which are on the x'- and y'-axes. With current in winding sa', a flux Ψ'_{sa} is generated and the interaction with the 4-pole flux Ψ'_{ma} generates a radial force of F'_x. When there is current in winding sb', a 2-pole flux Ψ'_{sb} is generated which results in a radial force of F'_y. The resulting radial force F' is a vector sum of these two radial forces and has a direction which is delayed by an angle 2θ. Hence if interference occurs in the radial forces in two perpendicular axes, the response of the radial positioning loops becomes worse. A basic example was discussed previously in section 3.3. More examples of the practical implementation of the bearingless motor concept which highlight this sort of alignment issue are discussed in this section.

Figure 8.2 shows a method to check the radial force interference under steady-state conditions [1]. An external static radial force F_{ex} is applied to a rotating shaft. The amplitude and direction of the external radial force are fixed. This can be implemented by using a weight hanging from a thread on a pulley with the other end fixed to a mechanical bearing on the rotating shaft. In addition, with a horizontal shaft arrangement, the shaft and rotor weight acts as an external static force so that the shaft arrangement has to be suspended by magnetic force with radial position feedback loops. If the motor magnetic field corresponds to the sine and cosine components in the modulation block then the radial force command F_x^* is generated which has the same amplitude but opposite direction to the external force so that the net force F_x is zero.

However, the radial force commands are automatically varied if there is a misalignment in angular position detection of the motor magnetic field. If the mechanical phase lag angle is θ then a radial force command F^* is generated with a phase lead of 2θ so that the generated radial force is exactly opposite in order to satisfy the force equilibrium. We can check the misalignment of the

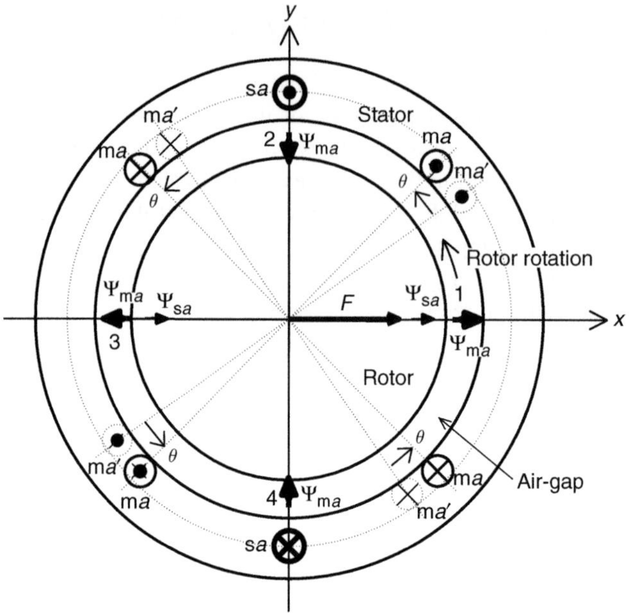

(a)

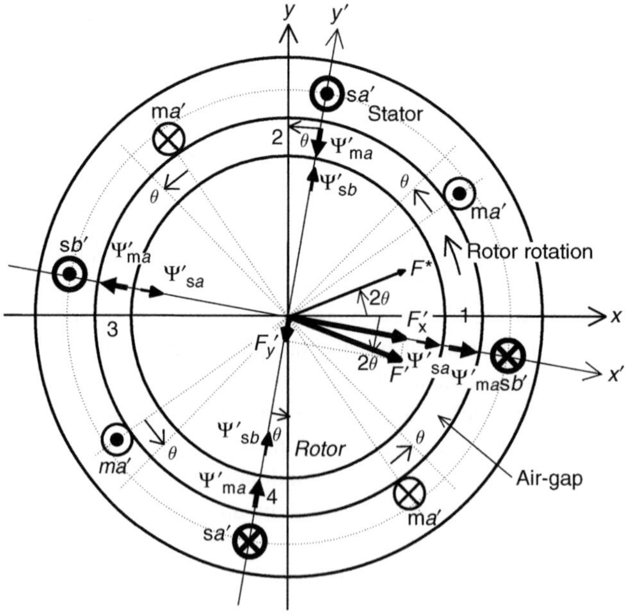

(b)

Figure 8.1 The variations in the radial force direction depending on the 4-pole flux orientation: (a) exact field flux orientation; (b) the 4-pole flux with a phase lag angle θ

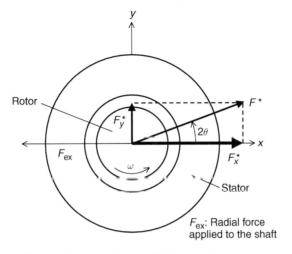

Figure 8.2   Radial force application to check radial force interference between x- and y-axes

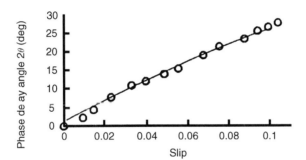

Figure 8.3   A phase shift angle

revolving motor magnetic field detection by observing the radial forces which are generated by the radial position controllers.

Figure 8.3 shows an example of the phase-shift angle for an induction-type bearingless motor. At slip = 0, the phase shift angle is almost zero, i.e., the line current and revolving magnetic field almost correspond. For example, the u-phase line current indicates a cosine component of flux linkage when the u-phase MMF and the x-axis are aligned. As the rotor slip is increased, the revolving magnetic field is delayed with respect to the line current (as would be expected in an induction motor). The amplitude of the revolving magnetic field also decreases when the current amplitude is kept constant. In the following examples, a field-oriented controller and a constant current controller are employed to see phase-shift influence.

Figure 8.4 shows the measured radial force references as a function of rotor slip for two cases. The first is with exact field orientation and the second is with the misalignment as shown in the previous figure. An external radial force is

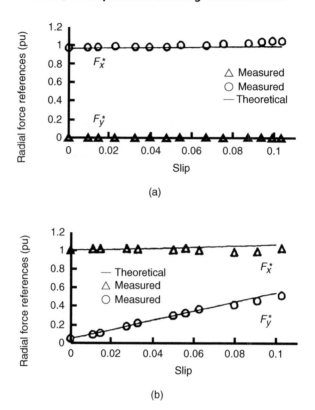

(a)

(b)

Figure 8.4 Radial force references: (a) exact field orientation; (b) misalignment in magnetic field

applied along the negative x-axis. In Figure 8.4(a), F_x^* and F_y^* are unity and zero respectively. In Figure 8.4(b), F_y^* increases as the slip increases and the revolving magnetic field is delayed with respect to the cosine and sine components of modulation. Since there is a negative phase shift in motor field orientation a positive F_y^* is generated automatically. At slip $= 0.1$, $F_y^* = 0.48$ and $F_x^* = 1$, which correspond to $2\theta = 26$ deg. This angle corresponds to the phase shift angle shown in Figure 8.3.

There are some practical points that can be checked during the measurement of the radial force references. The radial position loops may become unstable if the misalignment is too large so that the controller parameters in the radial position loops should be adjusted to allow for the phase shift angle. In some cases, decreasing the loop gains will provide some margin.

During measurements, the shaft is suspended by a magnetic force so that waveforms are produced from the radial force references. These include fluctuations caused by sensor noise, unbalance and mechanical misalignments. The applied external radial force may have synchronous variations caused by synchronous shaft displacements. Hence the ac component should be eliminated by a low pass filter to detect only the dc components.

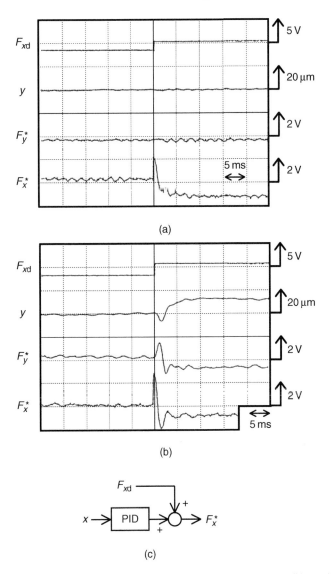

Figure 8.5 Step disturbance force suppression: (a) exact field orientation; (b) misalignment in magnetic field; (c) step disturbance summation

Figure 8.5 shows a comparison of step disturbance suppression in the radial position loops. A step disturbance signal is added to the output of the x-axis radial-position PID controller while the shaft is rotating at constant speed and suspended by a magnetic force. Hence the influence of the step disturbance results in x-axis shaft radial displacement so that the radial position controllers automatically generate a radial force reference F_x^* to suppress the disturbance signal. Because the speed of this process is determined by the rotor shaft inertia, there is a delay of several milliseconds. Figure 8.5(a,b) shows the behaviour of the disturbance signal F_{xd}, displacement y and radial force references F_y^* and F_x^*. The radial force reference F_x^* is the output signal from the step signal summation as shown in Figure 8.5(c). In Figure 8.5(a) the influence of step disturbance is seen in only F_x^* because of the exact field orientation. After the transient response, F_x^* decreases to cancel the step disturbance.

In Figure 8.5(b) the influence of the x-axis step disturbance is seen in the y-axis variables. A misalignment of field orientation generates radial force on the y-axis, hence the radial positioning loop automatically generates a signal F_y^* to suppress the disturbance. The radial displacement y cannot be totally suppressed. In addition, more vibration is seen in the response of F_x^* because of the inferior feedback loop characteristics caused by coupling between the two perpendicular axes (as previously shown in section 3.3).

Figure 8.6(a,b) shows a comparison of the waveforms during a small step in acceleration. The shaft speed command $2\omega^*$ is increased from 1280 to 1430 r/min as a step function. To follow the speed command the torque current increases to the maximum value so that the speed ramps up at a rate determined by the mechanical inertia. With an exact field orientation the influence of acceleration is not seen in radial displacements x and y. However, significant fluctuations are seen if there is a misalignment in field orientation. The misalignment angle appears just after the step change in the speed command so that it decreases as the shaft speed 2ω approaches the speed command. The misalignment angle is zero when shaft speed is following the speed command. Fluctuations in radial displacements are seen just after the maximum misalignment angle occurs (although misalignment angle is not shown here), which is during torque generation for the induction type of bearingless motor. The misalignment of the actual magnetic field components, and the modulating cosine and sine functions, should be minimized to prevent high radial disturbance.

8.2 VA requirement

In magnetic suspension, regulated ac current should be supplied to the suspension windings. These induce winding back-EMFs which are proportional to the winding inductance and supply frequency. To produce successful current regulation the rated voltage of the supply inverter must be larger than the induced EMFs. The product of the terminal voltage and line current is called the VA (unit: [volt-ampere]) requirement.

If the number of series turns N of a winding is increased then the required current needed to produce the same MMF decreases by $1/N$. However, the induced terminal voltage increases by N since the inductance increases by N^2. Hence the product of voltage and current is not a function of the number of winding turns although the VA requirement is found to be a unique value.

In this section the voltage and current requirements at the suspension winding terminals are derived, for example, for the induction-type bearingless motor.

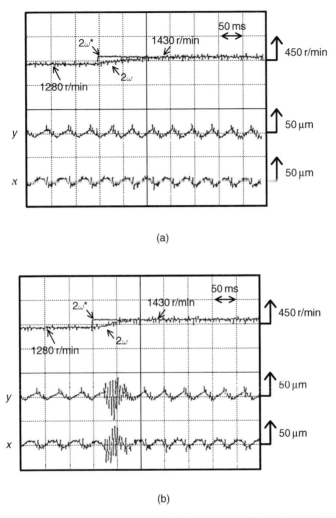

Figure 8.6 Small step acceleration: (a) exact field orientation; (b) misalignment in magnetic field

The required voltage and current for static radial force is derived and it is found that the VA requirement is proportional to the rotational speed. It is also shown that the no-load VA requirement at the motor winding terminals for motor operation magnetization is also proportional to rotational speed so that the ratio of these VA requirements is independent of speed. The VA ratio provides a simple index for a bearingless motor. In the following discussion of an induction type of bearingless motor, the VA requirement, its ratio and the power ratio are addressed. Finally the VA and efficiency definition for a permanent magnet bearingless motor is also described for completeness.

8.2.1 Induction-type bearingless motor

Let us suppose that a 3-phase 4-pole winding is the motor winding and a 3-phase 2-pole winding is the suspension winding. The rms value of a 3-phase line current in the 4-pole winding is I_m and the rms line-to-line voltage at the motor winding terminals V_{ml} is

$$V_{ml} = 2\sqrt{3}\omega L_4 I_m \qquad (8.1)$$

where ω is the mechanical angular speed of the shaft at no torque (synchronous speed) so that the current frequency is equal to 2ω (since it is a 4-pole winding). This assumes no rotor current. The total voltage and current requirement VA_m at the motor winding terminals is a product of the line-to-line voltage and current for all three phases and can be written as (assuming a balanced 3-phase set):

$$VA_m = \sqrt{3}V_{ml}I_m \qquad (8.2)$$

Substituting (8.1) into (8.2) yields

$$VA_m = 6\omega L_4 I_m^2 \qquad (8.3)$$

To produce a static radial force F the rms line current in the suspension windings I_s is

$$I_s = \frac{F}{3M'I_m} \qquad (8.4)$$

and has a frequency of 2ω. The line-to-line voltage V_{sl} at the suspension winding terminals is therefore

$$V_{sl} = 2\sqrt{3}\omega L_2 I_s \qquad (8.5)$$

so that the total VA requirement of the 2-pole winding for 3-phases is written as

$$VA_s = \sqrt{3}V_{sl}I_s \qquad (8.6)$$

Substituting (8.5) into (8.6) yields

$$VA_s = 6\omega L_2 I_s^2 \qquad (8.7)$$

Note that the VA requirements in (8.3) and (8.7) are proportional to the speed. Hence the ratio of the VA requirements is

$$\frac{VA_b}{VA_m} = \frac{L_2 I_s^2}{L_4 I_m^2} \qquad (8.8)$$

The VA ratio is a function of the inductances, motor exciting current and required suspension current. This ratio provides a simple index for the voltage and current

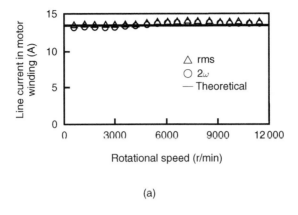

(a)

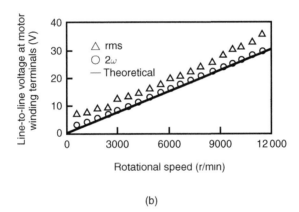

(b)

Figure 8.7 Voltage and current at motor terminals: (a) line current; (b) line-to-line voltage

requirements of the suspension inverter. For example, if the ratio is 0.01 then a motor excitation of 2 kVA requires a suspension inverter of 20 VA.

Figure 8.7 shows the measured rms values of line current and line-to-line voltage at the motor winding terminals. Over the whole speed range the line current is constant, providing reasonable excitation. The voltage is proportional to the rotational speed. Since the test machine is driven by a PWM inverter, the rms value is slightly more than the fundamental component, which has an angular frequency of 2ω (where ω is the rotational speed in rad/sec at no load). The fundamental component corresponds to the theoretical values.

Figure 8.8 shows waveforms for the line current i_{mu} and line-to-line voltage v_{muv} at a speed of 10 200 r/min. The current waveforms include the switching ripple caused by the PWM output voltage of the voltage-source inverter. Surge voltages are also seen at the switching instances.

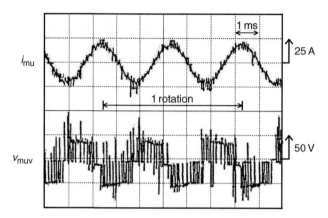

Figure 8.8 Current and voltage waveforms of motor terminals at 10 200 r/min

Figure 8.9 shows the rms current and voltage for the suspension winding. The current has several harmonic components in addition to the 2ω component necessary to produce the static suspension force. The major additional component has a frequency of ω which is dependent on the rotational speed and is maximum at the critical speed of 6000 r/min. This is because the vibration caused by mechanical unbalance is suppressed by the suspension winding current of this frequency. On the other hand the 2ω frequency component is almost constant since the static radial force is constant at 2.6 kgf across the whole speed range. In the voltage waveform the 2ω component is dominant and the voltage is proportional to the rotational speed, as expected.

Figure 8.10 shows the current and voltage waveforms for the suspension winding when a static force of 2.6 kgf is applied at a speed of 10 200 r/min. It can be observed that the 2ω voltage and current components are dominant even when the radial force suspension current is driven by a PWM inverter.

Table 8.1 shows a summary of the voltages, currents and VAs for the motor and suspension windings of the induction-type bearingless motor example. The rms values are measured by digital power meters that have a wide frequency range and the fundamental components are obtained by an FFT (fast Fourier transform) analyser. The measured (total rms and 2ω components) and theoretical values are shown for comparison. The measured VA ratio of the fundamental component is 0.0043 which is close to the theoretical value of 0.0039. The measured rms VA ratio is 0.0059, which is slightly high. The reasons for this discrepancy can be summarized as follows:

(a) The measured rms current includes a current component to suppress the mechanical unbalance. This has an angular frequency of ω.
(b) The measured rms current includes high frequency components caused by sensor noise.
(c) The measured rms voltage includes PWM harmonics.

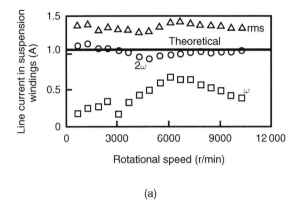

(a)

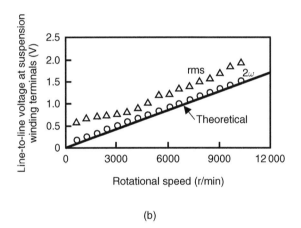

(b)

Figure 8.9 Current and voltage of suspension winding terminals with static force of 2.6 kgf: (a) line current; (b) line-to-line voltage

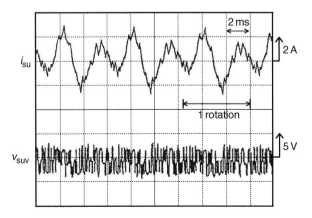

Figure 8.10 Current and voltage waveforms of suspension winding terminals at 10 200 r/min

Table 8.1   Voltage, current and VA at 10 200 r/min with 2.6 kgf

		Motor windings	Suspension windings
Measured 2ω component	V_f	25.7 V	1.49 V
	I_f	13.6 A	1.0 A
	$\sqrt{3}V_f I_f$	605 VA	2.6 VA
	VA ratio	0.0043	
Measured rms values	V_{rms}	31.0 V	1.94 V
	I_{rms}	14.0 V	1.3 A
	$\sqrt{3}V_{rms} I_{rms}$	752 VA	4.4 VA
	VA ratio	0.0059	
Theoretical	V_t	28.3 V	1.44 V
	I_t	13.6 A	1.0 A
	$\sqrt{3}V_t I_t$	666 VA	2.6 VA
	VA ratio	0.0039	

The example shown in this section is based on Reference [2]. For the test induction type of bearingless motor the VA ratio is quite low when compared to its shaft weight of 2.6 kg. The VA ratio increases as the required radial force increases. If a ninefold increase in the radial force is required in this machine then the VA ratio increases by 81 times (since it is a square relationship).

Several other machines have been tested. Experience suggests that there are several key issues that have to be addressed to obtain performance figures that are close to the theoretical. These are:

(a) A reduction of mechanical unbalance and misalignment of sensor target.
(b) Noise reduction of the sensor output signals.
(c) A reduction of the surface roughness of sensor target ring.
(d) Optimal voltage selection of the inverter for both the motor and radial force winding supply.

8.2.2 Permanent magnet type bearingless motor

An example of an inset permanent magnet bearingless motor is described in this section. In PM machines the motor winding terminal VA is zero at no-load (since there is no motor winding current, the excitation is via magnets). Hence the VA ratio has to be determined using the maximum rated torque VA requirement.

Table 8.2 shows the result of a load test at 10 000 r/min [3]. The voltage and current ratings of the motor winding terminals are 150 V and 11.1 A respectively. The input and shaft powers are 2.23, 2.06 kW. At the suspension winding terminals the current, voltage and power are 3.95 A, 39.3 V and 25.7 W. These values are required to suppress the synchronous disturbance caused by the mechanical shaft unbalance, the misalignment of shaft parts, sensor noise and so on. The VA ratio at the rated maximum torque point is 0.09. This VA ratio is

Table 8.2 Load test results of an inset permanent magnet bearingless motor

Rotational speed	10 000 r/min
Motor winding	
Line current I_m	11.1 A
Line-to-line voltage V_{ml}	150 V
Input VA, VA_m	2.88 kVA
Input power P_m	2.23 kW
Copper loss	95.2 W
Iron loss + other loss	75.0 W
Output power P_o	2.06 kW
Efficiency P_o/P_m	92.4%
Suspension winding	
Line current I_s	3.95 A
Line-to-line voltages V_{sl}	39.3 V
Input VA, VA_s	269 VA
Input power P_s	25.7 W
VA ratio	0.09
Efficiency $P_o/(P_m + P_s)$	91.3%

reasonably low. Further reduction is possible by improvement of the mechanical precision and sensor signal conditions; however, this adds additional cost.

8.3 Magnetic saturation

In this section the effect of magnetic steel saturation on the suspension force is discussed [4]. Results from the finite element analysis of a primitive bearingless machine are shown and an operating point is suggested. The concept of radial force density is then introduced.

Figure 8.11(a–c) shows flux plots for a primitive bearingless motor with suspension winding currents of $I_2 = 0.8$, 1.2 and 2.5 A. The cases for $I_2 = 0$ and 0.5 A were previously shown in Figure 6.3. At $I_2 = 0$ the flux distribution is symmetrical. As I_2 increases the 2-pole flux is superimposed. In Figure 8.11(a) the tooth flux density B_- is almost zero. In Figure 8.11(b) the radial force reaches the maximum value and after this point the radial force decreases as shown in Figure 8.11(c).

Figure 8.12 shows the variation of calculated tooth flux density. The current conditions a to e are for the flux plots in Figures 6.3(a,b) and 8.11(a–c) respectively. In addition, the measured flux densities in a test machine are shown. The flux densities B_+ and B_- are the same at $I_2 = 0$. Then B_+ increases, and the influence of magnetic saturation is seen as the maximum is reached. For silicon steel, which is used in almost all motors and generators, the saturated flux density is about 1.7 to 2.0 T. On the other hand, B_- keeps decreasing and becomes zero at around $I_2 = 0.8$ A. After this point B_- becomes negative and further decreases until saturation sets in and $B_- = -B_+$. Since the magnetic force is always attractive and is almost proportional to the square of airgap flux density, the effect of saturation on the radial force occurs from the operating

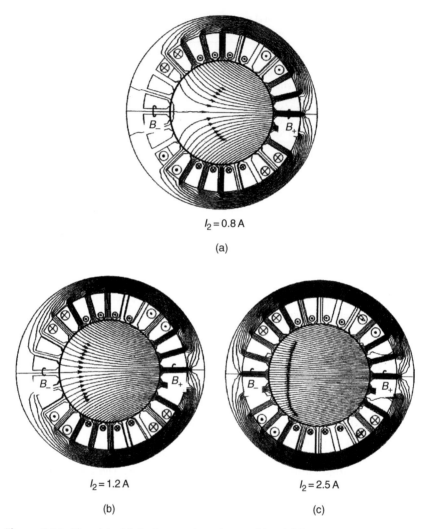

$I_2 = 0.8$ A

(a)

$I_2 = 1.2$ A

(b)

$I_2 = 2.5$ A

(c)

Figure 8.11 Flux plots: (a) B_- is approximately zero; (b) radial force is maximum; (c) radial force is decreased

point c; and by the time point e is reached, saturation has caused the airgap flux to become more evenly distributed, and hence the radial force to decrease.

Figure 8.13 shows the relationship between the radial force and suspension winding current. As illustrated by the finite element method (FEM) analysis, the radial force is maximum at point d, then it decreases as the steel saturation further increases. The derivative (slope) of the curve gradually decreases from around point c where saturation first occurs. Experiments were possible only up to just before point c because of a decrease in loop gain. It may be possible to operate further into the saturated region if a nonlinear controller is used to compensate radial force saturation.

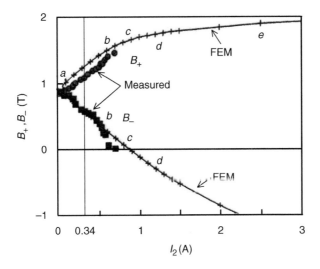

Figure 8.12 Flux density variation

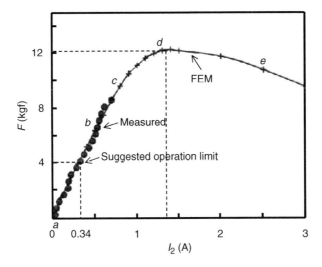

Figure 8.13 Radial force vs radial force current

A suggested operation limit point is about 4 kgf. One may feel that this operation point is too low for radial force generation. However, there are some merits to operation in the region between point *a* and this suggested point:

(a) Good linearity is obtained in the relationship between radial force and current.
(b) Flux density variation in the teeth is about 0.25 T for B_+ and B_-, which should not greatly affect the performance of the motor-drive characteristics.

In the case of a highly saturated electric machine, such as those used in aerospace applications, (b) may not apply. However, in most industrial applications there is some room for small flux variations.

The example shown in this section is based on specific dimensions. The rotor diameter D is 49 mm with an axial length l of 30.8 mm so that the rotor shadow area Dl is about 15 cm^2. A 4 kgf radial force has a radial force density of about 0.3 kgf/cm^2. This value may appear low compared to a radial magnetic bearing which has a practical radial force density of 1 kgf/cm^2. However, the area is usually much larger in a bearingless motor compared to a magnetic bearing, because the motor and magnetic bearing are combined, so that the suggested radial force density is acceptable in most cases.

References

[1] A. Chiba, R. Furuichi, Y. Aikawa, K. Shimada, Y. Takamoto and T. Fukao, "Stable Operation of Induction-Type Bearingless Motors Under Loaded Conditions", IEEE Trans. on IAS, Vol. 33, No. 4, July 1997, pp. 919–924.
[2] E. Ito, A. Chiba and T. Fukao, "A Measurement of VA Requirement in an Induction-type Bearingless Motor", IV International Symposium on Magnetic Suspension Technology, Gifu-Japan, May 1998, pp. 125–137.
[3] K. Inagaki, A. Chiba, M. A. Rahman and T. Fukao, "Performance Characteristics of Inset-Type Permanent Magnet Bearingless Motor Drives", IEEE PES, WMC, CDROM Singapore, January 2000.
[4] A. Chiba and T. Fukao, "The Maximum Radial Force of Induction Machine-Type Bearingless Motor Using Finite Element Analysis", ISMB, August 1994, pp. 333–338.

9

Cylindrical permanent magnet synchronous bearingless motors

Masahide Oshima

The advantages of the permanent magnet type of bearingless motor can be summarized as (1) small size and lightweight, (2) high power-factor and high efficiency and (3) suspension forces that can be generated without excitation current in the main winding. In this chapter, some structures and a control system suitable for cylindrical permanent magnet synchronous bearingless motors are shown. Armature reaction flux is considered in this type of machine in order to regulate the rotor radial position accurately. The rotor radial position control strategy uses this information to calculate the magnitude and the direction of the airgap flux and this is described together with the calculation of the resultant flux from the permanent magnet field excitation and the q axis flux. It is then shown how the permanent magnet thickness and the airgap length are determined for an optimal design. The influence of the stator winding MMF on the permanent magnet demagnetization is also described. The test results from a prototype of a surface permanent magnet synchronous bearingless motor are included as examples.

9.1 Structure of surface permanent magnet (SPM) rotor

Figure 9.1 shows a typical SPM rotor structure [1–4]. A 4-pole structure is described here as an example. Four permanent magnets are mounted on the iron surface. These are magnetized to produce four magnetic poles. A steel can or carbon-fibre casing usually wraps the permanent magnets in order to fix them to the rotor. It is important to design the permanent magnets to avoid irreversible

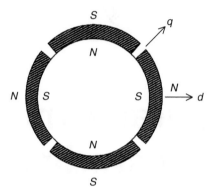

Figure 9.1 SPM type rotor structure

demagnetization because the suspension flux is superimposed on to the magnetic field and can act in the opposite direction to the magnet polarity. Conversely, however, thin permanent magnets result in more effective radial suspension force generation; this will be discussed in section 9.4.

9.2 Radial suspension force and suspension winding current

In this section, the theoretical relationship between the radial suspension force and the suspension winding current is derived. The permanent magnet MMF is replaced by an equivalent motor winding current component I_{mp} [5,6].

The equivalent 4-pole current $\bar{I}_{me}$ for the motor winding and magnets is represented as

$$\bar{I}_{me} = I_{mp} + jI_{mq} \tag{9.1}$$

where $\bar{I}_{me}$ denotes a phasor in a 2-phase coordinate system. I_{mq} is the q-axis component, which is the actual motor winding current. The d-axis main winding current should be zero to realize high torque. I_{mp} is the equivalent d-axis component of the permanent magnet MMF and $I_{mp} = \sqrt{3/2}E_0/X_m$, where E_0 is a synchronous internal voltage and X_m is an armature reaction reactance, so that I_{mp} is hereafter referred to as the equivalent permanent magnet current.

Stationary and synchronously rotating coordinate systems are drawn in Figure 9.2, where the perpendicular axes in the stationary coordinate are defined as the a- and b-axes. The counter-clockwise direction is defined as the positive angular and rotational direction. In addition, the motor windings N_{ma} and N_{mb}

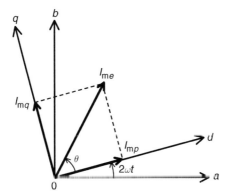

Figure 9.2   Coordinate transformation

generate MMFs on the a- and b-axes respectively. Hence the equivalent currents i_{mae} and i_{mbe} in motor windings N_{ma} and N_{mb} are represented as

$$i_{mae} = I_{me} \cos 2(\omega t + \theta/2) \tag{9.2}$$

$$i_{mbe} = I_{me} \sin 2(\omega t + \theta/2) \tag{9.3}$$

where

$$I_{me} = \sqrt{I_{mp}^2 + I_{mq}^2} \tag{9.4}$$

$$\theta = \tan^{-1}\left(\frac{I_{mq}}{I_{mp}}\right) \tag{9.5}$$

and ω is the mechanical angular speed. The resultant flux of the permanent magnet field excitation and the q-axis winding current can be considered as the 4-pole flux generated by the equivalent currents i_{mae} and i_{mbe}. The magnitude of the resultant flux is represented by a product of the self-inductance and the current amplitude I_{me}. The phase-lead mechanical angle $\theta/2$ is in the rotational direction of the revolving field, and it is caused by the armature reaction.

Let us define the currents flowing in the suspension windings N_{sa} and N_{sb} in the stator core as i_{sa} and i_{sb} respectively. The directions of the MMFs generated by the suspension winding currents i_{sa} and i_{sb} correspond to x- and y-axes respectively. Replacing i_{4a}, i_{4b}, i_{2a} and i_{2b} in (6.39) with i_{mae}, i_{mbe}, i_{sa} and i_{sb} and substituting (9.2) and (9.3) into i_{mae} and i_{mbe} produces the radial forces F_x and F_y:

$$\begin{bmatrix} F_x \\ F_y \end{bmatrix} = M' I_{me} \begin{bmatrix} \cos 2\left(\omega t + \frac{\theta}{2}\right) & \sin 2\left(\omega t + \frac{\theta}{2}\right) \\ \sin 2\left(\omega t + \frac{\theta}{2}\right) & -\cos 2\left(\omega t + \frac{\theta}{2}\right) \end{bmatrix} \begin{bmatrix} i_{sa} \\ i_{sb} \end{bmatrix} \tag{9.6}$$

Equation (9.6) shows the relationship between the perpendicular radial suspension force components F_x and F_y and the suspension winding currents i_{sa} and i_{sb}. Solving (9.6) for i_{sa} and i_{sb} yields

$$\begin{bmatrix} i_{sa} \\ i_{sb} \end{bmatrix} = \frac{1}{M'I_{me}} \begin{bmatrix} \cos 2\left(\omega t + \frac{\theta}{2}\right) & \sin 2\left(\omega t + \frac{\theta}{2}\right) \\ \sin 2\left(\omega t + \frac{\theta}{2}\right) & -\cos 2\left(\omega t + \frac{\theta}{2}\right) \end{bmatrix} \begin{bmatrix} F_x \\ F_y \end{bmatrix} \qquad (9.7)$$

From (9.4) and (9.5), the amplitude I_{me} and the phase-lead angle θ of the equivalent current in the motor winding are calculated using the q-axis current I_{mq}. Then the amplitude and phase of the suspension winding currents are obtained using (9.7) in accordance with I_{me} and θ. Hence, by considering the armature reaction, the cylindrical permanent magnet synchronous bearingless motor can be successfully operated without mutual interference between the radial suspension force components in the two perpendicular axes.

9.3 Equations of voltage and current

In this section, the voltage and current equations of the motor and the suspension windings are derived using the equivalent current in the motor windings shown in the previous section [7].

For a 4-pole machine, the amplitude V_{m2} of sinusoidal voltage at the motor winding terminals in a 2-phase coordinate system can be written as

$$V_{m2} = 2\omega L_m I_{me} \qquad (9.8)$$

where L_m is the self-inductance of the motor winding. Similarly, the fundamental voltage amplitude V_{s2} at the suspension winding terminals in the 2-phase coordinate system is also written as

$$V_{s2} = 2\omega L_s I_{s2} \qquad (9.9)$$

where L_s is the self-inductance of the suspension winding and I_{s2} is the fundamental amplitude of the suspension winding current. From (9.7), I_{s2} can be obtained from

$$I_{s2} = \frac{\sqrt{F_x^2 + F_y^2}}{M'I_{me}} \qquad (9.10)$$

Substituting (9.10) into (9.9) yields

$$V_{s2} = \frac{2\omega L_s}{M'I_{me}}\sqrt{F_x^2 + F_y^2} \qquad (9.11)$$

Example 9.1   When a cylindrical permanent magnet synchronous bearing-less motor operates at 3000 r/min and the q-axis current I_{mq} is 7.8 A, the equivalent permanent magnet current I_{mp} is 20 A and the self-inductances L_m and L_s of motor and suspension windings are 4.5 and 0.7 mH respectively. The derivative of mutual inductance M' is 0.32 H/m.

(a) Derive the equivalent current amplitude I_{me} in the motor winding.
(b) When a suspension force of 49 N is generated, derive the fundamental amplitude I_{s2} of the suspension winding current and also derive the rms value of the suspension winding current in the 3-phase coordinate system.
(c) Derive the voltage amplitudes V_{m2} and V_{s2} at the motor and suspension winding terminals.

Answer
(a) Substituting $I_{mp} = 20$ A, $I_{mq} = 7.8$ A into (9.4) yields $I_{me} = 21.5$ A.
(b) Substituting $F = \sqrt{F_x^2 + F_y^2} = 49$ N, $M' = 0.32$ H/m, $I_{me} = 21.5$ A
   into (9.10) yields $I_{s2} = 7.12$ A so that the rms value is $7.12 \times \sqrt{(2/3)} \times (1/\sqrt{2}) = 4.11$ A.
(c) Substituting $\omega = 2\pi \times 3000/60$ rad/s, $L_m = 4.5$ mH and $I_{me} = 21.5$ A into (9.8) yields $V_{m2} = 60.8$ V. Also substituting $\omega = 2\pi \times 3000/60$ rad/s, $L_s = 0.7$ mH and $I_{s2} = 7.12$ A into (9.9) yields $V_{s2} = 3.13$ V.

9.4 Guideline for permanent magnet thickness and airgap length

In this section the airgap flux density and radial suspension force for unity current are analytically derived. Using this analysis, the optimum design points of the permanent magnet thickness and airgap length are obtained [8,9].

9.4.1 Airgap flux density

Figure 9.3(a) shows the cross section of a cylindrical permanent magnet synchronous bearingless motor. As discussed later, thin permanent magnets are more effective in the generation of the radial suspension force. Therefore small permanent magnets are mounted on the rotor iron core. The radii of the stator inner surface and the rotor iron core are expressed as r and R respectively and the permanent magnet thickness as l_m. There is a guard ring to fix the permanent magnets and the thickness of this nonmagnetic ring is included in the airgap length l_g. The angular coordinates in stationary and synchronously rotating frames are ϕ_s and ϕ_r respectively, and the rotor angular position with

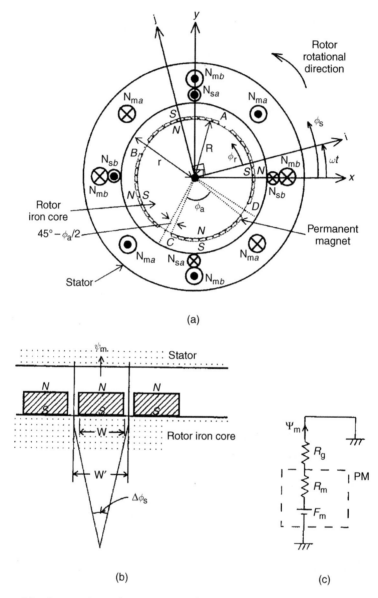

Figure 9.3 Cross section and magnetic equivalent circuit: (a) cross section; (b) enlarged rotor surface; (c) magnetic equivalent circuit

respect to time is ωt. Figure 9.3(b) shows a segment of the rotor iron surface. The permanent magnet width and the segment width per permanent magnet are w and w' respectively. Therefore the permanent magnet area S and the segment area S' are wl and $w'l$, where l is the axial length of the rotor. Figure 9.3(c) shows the corresponding magnetic equivalent circuit so that the permanent magnet MMF can be written as

$$F_{\mathrm{m}} = \frac{l_{\mathrm{m}}}{\mu_0} B_{\mathrm{r}} \qquad (9.12)$$

where B_{r} is the remanent flux density of the permanent magnet and μ_0 is the permeability of free space. The reluctances R_{g} and R_{m} of the airgap and the permanent magnets are therefore

$$R_{\varphi} = \frac{l_{\mathrm{g}}}{\mu_0 S} \qquad (9.13)$$

and

$$R_{\mathrm{m}} = \frac{l_{\mathrm{m}}}{\mu_0 S} \qquad (9.14)$$

Neglecting magnetic saturation and slot ripple, the flux in a segment can be obtained from

$$\Psi_{\mathrm{m}} = \frac{F_{\mathrm{m}}}{R_{\mathrm{m}} + R_{\mathrm{g}}} = \frac{\dfrac{l_{\mathrm{m}}}{\mu_0} B_{\mathrm{r}}}{\dfrac{l_{\mathrm{m}}}{\mu_0 S} + \dfrac{l_{\mathrm{g}}}{\mu_0 S}} = \frac{l_{\mathrm{m}}}{l_{\mathrm{m}} + l_{\mathrm{g}}} B_{\mathrm{r}} S \qquad (9.15)$$

Hence the airgap flux density B_{gp} can be written as

$$B_{\mathrm{gp}} = \frac{\Psi_{\mathrm{m}}}{S'} = \frac{l_{\mathrm{m}}}{l_{\mathrm{m}} + l_{\mathrm{g}}} \frac{S}{S'} B_{\mathrm{r}} \qquad (9.16)$$

The normalized permanent magnet thickness l_{m}/r and airgap length l_{g}/r with respect to the radius r of the stator inner surface are defined as l_{mn} and l_{gn}, respectively, so that the airgap flux density B_{gp} can be rewritten:

$$B_{\mathrm{gp}} = \frac{\dfrac{l_{\mathrm{m}}}{r}}{\dfrac{l_{\mathrm{m}}}{r} + \dfrac{l_{\mathrm{g}}}{r}} \frac{S}{S'} B_{\mathrm{r}} = \frac{l_{\mathrm{mn}}}{l_{\mathrm{mn}} + l_{\mathrm{gn}}} \frac{S}{S'} B_{\mathrm{r}} \qquad (9.17)$$

Figure 9.4 shows the relationship between the airgap flux density B_{gp} and the normalized permanent magnet thickness l_{mn} for lines of constant airgap length l_{gn}. In general the airgap flux density in an electric motor dictates the motor parameters and performance, most notably the voltage for a given rotational speed (the voltage constant in a dc machine), the efficiency and the utilization of the magnetic steel. The parameters are also governed by the motor geometry, especially the magnet reluctance to effective airgap reluctance ratio. For example,

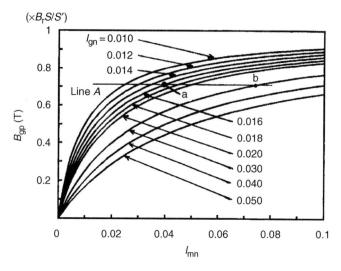

Figure 9.4 Airgap flux density and PM thickness

the design point has to be on the line A for an airgap flux density B_{gp} of 0.6 T (which is $0.6/0.9 = 0.71$ T in Figure 9.4) if the permanent magnets have a remanent flux density B_r of 1 T with an area ratio S/S' of 0.9. When the normalized airgap length l_{gn} is 0.016, point a is the design point and therefore the normalized permanent magnet thickness is 0.04. When $l_{gn} = 0.03$, point b is the design point. Thus, the desired flux density can be obtained for a given airgap length. This design method is correct for a conventional synchronous permanent magnet machine. However, in the bearingless motor it is necessary to generate a radial suspension force effectively. In the next section, the optimum permanent magnet thickness and optimum airgap length will be analytically derived.

9.4.2 Radial suspension force for unity current

From (9.10) the radial suspension force for unity suspension current under the no-load condition is

$$\frac{F}{I_{s2}} = M' I_{mp} \tag{9.18}$$

where the suspension force $F = \sqrt{F_x^2 + F_y^2}$. Note that I_{me} is equal to I_{mp} at no-load because $I_{mq} = 0$ in (9.4). An increase in the q-axis motor flux results in an increase in the radial suspension force for unity current, which is desirable. In (9.18) the derivative M' can be obtained from the mutual inductance matrix shown in (6.32). Differentiating the mutual inductance with respect to the rotor displacement yields

$$M' = \frac{\mu_0 \pi l n_2 n_4}{8} \frac{r - (l_m + l_g)}{(l_m + l_g)^2} \tag{9.19}$$

where n_4 and n_2 are the effective number of turns of the motor and suspension windings respectively, and $(l_m + l_g)$ is the effective airgap length in the permanent magnet type machine. From Figure 9.3(a) and equation (9.16) the fundamental component $B_{gp}(\phi_r)$ of the airgap flux density due to the permanent magnet field excitation is

$$B_{gp}(\phi_r) = \frac{4}{\pi} \frac{l_m}{l_m + l_g} \frac{S}{S'} B_r \cos 2\phi_r \tag{9.20}$$

For unity current in the motor winding, the MMFs A_{ma} and A_{mb} can be represented as

$$A_{ma} = n_4 \cos 2\phi_s \tag{9.21}$$

$$A_{mb} = n_4 \sin 2\phi_s \tag{9.22}$$

If the q-axis current I_{mq} is equal to zero then the equivalent current $I_{me} = I_{mp}$. So the equivalent currents i_{mae} and i_{mbe} in (9.2) and (9.3) are

$$i_{mae} = I_{mp} \cos 2\omega t \tag{9.23}$$

$$i_{mbe} = I_{mp} \sin 2\omega t \tag{9.24}$$

From (9.21) to (9.24) the resultant MMF A_{mp} is

$$A_{mp} = A_{ma} i_{mae} + A_{mb} i_{mbe} = n_4 I_{mp} \cos 2\phi_r \tag{9.25}$$

Since the reluctance in the effective airgap between the stator and rotor iron cores can be obtained from

$$\frac{1}{2\pi\mu_0} \int_R^r \frac{1}{x} dx$$

the airgap permeance P per segment arc angle $\Delta\phi_s$ can be derived using

$$P = l \frac{\Delta\phi_s}{2\pi} \frac{1}{\frac{1}{2\pi\mu_0} \int_R^r \frac{1}{x} dx} = \frac{\mu_0 l \Delta\phi_s}{\ln\left(1 + \frac{l_m + l_g}{R}\right)} \tag{9.26}$$

From (9.25) and (9.26) the airgap flux Ψ_{gM} per $\Delta\phi_s$ becomes

$$\Psi_{gM} = \frac{A_{mp}}{2} P = \frac{\mu_0 l \Delta\phi_s n_4 I_{mp}}{2\ln\left(1 + \frac{l_m + l_g}{R}\right)} \cos 2\phi_r \tag{9.27}$$

Hence the airgap flux density $B_{gM}(\phi_r)$ can be derived by dividing the Ψ_{gM} by the permanent magnet area S. Since $S = (R + l_m)\Delta\phi_s l$ then $B_{gM}(\phi_r)$ can be written as

$$B_{gM}(\phi_r) = \frac{\mu_0 n_4 I_{mp}}{2(R + l_m)\ln\left(1 + \dfrac{l_m + l_g}{R}\right)} \cos 2\phi_r \tag{9.28}$$

Equating (9.20) and (9.28) and then solving for I_{mp} yields

$$I_{mp} = \frac{8B_r S}{\pi\mu_0 n_4 S'} l_m k_{ip} \tag{9.29}$$

where the k_{ip} is given by

$$k_{ip} = \frac{r - l_g}{l_m + l_g} \ln \frac{r}{r - l_m - l_g} \tag{9.30}$$

If l_m and l_g are much less than r, and k_{ip} is nearly equal to 1, then I_{mp} can be approximated to

$$I_{mp} = \frac{8B_r S}{\pi\mu_0 n_4 S'} l_m \tag{9.31}$$

Substituting (9.19) and (9.31) into (9.18) yields the radial suspension force for unity current as

$$\frac{F}{I_{s2}} = \frac{l n_2 B_r S}{S'} \times \frac{l_{mn}(1 - l_{mn} - l_{gn})}{(l_{mn} + l_{gn})^2} \tag{9.32}$$

The first term of (9.32) cannot be increased without an increase in motor dimensions or an improvement of the permanent magnet material. The second term depends significantly on the ratios of the permanent magnet thickness and the airgap length to the radius of stator inner surface. If the second term reaches a maximum then the radial suspension force is also at a peak.

Figure 9.5 shows the characteristics for the relationship between the radial suspension force for unity rms suspension current and the normalized permanent magnet thickness for varying airgap length. It is seen that the radial suspension force for unity current is maximum for the condition of $l_{mn} = l_{gn}$. This maximum value increases with decreasing l_{gn}. Therefore it is recommended that the permanent magnet thickness is set to be equal to the airgap length and that the airgap length is as small as possible. However, if the permanent magnet thickness is set to be equal to the airgap length, the airgap flux density of the permanent magnet field excitation is expressed as $B_{gp} = 0.5\, B_r S/S'$ by (9.17) so that the flux density B_{gp} is no more than 0.6 T, even if rare-earth permanent magnets with a cylindrical structure are employed where $B_r = 1.2$ T and $S/S' = 1$. It will be

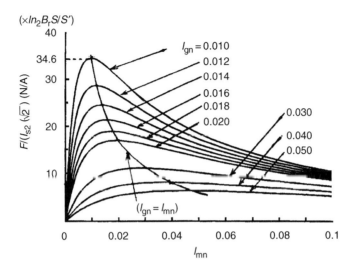

Figure 9.5 Radial suspension force for unity current and PM thickness

less than 0.6 T if a segmented permanent magnet rotor structure where $S/S' < 1$ is employed. In general, a high airgap flux density is preferred to maximize the rotational torque capability (except with high speed motors). In addition, thin permanent magnets may be irreversibly (or permanently) demagnetized. Therefore, it is necessary that the permanent magnet thickness and the airgap length should be designed so that the airgap flux density and the permanent magnet permanent demagnetization issues are addressed. This leads to a compromise between these issues and the need for a high radial suspension force per unity suspension current. The issue of irreversible demagnetization is discussed in section 9.5.

Example 9.2 The specification of a cylindrical permanent magnet synchronous bearingless motor is shown in Table 9.1. Derive the following:

(a) The airgap flux density B_{gp}.
(b) The derivative M' for the mutual inductance between the motor and suspension windings.
(c) The equivalent permanent magnet current I_{mp}.
(d) The radial suspension force for unity current F/I_{s2}.

Answer
(a) From (9.16), $B_{gp} = 0.43\,\text{T}$.
(b) From (9.19), $M' = 0.32\,\text{H/m}$.
(c) From (9.31), $I_{mp} = 24.0\,\text{A}$.
(d) From (9.32), $F/I_{s2} = 7.57\,\text{N/A}$.

Table 9.1 Specification of a cylindrical PM synchronous bearingless motor

Parameter	Symbol	Value
Radius of stator inner surface	r	25 mm
Radius of rotor iron core	R	23 mm
PM thickness	l_m	1 mm
Airgap length	l_g	1 mm
Rotor axial length	l	50 mm
Pole arc angle of rotor	ϕ_a	76.2 deg
PM area density	S/S'	0.86
PM residual flux density	B_r	1 T
Effective number of turns of		
Motor winding	n_4	72.7 turns
Suspension winding	n_2	30.6 turns

9.5 Irreversible permanent magnet demagnetization and MMF limitations of stator windings

In surface permanent magnet bearingless motors, thin permanent magnets are preferable to generate the radial suspension force with less current. Both motor and suspension fluxes are superimposed on the rotor permanent magnets. Therefore, irreversible (or permanent) demagnetization of the permanent magnets is a more serious problem in these motors than in conventional electric motors.

In order to protect the permanent magnets against irreversible demagnetization, a rotor structure with small laminated projections from the rotor iron core between the magnets (as shown in Figure 9.6) was proposed [7] so that the leakage flux goes through the permanent magnet, air, the projection and the rotor iron core before returning to the permanent magnet. Because of these small projections (that are fabricated into the rotor lamination stamping), the permanent magnets can also be easily fixed to the rotor. The protection of the magnet against irreversible demagnetization is slightly increased when compared to a rotor structure without the small inter-magnet projections. However, the magnet leakage flux is also increased.

It is therefore necessary that the maximum current ratings of the motor and suspension windings are properly determined so that the permanent magnets are not irreversibly demagnetized. In the next section the positioning of the permanent magnets which produces high demagnetization is assessed [10] and the maximum MMFs of the stator windings at the demagnetization limit are derived [11].

9.5.1 Permanent magnet positioning for high demagnetization

Figure 9.7 shows an example with q-axis motor current and suspension current flux paths. The rotor angular position ωt is 0 deg. When the q-axis motor flux Ψ_{mq}

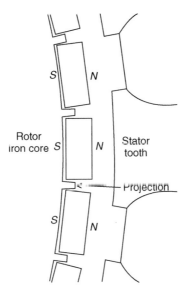

Figure 9.6 SPM rotor surface with small projections

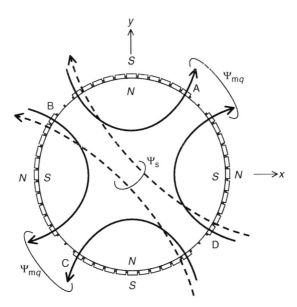

Figure 9.7 Position of PM with possibility of irreversible demagnetization

is generated as shown, torque is generated in the counter-clockwise direction. In this case, Ψ_{mq} goes through the permanent magnets denoted as A, B, C and D in the opposite direction to their magnetization. In addition, the suspension flux Ψ_s is shown. Note that both Ψ_{mq} and Ψ_s go through permanent magnet D against the magnetization so that D is the most critical permanent magnet. The q-axis motor flux Ψ_{mq} is synchronously rotating with the rotor but the suspension flux Ψ_s is rotating with a frequency twice that of the rotor, i.e., Ψ_s is not synchronized to the revolving rotor magnetic field. Therefore, not only D but also A, B and C will experience the same degree of demagnetization at some point with the possibility of irreversible demagnetization.

9.5.2 Limits of stator winding MMFs

In this section the MMF limits of the motor and suspension winding currents due to permanent magnet demagnetization are derived.

The flux density in the most critically demagnetized permanent magnet, including the effects of the motor and suspension MMFs as well as the permanent magnet magnetization itself, can be calculated. The permanent magnet field excitation has already been obtained in a previous section, so let us derive the flux due to the motor and suspension winding MMFs. When there is only a q-axis component in the motor winding current, from (9.2) to (9.5):

$$i_{ma} = -I_{mq} \sin 2\omega t \tag{9.33}$$

$$i_{mb} = I_{mq} \cos 2\omega t \tag{9.34}$$

From (9.21), (9.22), (9.33) and (9.34), the resultant MMF A_m is

$$A_m = A_{ma}i_{ma} + A_{mb}i_{mb} = n_4 I_{mq} \sin 2\phi_r \tag{9.35}$$

Hence, from (9.26) and (9.35), the airgap flux due to the motor winding current Ψ_{Mm} per segment and angle $\Delta\phi_s$ becomes

$$\Psi_{Mm} = \frac{A_m}{2} P = \frac{\mu_0 l \Delta\phi_s n_4 I_{mq}}{2 \ln\left(1 + \dfrac{l_m + l_g}{R}\right)} \sin 2\phi_r \tag{9.36}$$

The total resultant airgap flux Ψ_{gm} per segment due to the permanent magnet flux Ψ_m and the motor winding MMF flux Ψ_{Mm} can be obtained by (9.15) and (9.36). Let us consider the airgap region under the pole face. We can define the mechanical arc of the permanent magnets for one pole as ϕ_a, this is about 80 deg for the example in Figure 9.7.

The critical value of Ψ_{gm} will occur at the angular positions of $(90 - \phi_a/2), (180 - \phi_a/2), (270 - \phi_a/2)$ and $(-\phi_a/2)$, as denoted by A, B, C and D in Figures 9.3(a) and 9.7. The surface of the permanent magnets will be most susceptible to demagnetization, so substituting $\phi_r = (90 - \phi_a/2), (180 - \phi_a/2), (270 - \phi_a/2)$ and $(-\phi_a/2)$ into (9.36) gives the minimum flux per segment through the magnet surface, which can be expressed as

$$\Psi_{gm} = \Psi_m - |\Psi_{Mm}| = \frac{l_m}{l_m + l_g} B_r S - \frac{\mu_0 l \Delta\phi_s \sin\phi_a}{2 \ln\left(1 + \dfrac{l_m + l_g}{R}\right)} n_4 I_{mq} \tag{9.37}$$

Now let us consider the influence of the suspension winding MMF. The maximum airgap flux Ψ_{Mr} per segment due to the suspension winding MMFs is similar to (9.36) so that

$$\Psi_{Mr} = \frac{\mu_0 l \Delta \phi_s n_2 I_{s2}}{2 \ln \left(1 + \frac{l_m + l_g}{R} \right)} \tag{9.38}$$

Hence, the minimum value Ψ_{min} on the surface of permanent magnets A, B, C and D is

$$\Psi_{min} = \Psi_{gm} - \Psi_{M_1} \tag{9.39}$$

From (9.37) to (9.39), the minimum flux density B_{min} is therefore

$$B_{min} = \frac{\Psi_{min}}{S} = \frac{l_m}{l_m + l_g} B_r - \frac{\mu_0 \sin \phi_a}{2(R + l_m) \ln \left(1 + \frac{l_m + l_g}{R} \right)} n_4 I_{mq}$$

$$- \frac{\mu_0}{2(R + l_m) \ln \left(1 + \frac{l_m + l_g}{R} \right)} n_2 I_{s2} \tag{9.40}$$

As the MMF $n_4 I_{mq}$ of the motor winding and the MMF $n_2 I_{s2}$ of the suspension winding increase, the flux density B_{min} decreases. The minimum flux density occurs at the current limits.

Figure 9.8 shows B–H curves for a rare-earth permanent magnet. When the operating point is between points b and k on curve c, the permanent magnet is not irreversibly demagnetized. However, if the reverse magnetic field is increased further then the operating point will be moved to the left of point k so that the permanent magnet is irreversibly demagnetized. The flux density at point k is the minimum value B_{min}. Substituting this minimum value B_{min} and the motor dimensions into (9.40) and solving for the MMF $n_4 I_{mq}$ and $n_2 I_{s2}$, one can obtain the maximum MMF values. When the permanent magnet temperature rises, the B–H curve moves in the direction of point 0 from curve c to curve c'. Therefore, the permanent magnets can be permanently demagnetized at a lower magnetic field if the temperature is raised.

Example 9.3 For a machine with the specification shown in Table 9.1:

(a) Derive the flux density B_{PM} on the permanent magnet surface using the expression for the permanent magnet MMF.
(b) Derive the flux densities B_m and B_s due to the motor and suspension winding MMFs at a current rating of 7.8 A for both.
(c) Derive the minimum flux density B_{min} on the permanent magnet surface at the current ratings.
(d) Derive B_{min} when the motor and suspension winding currents are twice the rated values.

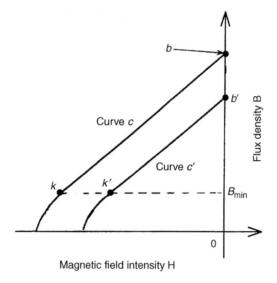

Figure 9.8 PM B–H curves

Answer

(a) The first term $\dfrac{l_m}{l_m + l_g} B_r$ on the right-hand side of (9.40) is the flux density B_{PM} on the permanent magnet surface. Substituting $l_m = l_g = 1\,\text{mm}$ and $B_r = 1\,\text{T}$ yields $B_{PM} = 0.5\,\text{T}$.

(b) The second and third terms on the right-hand side of (9.40) are the flux components due to the motor and suspension winding MMFs. Therefore, $B_m = 0.17\,\text{T}$ and $B_s = 0.07\,\text{T}$ at a current rating of 7.8 A.

(c) From the above results, $B_{min} = 0.26\,\text{T}$.

(d) Substituting $I_{mq} = I_{s2} = 7.8 \times 2 = 15.6\,\text{A}$ into (9.40) yields $B_{min} = 0\,\text{T}$. If the acceptable flux density of the employed permanent magnet is $0\,\text{T}$ then the maximum currents in the motor and suspension windings are 15.6 A, i.e., twice the rated current values.

9.6 Control system configuration

Accurate information on both the magnitude and direction of the airgap flux is required to regulate the rotor radial position precisely. In this section, a control system configuration which considers the armature reaction is introduced. Figure 9.9 shows the control system block diagram of a cylindrical permanent magnet synchronous bearingless motor [5,6]. In the rotor radial position controller, the rotor radial displacements x and y in x- and y-directions are detected by gap sensors. The differences between the detected displacements and

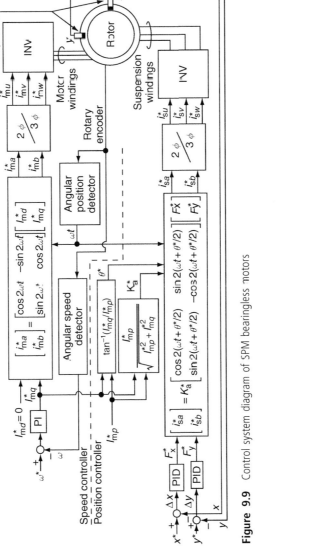

Figure 9.9 Control system diagram of SPM bearingless motors

the commands x^* and y^* are amplified by proportional-integral-derivative (PID) controllers so that the radial suspension force commands F_x^* and F_y^* in x- and y-directions are correctly determined. The rotor radial position commands are set to be the center position in x- and y-axes, i.e., $x^* = 0$, $y^* = 0$. As derived in (9.7), the radial suspension force commands are transformed into the radial suspension winding current commands i_{sa}^* and i_{sb}^* for the sa- and sb-phases so that

$$\begin{bmatrix} i_{sa}^* \\ i_{sb}^* \end{bmatrix} = \frac{1}{M'I_{me}^*} \begin{bmatrix} \cos 2\left(\omega t + \frac{\theta^*}{2}\right) & \sin 2\left(\omega t + \frac{\theta^*}{2}\right) \\ \sin 2\left(\omega t + \frac{\theta^*}{2}\right) & -\cos 2\left(\omega t + \frac{\theta^*}{2}\right) \end{bmatrix} \begin{bmatrix} F_x^* \\ F_y^* \end{bmatrix} \quad (9.41)$$

I_{me}^* and θ^* are

$$I_{me}^* = \sqrt{I_{mp}^{*\,2} + I_{mq}^{*\,2}} \quad (9.42)$$

$$\theta^* = \tan^{-1}\left(\frac{I_{mq}^*}{I_{mp}^*}\right) \quad (9.43)$$

where I_{mq}^* is the q-axis motor current command and I_{mp}^* is the equivalent current command for the permanent magnets. The current command I_{mq}^* is determined by amplifying a speed error in a proportional-integral (PI) controller. The error is calculated using the detected angular speed ω (from a rotary encoder) and the speed command ω^*. The current command $I_{mp}^* = E_o/X_m$, where E_o is the measured open-circuit voltage and X_m is the measured armature reactance. Substituting (9.42) into the right-hand side coefficient of (9.41) and rearranging yields

$$\frac{1}{M'I_{me}^*} = \frac{1}{M'I_{mp}^*} \cdot \frac{I_{mp}^*}{\sqrt{I_{mp}^{*\,2} + I_{mq}^{*\,2}}} \quad (9.44)$$

The first term $1/M'I_{mp}^*$ on the right-hand side is a constant and the gains of the PID controllers are adjusted to be equal to $1/M'I_{mp}^*$. Let us now replace the second term as K_a^* so that

$$K_a^* = \frac{I_{mp}^*}{\sqrt{I_{mp}^{*\,2} + I_{mq}^{*\,2}}} \quad (9.45)$$

K_a^* is the ratio of the permanent magnet excitation to the resultant flux from both the permanent magnet excitation and the q-axis current. In the control system, the radial suspension force commands F_x^* and F_y^* are multiplied by K_a^*. This varies in accordance with the q-axis motor current command I_{mq}^* so that the loop gain of the rotor radial position controller can be successfully maintained at a constant value. The phase command θ^* of the radial suspension winding current is determined by the calculation of $\tan^{-1}(I_{mq}^*/I_{mp}^*)$, where I_{mq}^*/I_{mp}^* is the ratio of the q-axis current to the equivalent current of the permanent magnet field excitation.

The amplitude and phase of the radial suspension winding current commands i_{sa}^* and i_{sb}^* are controlled by the q-axis motor current command I_{mq}^*. As a result, by consideration of armature reaction, the shaft can be successfully suspended without mutual interference between the radial suspension force components in two perpendicular axes.

References

[1] J. Bichsel, "The Bearingless Electrical Machine", NASA-CP-3152-PT-2, 1992, pp. 561–573.

[2] Y. Okada, S. Shimura and T. Ohishi, "Horizontal Experiments on a Permanent Magnet Synchronous Type and Induction Type Levitated Rotating Motor", Proc. of IPEC-Yokohama, 1995, pp. 340–345.

[3] B.A. Steele and L.S. Stephens, "A Test Rig for Measuring Force and Torque Production in a Lorentz, Slotless Self Bearing Motor", Proc. of 7th Int. Symposium on Magnetic Bearings, 2000, pp. 407–412.

[4] K. Nenninger, W. Amrhein, S. Silber, G. Trauner and M. Reisinger, "Magnetic Circuit Design of a Bearingless Single-Phase Slice Motor", Proc. of 8th Int. Symposium on Magnetic Bearings, 2002, pp. 265–270.

[5] M. Ooshima, S. Miyazawa, A. Chiba, F. Nakamura and T. Fukao, "Parameter Measurements and Radial Position Control Characteristics of a Permanent Magnet-Type Bearingless Motor Under Loaded Conditions", Trans. IEE, Japan, Vol. 120-D, No. 8/9, 2000, pp. 1015–1023 (*in Japanese*).

[6] M. Ooshima, S. Miyazawa, A. Chiba, F. Nakamura and T. Fukao, "Performance Evaluation and Test Results of a 11 000 r/min, 4 kW Surface-Mounted Permanent Magnet-Type Bearingless Motor", Proc. of 7th Int. Symposium on Magnetic Bearings, 2000, pp. 377–382.

[7] M. Ooshima, S. Miyazawa, T. Deido, A. Chiba, F. Nakamura and T. Fukao, "Characteristics of a Permanent Magnet-Type Bearingless Motor", IEEE Transactions on Industry Applications, Vol. 32, No. 2, 1996, pp. 363–370.

[8] M. Ooshima, A. Chiba, T. Fukao and M.A. Rahman, "Design and Analysis of Permanent Magnet-Type Bearingless Motors", IEEE Transactions on Industrial Electronics, Vol. 43, No. 2, 1996, pp. 292–299.

[9] M. Ooshima, S. Miyazawa, T. Deido, A. Chiba, F. Nakamura and T. Fukao, "An Analysis and Characteristics of a Permanent Magnet-Type Bearingless Motor", Trans. IEE Japan, Vol. 115-D, No. 9, 1995, pp. 1131–1139 (*in Japanese*).

[10] A. Chiba, S. Onoya, T. Kikuchi, M. Ooshima, S. Miyazawa, F. Nakamura and T. Fukao, "An Analysis of a Prototype Permanent-Magnet Bearingless Motor Using Finite Element Method", Proc. of 5th Inc. Symposium on Magnetic Bearings, 1996, pp. 351–356.

[11] M. Ooshima, S. Miyazawa, A. Chiba, F. Nakamura and T. Fukao, "A Rotor Design of a Permanent Magnet-Type Bearingless Motor Considering Demagnetization", IEEE Proceedings of the power conversion conference-NAGAOKA, Vol. 2, 1997, pp. 655–660.

10

Inset types of permanent magnet bearingless motor

Masahide Oshima

In the previous chapter a method for controlling radial force generation, when considering the armature reaction, was described. The radial suspension force can be controlled without interference between the components in two perpendicular axes. With the correct design of the decoupling controller and rotor, suspension force generation under torque-generating conditions can be further enhanced.

In the inset type of permanent magnet motor, the armature reaction is quite high and advantage can be taken of this to enhance the suspension force. The key is the construction of a precise and accurate decoupling controller with the identification of radial force parameters.

In this chapter an inset type of permanent magnet rotor is put forward, which has a typical salient-pole rotor design as found in permanent magnet motors, and the structure, features and a position control strategy for this motor, as well as an off-line parameter identification method, are described.

10.1 Structure and features of an inset type of permanent magnet rotor

Figure 10.1 shows an inset type of permanent magnet rotor structure. This type of rotor has q-axis saliency and the permanent magnets are mounted on the rotor iron core between the salient poles. Hence the motor q-axis magnetic

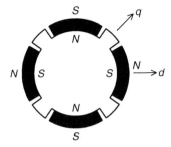

Figure 10.1   Inset type PM rotor

permeance is larger than that of the d-axis and the features of this type rotor can be summarized as follows:

1. The magnetic permeance in the motor q-axis flux path is large; therefore, the radial suspension force can be increased by taking advantage of the armature reaction flux.
2. A reluctance torque is also generated when field-weakening the motor winding current (i.e., when the motor winding current is phase-advanced so that it is no longer on the q-axis), resulting in an increase in the torque or extended operation above the base speed.

10.2 Mutual interference between radial suspension forces

Figure 10.2(a) shows the cross section of an inset type of permanent magnet bearingless motor and illustrates the principle of radial force generation with no torque [1]. Figure 10.2(b) shows radial force generation when there is torque. If the rotor rotational position is zero then the permanent magnet field excitation Ψ_m is oriented as shown in Figure 10.2(a). In this instance, when suspension winding current flows in winding N_{sa} in the positive direction, there are two representative suspension flux paths. The first goes through the permanent magnets and the second goes through the salient rotor poles. These flux paths are denoted by Ψ_{sa1} and Ψ_{sa2}; Ψ_{sa1} is in the same direction across airgap section 1 as Ψ_m, while it is in the opposite direction across airgap section 2. Therefore a suspension force is generated in the positive x-axis direction. If there is torque generation then the q-axis motor flux Ψ_{mq} is generated so that torque is produced in a counter-clockwise direction as shown in Figure 10.2(b). During this condition, Ψ_{mq} and Ψ_{sa2} have the same direction across airgap sections 3 and 4. However, the direction of Ψ_{mq} is reversed across airgap sections 5 and 6. As a result, a suspension force is generated in the positive y-axis direction. Hence it can be seen that suspension forces are generated in the y direction as well as the

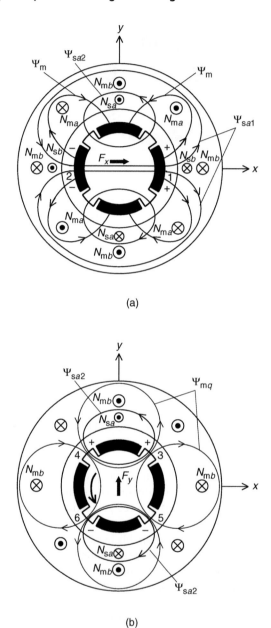

(a)

(b)

Figure 10.2 Principle of radial force generation: (a) at no load; (b) under torque-loaded condition

x direction when there is torque. The suspension force in the y direction depends on the magnitude of the q-axis motor flux, which varies with the motor torque condition. Therefore, it is necessary for the suspension forces to be decoupled so that the radial position regulation is insensitive to the motor torque variation.

10.3 Rotor position control strategy

The inset type of permanent magnet motor has q-axis salient poles so that the operation strategy is based upon the relationship between the suspension force and the suspension winding current in a salient-pole permanent magnet bearingless motor, which ensures that the rotor radial position control can be performed [2]. The relationship between the suspension force and the current was shown in (7.21) and the system control diagram was illustrated in Figure 7.17. In addition, the voltages at the motor winding terminals can be expressed using (7.3) in synchronously rotating coordinates, and the rotational torque is given by (7.8)

10.4 Identification of suspension force parameters

In the inset type of permanent magnet bearingless motor, the suspension force parameters λ'_m, M'_q and M'_d are used in a decoupling controller as shown previously in Figure 7.17. In this section, an off-line measurement method for these parameters is introduced [2–5]. This method does not require magnetic suspension while measuring the parameters so that the rotor should be mechanically held and fixed at the centre position.

From the relationship between the flux linkages and the winding currents in (7.11), when the suspension d- and q-windings are open-circuit (so that the currents i_{sd} and i_{sq} are zero), the flux linkages λ_{sd} and λ_{sq} of suspension d- and q-windings become

$$
\begin{bmatrix} \lambda_{sd} \\ \lambda_{sq} \end{bmatrix} = \begin{bmatrix} M'_d i & M'_q j \\ -M'_d j & M'_q i \end{bmatrix} \begin{bmatrix} i_{md} \\ i_{mq} \end{bmatrix} + \begin{bmatrix} \lambda'_m i \\ -\lambda'_m j \end{bmatrix}
$$

$$
= \begin{bmatrix} \lambda'_m + M'_d i_{md} & M'_q i_{mq} \\ M'_q i_{mq} & -\lambda'_m - M'_d i_{md} \end{bmatrix} \begin{bmatrix} i \\ j \end{bmatrix} \tag{10.1}
$$

The radial position relationship between the i- and j-axes and the x- and y-axes was written in (7.17). A similar relationship can also be defined for the flux linkages as

$$
\begin{bmatrix} \lambda_{sd} \\ \lambda_{sq} \end{bmatrix} = \begin{bmatrix} \cos \phi & \sin \phi \\ -\sin \phi & \cos \phi \end{bmatrix} \begin{bmatrix} \lambda_{sa} \\ \lambda_{sb} \end{bmatrix} \tag{10.2}
$$

where λ_{sa} and λ_{sb} are the flux linkages of the suspension windings sa and sb which are fixed to the stationary coordinates. Substituting (7.17) and (10.2) into (10.1) yields

$$
\begin{bmatrix} \lambda_{sa} \\ \lambda_{sb} \end{bmatrix} = K_f \begin{bmatrix} \cos(2\phi + \theta_f) & \sin(2\phi + \theta_f) \\ \sin(2\phi + \theta_f) & -\cos(2\phi + \theta_f) \end{bmatrix} \begin{bmatrix} x \\ y \end{bmatrix} \tag{10.3}
$$

where K_f and θ_f were previously shown in (7.25) and (7.26) as

$$K_f = \sqrt{(\lambda'_m + M'_d i_{md})^2 + (M'_q i_{mq})^2} \qquad (10.4)$$

$$\theta_f = \tan^{-1}\left(\frac{M'_q i_{mq}}{\lambda'_m + M'_d i_{md}}\right) \qquad (10.5)$$

Hence, when the shaft is driven at constant speed, the induced voltages v_{sa} and v_{sb} at the terminals of the suspension windings N_{sa} and N_{sb} are

$$\begin{bmatrix} v_{sa} \\ v_{sb} \end{bmatrix} = \frac{d}{dt}\begin{bmatrix} \lambda_{sa} \\ \lambda_{sb} \end{bmatrix} = 2\omega K_f \begin{bmatrix} -\sin(2\phi+\theta_f) & \cos(2\phi+\theta_f) \\ \cos(2\phi+\theta_f) & \sin(2\phi+\theta_f) \end{bmatrix}\begin{bmatrix} x \\ y \end{bmatrix} \qquad (10.6)$$

where ω is the shaft angular speed. The rotor angular position $\phi = \omega t$, so transforming v_{sa} and v_{sb} into the induced voltages v_{su}, v_{sv} and v_{sw} of the 3-phase suspension windings N_{su}, N_{sv} and N_{sw} yields

$$\begin{bmatrix} v_{su} \\ v_{sv} \\ v_{sw} \end{bmatrix} = \sqrt{\frac{2}{3}}\begin{bmatrix} 1 & 0 \\ \frac{-1}{2} & \frac{\sqrt{3}}{2} \\ \frac{-1}{2} & \frac{-\sqrt{3}}{2} \end{bmatrix}\begin{bmatrix} v_{sa} \\ v_{sb} \end{bmatrix}$$

$$= 2\sqrt{\frac{2}{3}}\omega K_f \begin{bmatrix} -\sin(2\phi+\theta_f) & \cos(2\phi+\theta_f) \\ -\sin\left(2\phi+\theta_f-\frac{2}{3}\pi\right) & \cos\left(2\phi+\theta_f-\frac{2}{3}\pi\right) \\ -\sin\left(2\phi+\theta_f-\frac{4}{3}\pi\right) & \cos\left(2\phi+\theta_f-\frac{4}{3}\pi\right) \end{bmatrix}\begin{bmatrix} x \\ y \end{bmatrix} \qquad (10.7)$$

Therefore the line-to-line voltages v_{suv}, v_{svw} and v_{swu} can be expressed as

$$\begin{bmatrix} v_{suv} \\ v_{svw} \\ v_{swu} \end{bmatrix} = \begin{bmatrix} v_{su} - v_{sv} \\ v_{sv} - v_{sw} \\ v_{sw} - v_{su} \end{bmatrix} = 2\sqrt{2}\omega K_f \begin{bmatrix} \cos\left(2\phi+\theta_f+\frac{2}{3}\pi\right) & \sin\left(2\phi+\theta_f+\frac{2}{3}\pi\right) \\ \cos(2\phi+\theta_f) & \sin(2\phi+\theta_f) \\ \cos\left(2\phi+\theta_f-\frac{2}{3}\pi\right) & \sin\left(2\phi+\theta_f-\frac{2}{3}\pi\right) \end{bmatrix}\begin{bmatrix} x \\ y \end{bmatrix}$$

$$(10.8)$$

It is obvious from (10.8) that the induced voltages are proportional to the angular speed and rotor radial displacement. One can obtain K_f by differentiating the induced voltage with respect to the rotor radial displacement so that

$$K_f = \frac{1}{2\sqrt{2}\omega}\frac{\partial V_s}{\partial x}\Big|_{y=0} \quad \text{or} \quad \frac{1}{2\sqrt{2}\omega}\frac{\partial V_s}{\partial y}\Big|_{x=0} \qquad (10.9)$$

where V_s is the amplitude of the induced line-to-line voltage at the suspension windings when the rotor has x displacement while the displacement in the

y direction is kept zero, or vice-versa. The partial derivative equation is used for the following reasons:

1. The centre position is not easy to obtain in experiments.
2. The induced voltage is proportional to the displacement around the centre.

The force parameters λ'_m, M'_q and M'_d can be obtained using the measurement techniques described below.

1. When $i_{md} = i_{mq} = 0$, λ'_m is given by (10.4):

$$\lambda'_m = K_f \tag{10.10}$$

When the motor is rotated with $i_{md} = i_{mq} = 0$, the induced line-to-line voltages at the suspension winding terminals are measured with rotor eccentricity. Substituting the induced voltage derivative with respect to the rotor radial displacement and the angular speed into (10.9), the parameter λ'_m can be obtained from (10.10). If the external driver is not suitable then a small motor current should be supplied.
2. When $i_{md} = 0$, K_f is written in (10.4) as

$$K_f = \sqrt{\lambda'^2_m + \left(M'_q i_{mq}\right)^2} \tag{10.11}$$

Equating (10.9) and (10.11) and solving for M'_q yields

$$M'_q = \frac{1}{i_{mq}} \sqrt{\left(\frac{1}{2\sqrt{2}\omega} \frac{\partial V_s}{\partial x}\bigg|_{y-0}\right)^2 - \lambda'^2_m} \tag{10.12}$$

When the motor is operated under torque load while $i_{md} = 0$, the induced line-to-line voltages are measured with an eccentric rotor displacement. Then, substituting the induced voltage derivative, the angular speed, the q-axis current and the calculated λ'_m in 1 into (10.12), one can obtain the parameter M'_q.
3. Substituting (10.4) into (10.9) and solving for M'_d yields

$$M'_d = \frac{1}{i_{md}} \left\{ \sqrt{\left(\frac{1}{2\sqrt{2}\omega} \frac{\partial V_s}{\partial x}\bigg|_{y-0}\right)^2 - \left(M'_q i_{mq}\right)^2} - \lambda'_m \right\} \tag{10.13}$$

The d- and q-axis currents are provided by the controller and the induced line-to-line voltages are measured. Again, substituting the induced voltage derivative, the angular speed, the d- and q-axis currents and the calculated λ'_m and M'_q in 1 and 2 into (10.13), one can obtain the parameter M'_d.

Example 10.1   Figure 10.3(a,b) shows the variations in the measured induced voltage amplitudes at the suspension winding terminals with respect to the rotor radial displacement in the x direction at no torque and rated torque. The specification of a prototype machine is given in Table 10.1. The gradients of the characteristics are 1842 V/m in Figure 10.3(a) and 3279 V/m in Figure 10.3(b). The rotational speed is 3000 r/min and the d-axis current is zero. The q-axis current i_{mq} is zero in Figure 10.3(a) and 13.86 A in Figure 10.3(b). The shaft is supported by a ball bearing located in a housing so that the radial position can be fixed at any desired position. The y position is kept zero by monitoring the voltage waveforms on an oscilloscope. Derive the force parameters λ'_m and M'_q.

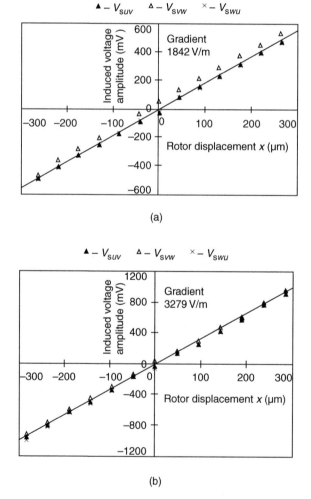

(a)

(b)

Figure 10.3   Induced voltage amplitudes at suspension winding terminals: (a) at no load; (b) under torque-loaded condition

Table 10.1 Specification of an inset type machine

Parameter	Value
Stator	
Outer diameter	100 mm
Inner diameter	50 mm
Rotor	
Outer diameter	47 mm
PM	Sm–Co
PM thickness	5 mm
Airgap length	1.5 mm
Stack length	50 mm

Answer

From (10.9) and (10.10) and with no load

$$\lambda'_m = \frac{1}{2\sqrt{2}} \times \frac{1}{\dfrac{2\pi}{60} \times 3000} \times 1842 = 2.07 \, \text{Wb/m} \tag{10.14}$$

With a load condition, and from (10.12):

$$M'_q = \frac{1}{13\,86} \sqrt{\left(\frac{1}{2\sqrt{2} \times \dfrac{2\pi}{60} \times 3000} \times 3279 \right)^2 - 2.07^2} = 0.22 \, \text{H/m} \tag{10.15}$$

Example 10.2 Figure 10.4 shows the relationship between the measured radial force and the suspension winding current operating at 1000 r/min,

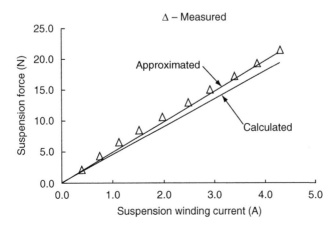

Figure 10.4 Radial suspension force

$i_{md} = 0\,\text{A}$ and $i_{mq} = 13.86\,\text{A}$. The horizontal axis is the amplitude of the suspension winding current in the 3-phase coordinate system. Derive the suspension force for unity current and compare with the measured values.

Answer
From the approximated line in Figure 10.4, the measured suspension force for unity current is 4.83 N/A. From (7.27), the theoretical suspension force F/I_{s3} for unity current in 3-phase coordinates is

$$\frac{F}{I_{s3}} = \sqrt{\frac{3}{2}} K_{\text{f}} \tag{10.16}$$

Substituting (10.11) into (10.16) yields

$$\frac{F}{I_{s3}} = \sqrt{\frac{3}{2}} \times \sqrt{\lambda_{\text{m}}'^2 + \left(M_q' i_{mq}\right)^2} \tag{10.17}$$

Substituting the calculated parameters λ_{m}' and M_q' in Example 10.1 and $i_{mq} = 13.86\,\text{A}$ into (10.17) gives $F/I_{s3} = 4.51\,\text{N/A}$. Therefore the calculated value is within 7% error. This is acceptable for an off-line parameter measurement method. For more accurate parameter measurements, on-line measurements are necessary which require successful magnetic suspension.

References

[1] S. Hara, N. Sugitani, A. Chiba and T. Fukao, "Radial Forces of Salient Pole Permanent Magnet Type Bearingless Motors", Proceedings of the 9th Symposium on Electromagnetics and Dynamics, 1997, pp. 541–546 (*in Japanese*).

[2] K. Inagaki, A. Chiba, M. A. Rahman and T. Fukao, "Performance Characteristics of Inset-Type Permanent Magnet Bearingless Motor Drives", IEEE Power Engineering Society Winter Meeting WM2000 Conference Record CDROM, Singapore, January, 2000.

[3] K. Hiraguri, A. Chiba and T. Fukao, "Influence of Space Harmonics on Induced Voltages in Bearingless Motors", Papers of Technical Meeting on Semiconductor Power Converters SPC-97-51, May 16, 1997, pp. 91–98 (*in Japanese*).

[4] K. Inagaki, A. Chiba and T. Fukao, "A Measurement of Radial Force Coefficient in Permanent Magnet Type Bearingless Motors", IEE Japan, The Papers of Technical Meeting on Linear Drives LD-98-78, 1998, pp. 1–6 (*in Japanese*).

[5] K. Inagaki, N. Fujie, A. Chiba and T. Fukao, "A Measurement of Machine Parameters and Decoupling Control in Inset Permanent Magnet Type Bearingless Motors", The Papers of Technical Meeting on Semiconductor Power Converters, SPC-99-48, May 17, 1999, pp. 49–56 (*in Japanese*).

11

Buried permanent magnet bearingless motors

Masahide Oshima

Recently, buried permanent magnet (BPM) motors, often referred to as interior permanent magnet (IPM) motors, have become more commonplace. By employing a field-weakening control, adjustable speed operation can be realized over a wide speed range. In addition, a reluctance torque is generated which increases the motor efficiency. In this chapter, the BPM bearingless motor is introduced. The linear position control strategy employed in cylindrical and inset permanent magnet types of motor is not applicable because partial magnetic saturation occurs in the BPM motor so that the suspension force parameters λ'_{m}, M'_q and M'_d are influenced by the operating currents. Therefore a successful position control strategy is introduced to solve this problem with off-line and on-line identification methods of parameter variations. The rotor structure and features are described. One of the advantages of this machine is that enhanced radial force can be obtained.

11.1 BPM rotor structures

Figure 11.1(a–c) shows the BPM rotor structures. The advantages of the BPM rotors are as follows:

1. The airgap length can be small compared with those in the other types of permanent magnet machines, resulting in an increase in the suspension force for unity suspension winding MMF.
2. The leakage flux goes around the permanent magnets in the rotor core, helping to prevent the irreversible demagnetization of the permanent magnets.
3. Since the permanent magnets are buried in the rotor core, they are protected against centrifugal force at high speed operation.

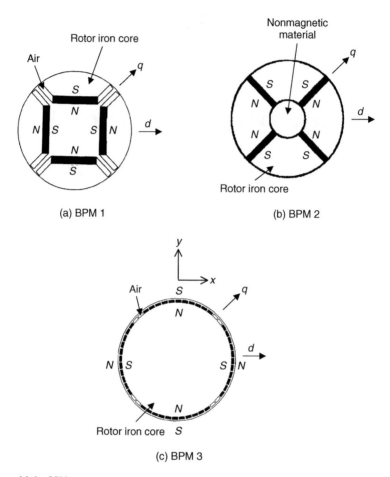

Figure 11.1 BPM rotor structures

The structures BPM 1 and BPM 2, shown in Figure 11.1(a,b), are typical interior permanent magnet and spoke types of rotor [1]. The q-axis flux paths are sufficient to increase the saliency ratio. However, these types of rotor are not very suitable for a bearingless motor arrangement. Since the suspension forces are generated by the superimposition of the permanent magnet field and the suspension flux, and the suspension flux goes through the thick permanent magnets and the nonmagnetic material, the suspension forces for a unity suspension winding MMF are small. On the other hand, the rotor structure BPM 3 in Figure 11.1(c) has thin permanent magnets buried just below the rotor surface, which improves radial suspension force generation because of the low permanent magnet magnetic reluctance [2–4]. Therefore, this type of rotor is suitable for a BPM bearingless motor; and the suspension force, the permanent magnet demagnetization and the control strategy for this machine are introduced in this chapter.

11.2 Suspension force for unity current and permanent magnet demagnetization

The BPM 3 rotor shown in Figure 11.1(c) has a ratio of suspension force per unity suspension winding current which is higher than the other types of permanent magnet bearingless motors described in previous chapters. This is due to the small airgap length and the thin permanent magnets. In this section, the suspension force per unity current is compared in order to emphasize that the BPM 3 rotor structure is suitable in radial force generation. The ability to withstand irreversible demagnetization is also described.

Example 11.1 Figure 11.2 shows the suspension forces analysed by the finite element method (FEM) with no torque. The horizontal axis is the amplitude of the suspension winding current in 3-phase coordinates. The analysis is carried out for the SPM and inset types of motor as well as the BPM types 1 and 3 as shown Figure 11.1(a,c). The SPM and inset types are analysed for the structures described in Figures 9.6 and 10.1 and the specifications of these rotors are summarized in Table 11.1. In the SPM and inset types, the equivalent airgap length is large because these motors require a cylindrical cover to fix the permanent magnets to the rotor surface. The stator structure and dimensions are the same for all motors. If the suspension winding current is above 15 A, the suspension forces are saturated in the SPM and BPM 3 motors. This is caused by magnetic saturation in the stator

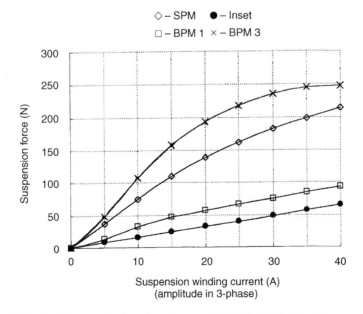

Figure **11.2** Radial suspension force for unity current in PM bearingless motors

Table 11.1　Specification of PM bearingless motors

	SPM	Inset	BPM 1	BPM 3
Rotor outer diameter (mm)	48	47	49.2	49
Equivalent airgap length (mm)	1	1.5	0.4	0.5
PM thickness (mm)	1	5	3	1
Force/Current (N/A)	7.5	2	3	11

teeth. Table 11.1 also shows the suspension force divided by the suspension winding current ratio (F/I_{s3}). F/I_{s3} for the BPM 1 and inset types are lower due to the thick permanent magnets whereas F/I_{s3} for BPM 3 is higher due to the small airgap length and thin permanent magnets (F/I_{s3} for BPM 3 is 5.5 times of that of the inset-type).

As discussed in Chapter 9, permanent magnet demagnetization is an important issue in permanent magnet bearingless motors. The torque and suspension force are restricted by irreversible demagnetization of the permanent magnets. By burying the permanent magnets in the rotor iron core a uniform flux distribution across the magnet surface is obtained so that the problem of partial irreversible demagnetization is eased. As a result, the resistance to irreversible demagnetization of the BPM type of rotor is much better [2].

11.3 Rotor position control strategy

11.3.1 Influence of suspension force parameter variations

The BPM bearingless motor is a salient-pole motor so that, using the control system shown in Figure 7.17, the rotor radial position can be, in principle, stably regulated. However, the suspension force becomes saturated due to partial magnetic saturation caused by the q-axis motor flux since the airgap length is smaller than the SPM and inset types of machine. Hence the suspension force parameters λ'_m, M'_q and M'_d all have load variation. If one assumes that these parameters are constant then the magnetic suspension cannot be stable. Therefore the identification of the parameter variations and the controller integration of these variables are key issues for successful suspension.

Example 11.2　Figure 11.3 shows an example of the suspension force analysed by FEM using the BPM machine shown in Table 11.2 and Figure 11.1(c) [3,4]. The suspension windings which generate the MMFs in the x and y directions are N_{sa} and N_{sb}. In this analysis, the currents i_{sa} and i_{sb}

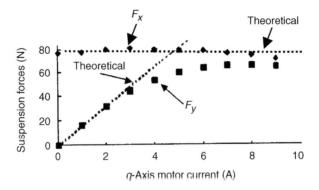

Figure 11.3 Suspension force components in the perpendicular axes *x* and *y*

Table 11.2 Specification of a BPM 3

Parameter	Symbol	Value
Radius of stator inner surface	r	25 mm
Radius of rotor iron core	R	24.5 mm
PM thickness	l_m	1 mm
Buried PM depth	l_d	0.5 mm
Airgap length	l_g	0.5 mm
Rotor axial length	l	50 mm
PM residual flux density	B_r	1 T
Effective number of turns of		
Motor winding	n_4	72.7 turns
Suspension winding	n_2	30.6 turns

are set to 4 and 0 A in these windings. The suspension force in the *x* direction is almost constant even though the *q*-axis current is increasing because it is generated by superposition of the suspension flux onto the permanent magnet field. However, the suspension force in the *y* direction becomes saturated above a *q*-axis current of 3 A. The *y*-axis force is caused by the superposition of the suspension flux onto the *q*-axis motor flux. Since the stator teeth are magnetically saturated above 3 A, the *y*-axis force likewise saturates.

Figure 11.4 shows an example of the measured suspension force commands using constant parameters [3,4]. When the suspension force is generated only in the *x* direction, the theoretical suspension force commands F_x^* and F_y^* are 1 and 0 pu respectively. However, the measured suspension force commands F_x^* and F_y^* are not equal to theoretical values when the *q*-axis motor current is above 2 A and the suspension forces have mutual interference in two perpendicular axes. Hence the control system has less damping resulting in a large rotor position fluctuation, and the rotor shaft touches down at high current.

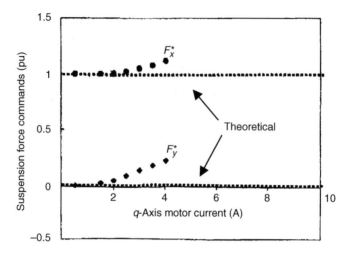

Figure 11.4 Mutual interference of suspension forces

11.3.2 Successful suspension with nonlinear magnetic characteristics

In order to regulate the rotor position accurately, it is important to consider the magnetic saturation caused by the q-axis motor current. The suspension winding currents can only be successfully controlled by considering the parameter variations. Hence it is necessary to express the suspension force parameters as functions of the motor winding currents, and in this section, therefore, an identification method for the suspension force parameters is introduced and an example of suspension force decoupling is described [3,4].

Example 11.3 Figure 11.5 shows the variation of the suspension force parameter M'_q (as an example of the identification of the suspension force parameters). Initially substituting $i_{sq} = 0$ into (7.16) and solving for M'_q yields

$$M'_q = \frac{F_j}{i_{mq}i_{sd}} \tag{11.1}$$

For some values of i_{mq} and i_{sd}, the suspension force F_j in the j direction is calculated by FEM analysis and then the calculated F_j and the currents i_{mq} and i_{sd} are substituted into (11.1); Figure 11.5 shows the results. M'_q decreases with increasing q-axis motor current and it is therefore necessary to approximate M'_q to a line, in this particular machine case, in order to keep the suspension loop gain constant.

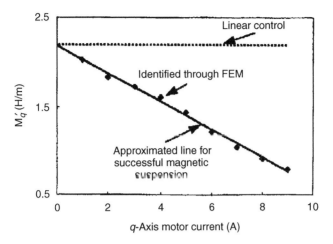

Figure 11.5 Suspension force parameter

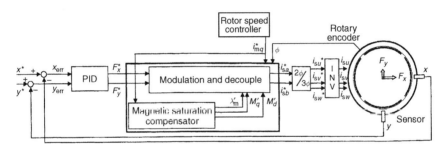

Figure 11.6 Successful position control system

Figure 11.6 shows a successful control system for the BPM type bearingless motor. In the magnetic saturation compensator, M'_q is a function of the q-axis current command i^*_{mq}. λ'_m and M'_d can also be functions of i^*_{mq}. However, this is not the case in this example; although in some bearingless motor controls, saturation compensators would be required for λ'_m and M'_d.

In this example, M'_q is derived using a series of static FEM analyses. This method is useful for obtaining rough parameter variation estimates off-line. However, FEM results always have a discrepancy from the actual machine parameters because of mechanical tolerances, end effects and other effects. Hence on-line identification is necessary; this can be done using the static force and torque generation tests shown previously in Figure 11.4. The discrepancies of the radial force commands do include radial force parameter errors [3]. Identification is possible up to 4 A, and parameter estimates at higher current values can be obtained by extrapolation for implementation in the controller. More accurate parameter identification is possible if the measured range is extended above 4 A with the parameters again extrapolated.

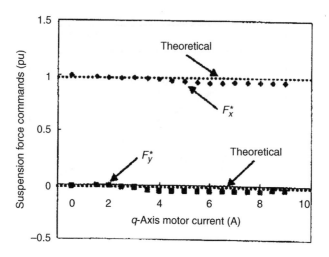

Figure 11.7 Suspension force commands with F_x considering parameter variation

---

Example 11.4 Examples of some experimental results for the suspension force decoupling, and also the suppression response to a suspension force disturbance and rotor acceleration, are described below [3,4].

Figure 11.7 shows the test results of the suspension force decoupling. The suspension force commands F_x^* and F_y^* remain at 1 and 0 pu even though the q-axis motor current is increased.

Figure 11.8 shows the suppression response of a x-axis suspension force disturbance. The y-axis suspension force command and the displacement are not influenced due to good decoupling.

Finally, Figure 11.9 shows the rotor acceleration test result. The rotor speed ω is accelerated from 1000 to 2140 r/min using a reference step function. It

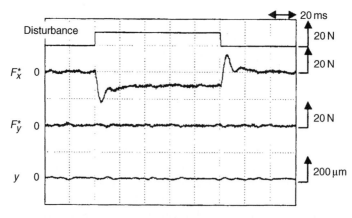

Figure 11.8 Suppression response of suspension force disturbance

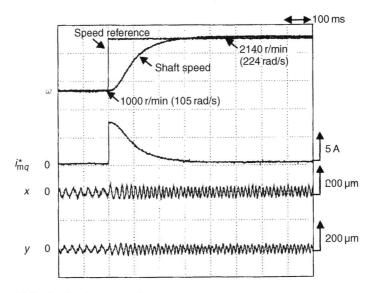

Figure 11.9   Acceleration test result

can be seen that the rotor radial vibrations are suppressed during the rotor acceleration when the motor currents are maximum.

Figures 11.7–11.9 illustrate that the decoupling of the suspension forces in two perpendicular axes can be achieved. Hence the rotor shaft is stably supported even though partial magnetic saturation occurs.

References

[1] T. Ohishi, Y. Okada and S. Miyamoto, "Levitation Control of IPM Type Rotating Motor", Proceedings of the Fifth International Symposium on Magnetic Bearings, 1996, pp. 327–332.

[2] M. Ooshima, S. Miyazawa, Y. Shima, A. Chiba, F. Nakamura and T. Fukao, "Increase in Radial Forces of a Bearingless Motor with Buried Permanent Magnet-Type Rotor", The 4th International Conference on Movic Proceedings, Vol. 3, 1998, pp. 1077–1082.

[3] N. Fujie, R. Yoshimatsu, A. Chiba, M. Ooshima, M. A. Rahman and T. Fukao, "A Decoupling Control Method of Buried Permanent Magnet Bearingless Motors Considering Magnetic Saturation", Proceedings of IPEC-Tokyo 2000, Vol. 1, 2000, pp. 395–400.

[4] N. Fujie, R. Yoshimatsu, A. Chiba, M. Ooshima and T. Fukao, "Influence of Magnetic Saturation on Armature Reaction Flux in Buried Permanent Magnet Type Bearingless Motors", IEE Japan, The Papers of Technical Meeting on Linear Drives LD-99-153, pp. 33–38, 1999 (*in Japanese*).

12

Synchronous reluctance bearingless motors

Osamu Ichikawa

Since there are no rotor windings or permanent magnets the synchronous reluctance machine can be used for high-speed applications including operation at high temperature. Synchronous reluctance machines are also inherently low cost (for the same reasons), which makes them attractive in many applications.

In this chapter, synchronous reluctance bearingless motors are introduced and their principles, physical structure and a compensator structure are described. In these machines, the unbalanced magnetic pull due to rotor eccentricity has to be considered in order to realize stable and robust magnetic suspension. Hence compensation of the suspension force (which is proportional to the shaft displacement) is necessary. The described controller is also applicable to other bearingless motors which have a considerable force-displacement factor and require unbalanced magnetic pull compensation.

12.1 Torque characteristics

12.1.1 Rotor and stator

Figure 12.1 shows the cross section of a 4-pole synchronous reluctance bearingless motor with only the u-phase motor and suspension windings shown (of a 3-phase winding). Typically, both the rotor and the stator are constructed from laminated silicon steel and the stator is identical with that of a bearingless induction motor or a permanent magnet motor, i.e., the inner surface of the stator is cylindrical and the stator windings have general sinusoidal distribution. The stator has both n-pole motor windings and $(n \pm 2)$-pole suspension windings. Because the rotor has salient poles, reluctance torque is produced when the 4-pole motor winding is excited and there is a phase angle. This is the sole source of torque in this type of motor so that in order to produce torque effectively with

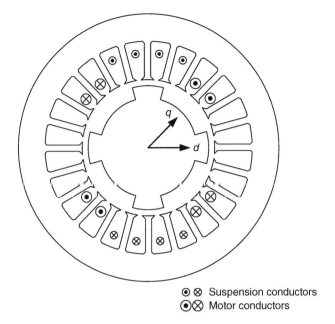

⊙ ⊗ Suspension conductors
⊙⊗ Motor conductors

Figure 12.1 Rotor and stator of a synchronous reluctance bearingless motor

low current, the airgap should be designed to be as small as possible and a high saliency ratio is necessary to achieve high torque and power-factor.

12.1.2 Voltage and current

The terminal voltages for the synchronous reluctance bearingless motor, as derived in Chapter 7, are

$$\begin{bmatrix} v_{md} \\ v_{mq} \end{bmatrix} = \begin{bmatrix} R_s & 0 \\ 0 & R_s \end{bmatrix} \begin{bmatrix} i_{md} \\ i_{mq} \end{bmatrix} + \begin{bmatrix} PL_d & -\omega L_q \\ \omega L_d & PL_q \end{bmatrix} \begin{bmatrix} i_{md} \\ i_{mq} \end{bmatrix} \qquad (12.1)$$

Since these motors do not have permanent magnets, d-axis excitation is provided only by the d-axis component of the motor winding current. Depending on the saliency of the rotor, the d-axis inductance L_d should be much higher than q-axis inductance L_q and the ratio of L_d/L_q is defined as the saliency ratio.

Figure 12.2 shows a phasor diagram for a synchronous reluctance motor having a saliency ratio L_d/L_q of 3, with the condition that the motor d-axis current is equal to the motor q-axis current. A ratio of 3 is typical for the type of rotor shown in Figure 12.1. In the phasor diagram, Ψ_m is the motor winding flux linkage, i_m is the motor current and v_m is the terminal voltage. Since d-axis current is required for excitation, the phase angle ϕ between current and terminal voltage is rather large. Hence typical synchronous reluctance motors do not have high power-factor. Another way to consider the phasor diagram (to assess the

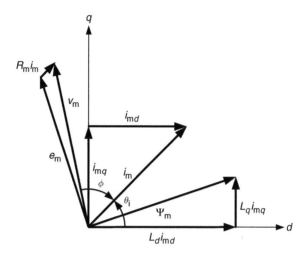

Figure 12.2 Phasor diagram

reluctance torque) is to neglect the winding resistance and put the current on the d-axis and then on the q-axis. In both cases the phase angle between the voltage and current will be 90 deg which means there is no input power and hence no torque. When the current is not on these axes then the difference between L_d and L_q leads to a phase angle which is less than 90 deg which means there is input power and hence a reluctance torque (or output power if the angle is greater than 90 deg and the machine is generating). For a certain current magnitude the peak torque will occur when the motor winding current is at 45 deg electrical deg.

It is possible to increase the saliency ratio and the power-factor by some modifications of the rotor. For example, a high saliency ratio can be obtained with flux barriers in the rotor.

12.1.3 Torque

The torque of synchronous reluctance machines can be written as:

$$T_r = p_p \left[(L_d - L_q) i_{md} i_{mq} \right] = p_p (L_d - L_q) i_m^2 \frac{\sin 2\theta_i}{2} \qquad (12.2)$$

where p_p is the number of pole-pairs of the motor windings. From (12.2), the torque can be controlled with both d-axis and q-axis motor winding current or, put another way, the magnitude and phase angle of the motor current. Hence the current phase angle θ_i is defined by

$$\theta_i = \arctan \frac{i_{mq}}{i_{md}} \qquad (12.3)$$

There are several methods for regulating the motor currents [1,2]. Suppose that current phase angle is regulated while the magnitude of the current is kept constant. In this operation, the copper losses are constant and the maximum torque is obtained at a phase angle $\theta_i = \pi/4$. Hence, the copper losses are a minimum at this phase angle for any torque. Next, suppose that the magnitude of the flux linkage is kept constant so that both the d- and q-axis currents are regulated to satisfy the following constraint:

$$\Psi_m = \sqrt{(L_d i_{md})^2 + (L_q i_{mq})^2} = \text{constant} \tag{12.4}$$

In this operating mode, the iron losses due to the motor flux are almost constant and the maximum torque is obtained at $\theta_i = \arctan(L_d/L_q)$. Hence the iron losses are minimized at this phase angle for any torque. Moreover, this phase angle is also optimal for a fast torque response under a limited terminal voltage condition. Therefore the optimal phase angle for minimizing the motor losses is between the above phase angles so that

$$\frac{\pi}{4} \le \theta_i \le \arctan \frac{L_d}{L_q}$$

For correct operation the phase angle should be determined by taking into account these losses, as well as the terminal voltage, the rotational speed and any variations of the motor constants. Torque can be regulated using the current amplitude.

12.2 Radial force characteristics

12.2.1 Radial forces

The synchronous reluctance bearingless motor produces a suspension force by the superposition of the n-pole motor flux and the $(n \pm 2)$-pole suspension flux. If a bearingless motor has a 2-pole suspension winding in addition to the 4-pole motor winding then the suspension force can be derived (as in equations (7.20) and (7.21))

$$\begin{bmatrix} F_{1x} \\ F_{1y} \end{bmatrix} = \begin{bmatrix} \cos\phi & -\sin\phi \\ \sin\phi & \cos\phi \end{bmatrix} \begin{bmatrix} M_d' i_{md} & M_q' i_{mq} \\ M_q' i_{mq} & -M_d' i_{md} \end{bmatrix} \begin{bmatrix} \cos\phi & \sin\phi \\ -\sin\phi & \cos\phi \end{bmatrix} \begin{bmatrix} i_{sa} \\ i_{sb} \end{bmatrix} \tag{12.5}$$

$$\begin{bmatrix} F_{1x} \\ F_{1y} \end{bmatrix} = \begin{bmatrix} M_d' i_{md} & -M_q' i_{mq} \\ M_q' i_{mq} & M_d' i_{md} \end{bmatrix} \begin{bmatrix} \cos 2\phi & \sin 2\phi \\ \sin 2\phi & -\cos 2\phi \end{bmatrix} \begin{bmatrix} i_{sa} \\ i_{sb} \end{bmatrix} \tag{12.6}$$

The radial forces F_{1x} and F_{1y} can be regulated using the suspension currents i_{sa} and i_{sb} so that the direction of the suspension force varies with both the rotor rotational angle and the ratio $(M_d' i_{md})/(M_q' i_{mq})$. The magnitude of the force is also a function of motor currents i_{md} and i_{mq}. Hence the suspension currents must be regulated in accordance with (12.6) in order to produce the required radial force.

12.2.2 Unbalanced magnetic pull

Figure 12.3 shows the measured d- and q-axis inductances for the motor wind-ings with respect to the normalized rotor radial displacement. These inductances can be derived from self and mutual inductances for a 2- or 3-phase stator winding. If we approximate the variation of the d- and q-axis inductances to first-order equations of radial displacements x and y from the centre so that these inductances are then constant, then no radial force due to rotor eccentricity would be generated according to the analysis, as described in Chapter 6 [3,4]. However, due to the small airgap of the synchronous reluctance machine, the induct-ances vary significantly with rotor displacement (whereas in thick permanent magnet machines they do not); hence the unbalanced magnetic pull should not be neglected. Therefore let us assume that the variations of the inductances are second-order equations [5] so that

$$L_d = a_d(x^2 + y^2) + c_d \tag{12.7}$$

$$L_q = a_q(x^2 + y^2) + c_q \tag{12.8}$$

The magnetic energy stored in the d- and q-axis inductances varies in accordance with the rotor displacement. From the expressions of inductance in (12.7) and (12.8), the radial forces can be derived, and it is found that they are proportional to the radial displacement, where

$$F_{2x} = (a_d i_{md}^2 + a_q i_{mq}^2)x \tag{12.9}$$

$$F_{2y} = (a_d i_{md}^2 + a_q i_{mq}^2)y \tag{12.10}$$

Figure 12.4 shows the resultant unbalanced magnetic pull. Since the inductances increase with rotor displacement, the generated unbalanced magnetic pull is in

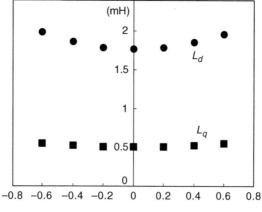

Normalized radial displacement x_n (pu)

Figure 12.3 The d- and q-axis inductances of motor windings

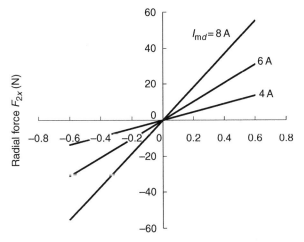

Figure 12.4 Unbalanced magnetic pull

the displaced direction (where the airgap flux is most concentrated), so that the displacement-originated force is proportional to the radial displacement. A proportional gain controller can regulate the suspension currents to produce a radial force which is also proportional to the radial displacement; however, this force acts towards the stator centre, i.e., in the opposite direction of the unbalanced magnetic pull. Thus, in a feedback controller, the proportional gain K_p can be modified to become

$$K_p - (a_d i_{md}^2 + a_q i_{mq}^2)$$

as a result of the unbalanced magnetic pull. The influence on the suspension system can be seen in the frequency range where the proportional gain is dominant.

Example 12.1   A test machine, having a coefficient of unbalanced magnetic pull $a_d = 4000\,\mathrm{Nm^{-1}\,A^2}$ has a motor d-axis current $i_{md} = 8\,\mathrm{A}$. The proportional gain of the suspension controller $K_p = 5 \times 10^5\,\mathrm{N/m}$, including a sensor gain. Derive the decrease in loop gain (in dB) at the frequency where the proportional gain is dominant. The decrease is caused by an unbalanced magnetic pull.

Answer
The unbalanced magnetic pull per meter of radial displacement is $4000 \times 8^2\,\mathrm{N/m}$. The sum of proportional gain and unbalanced magnetic pull is

$$5 \times 10^5 - 4000 \times 8^2\,\mathrm{N/m} \tag{12.11}$$

Therefore, the difference in loop gain becomes

$$20 \log_{10} \frac{5 \times 10^5 - 4000 \times 8^2}{5 \times 10^5} = -6.2 \, \text{dB} \qquad (12.12)$$

Hence the loop gain is decreased by 6.2 dB in the middle frequency range due to the unbalanced magnetic pull caused by the exciting current.

12.3 Drive systems

Synchronous reluctance bearingless motors have both a motor control system and a suspension control system. These have dependent relationships with each other due to the characteristics of the radial force. In this section, a drive system for a synchronous reluctance bearingless motor with a 4-pole motor winding and 2-pole suspension winding is introduced.

12.3.1 Suspension control system

Figure 12.5 shows the suspension control system where the radial positions are detected and compared with the references. These positions are also used for the calculation and cancellation of the unbalanced magnetic pull; this is done by subtracting the calculated unbalanced magnetic pulls F_{2x}^* and F_{2y}^* from the outputs of the suspension controllers. Hence, the suspension controllers can be designed without considering the unbalanced magnetic pull. Block A in

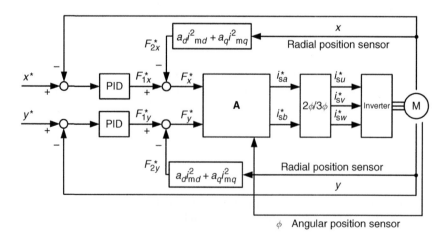

Figure 12.5 Suspension control system

Figure 12.5 converts the suspension force commands into appropriate suspension current commands using the following equation [6]:

$$\begin{bmatrix} i^*_{sa} \\ i^*_{sb} \end{bmatrix} = \begin{bmatrix} \cos 2\phi & \sin 2\phi \\ \sin 2\phi & -\cos 2\phi \end{bmatrix}^{-1} \begin{bmatrix} M'_d i_{md} & -M'_q i_{mq} \\ M'_q i_{mq} & M'_d i_{md} \end{bmatrix}^{-1} \begin{bmatrix} F^*_x \\ F^*_y \end{bmatrix} \tag{12.13}$$

$$\begin{bmatrix} i^*_{sa} \\ i^*_{sb} \end{bmatrix} = \frac{1}{(M'_d i_{md})^2 + (M'_q i_{mq})^2} \begin{bmatrix} \cos 2\phi & \sin 2\phi \\ \sin 2\phi & -\cos 2\phi \end{bmatrix} \begin{bmatrix} M'_d i_{md} & M'_q i_{mq} \\ -M'_q i_{mq} & M'_d i_{md} \end{bmatrix} \begin{bmatrix} F^*_x \\ F^*_y \end{bmatrix} \tag{12.14}$$

which is derived from (12.6). If the motor has 3-phase windings, the 2-phase current commands should be transformed into 3-phase values using the block $2\phi/3\phi$ where

$$\begin{bmatrix} i^*_{su} \\ i^*_{sv} \\ i^*_{sw} \end{bmatrix} = \sqrt{\frac{2}{3}} \begin{bmatrix} 1 & 0 \\ -\frac{1}{2} & \frac{\sqrt{3}}{2} \\ -\frac{1}{2} & \frac{-\sqrt{3}}{2} \end{bmatrix} \begin{bmatrix} i^*_{sa} \\ i^*_{sb} \end{bmatrix} \tag{12.15}$$

12.3.2 Motor control system

Figure 12.6 shows the motor control system. The rotor angular position is required for coordinate transformation from rotational to stationary coordinates and the derivative of the angular position is required to maintain the rotational speed. A torque command T^*_r is generated from the speed error in the speed regulator, which is a proportional-integral (PI) controller. In block B, the d- and q-axis current commands are determined from the torque command using

$$i^*_{md} i^*_{mq} = \frac{1}{2(L_d - L_q)} T^*_r \tag{12.16}$$

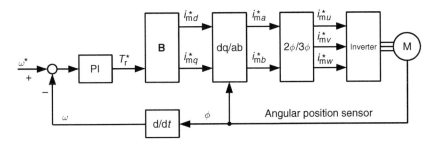

Figure 12.6 Motor control system

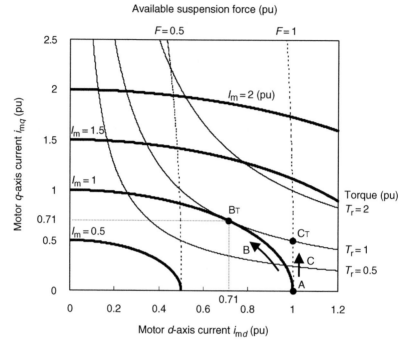

Figure 12.7 Available torque and suspension force at $M'_d : M'_q = 10:1$

which is derived from (12.2). There are several combinations of i^*_{md} and i^*_{mq} which can produce the desired torque. The d- and q-axis motor current commands are then transformed into stationary coordinates in block dq/ab where

$$\begin{bmatrix} i^*_{ma} \\ i^*_{mb} \end{bmatrix} = \begin{bmatrix} \cos 2\phi & -\sin 2\phi \\ \sin 2\phi & \cos 2\phi \end{bmatrix} \begin{bmatrix} i^*_{md} \\ i^*_{mq} \end{bmatrix} \tag{12.17}$$

And from this the 2-phase current commands are transformed into 3-phase current commands in block $2\phi/3\phi$ so that

$$\begin{bmatrix} i^*_{mu} \\ i^*_{mv} \\ i^*_{mw} \end{bmatrix} = \sqrt{\frac{2}{3}} \begin{bmatrix} 1 & 0 \\ -\frac{1}{2} & \frac{\sqrt{3}}{2} \\ -\frac{1}{2} & \frac{-\sqrt{3}}{2} \end{bmatrix} \begin{bmatrix} i^*_{ma} \\ i^*_{mb} \end{bmatrix} \tag{12.18}$$

Figure 12.7 shows the available torque and radial force in the motor current plane from a motor where the ratio M'_d/M'_q is 10. The winding currents, torque and suspension force are normalized values. If the torque of the motor is zero then the q-axis current should be set to zero and the d-axis current i_{md} to 1 in order to produce the maximum radial force $F = 1$. Hence, operating point A

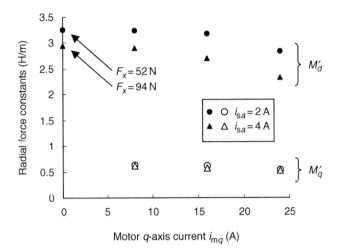

Figure 12.8 Radial force constants under magnetic saturation

provides the maximum suspension force with no load operation. If the torque current is increased along arrow B to produce torque with the motor current magnitude at the limit, then the available radial force decreases. The maximum torque is achieved at point B_T. Another control strategy which increases both the torque and suspension force is to operate along the line indicated by arrow C. Maximum torque is obtained at operating point C_T and both the torque and suspension force are 1 pu. The characteristics described by arrows B and C are determined by the combination of i^*_{md} and i^*_{mq}.

12.3.3 Magnetic saturation

When torque is required, and if the motor current is increased along arrow C in Figure 12.7, then the flux density of the motor increases. Moreover, when suspension force is required, the flux density increases in some parts of the motor [7,8]. This flux density increase may result in magnetic saturation, especially in the stator teeth. Figure 12.8 shows the analysed radial force constants of a test machine at $i_{md} = 8\,\text{A}$. The machine has the following dimensions: rotor diameter $= 49.2\,\text{mm}$, rotor stack length $= 50.0\,\text{mm}$ and stator inner diameter $= 50.0\,\text{mm}$. The maximum flux density in the stator teeth is $0.9\,\text{T}$ when only d-axis current $i_{md} = 8\,\text{A}$ is provided. The radial force constants decrease with an increase in torque current and suspension current. It is obvious from (12.6) that these variations of the radial force constants may cause the loop gain to decrease, resulting in force direction variation in the suspension control system. Therefore, the suspension control system performance may be poor when the motor produces high suspension force and torque. To avoid this, the parameters in the suspension control, as described by (12.13), should be calibrated in order

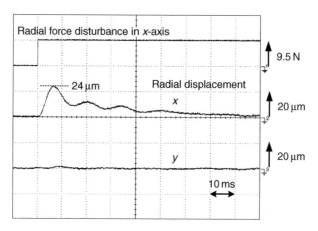

Figure 12.9 Step disturbance at $i_{md} = 6$ A and $i_{mq} = 11$ A

to cancel these variations if the motor is operated in the nonlinear region. One solution was described in the previous chapter; the parameter variations are also significant in synchronous reluctance machines. There are several ways to obtain estimates of the parameter variations:

1. On-line parameter tuning.
2. Parameter tables obtained by analysis or off-line measurements [9].
3. Approximated equations obtained by analysis or off-line measurements.

12.3.4 Characteristics of a test machine

Figure 12.9 shows the radial displacement of a test machine when $i_{md} = 6$ A and $i_{mq} = 11$ A [5]. A radial step force of 9.5 N is applied to the rotor by adding a step disturbance to the suspension force command. It is shown that the radial positions on the x- and y-axes are stably controlled and that there is no interaction between the two axes even at high torque where there are magnetic saturation conditions. This is because parameters M'_d and M'_q have been identified and calculated correctly.

Figures 12.10 and 12.11 show the open-loop bode plots for a suspension control system [5]. The results in Figure 12.10 were measured without compensators for the negative displacement-stiffness, i.e., without cancellation of the unbalanced magnetic pulls F^*_{2x} and F^*_{2y} as shown in Figure 12.5. Due to the variations of unbalanced magnetic pull, the loop gain of the suspension control system decreases with an increase of the d-axis current. On the other hand, in Figure 12.11, the results were obtained with the negative-displacement-stiffness compensators, and here it is shown that the loop gain of the system is successfully kept constant for all d-axis current values.

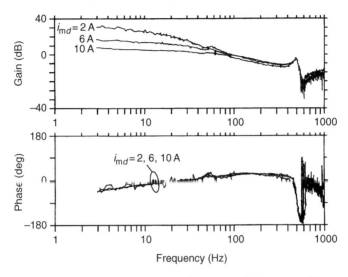

Figure 12.10 Bode plot without a compensator for negative displacement-stiffness

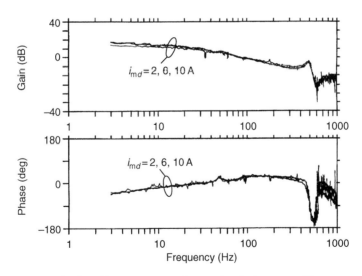

Figure 12.11 Bode plot with a compensator for negative displacement-stiffness

If the d-axis current is changed during operation, the negative-displacement-stiffness compensator is necessary to obtain the constant characteristics of the suspension control system for any d-axis current. It is advantageous to compensate for the unbalanced magnetic pull even if the motor is operated with a constant d-axis current in order to design the suspension controller.

References

[1] T. Fukao, A. Chiba and M. Matsui, "Test Results on a Super High Speed Amorphous Iron Reluctance Motor", IEEE Transactions on IA, Vol. 25, No. 1, 1989, pp. 119–125.

[2] A. Chiba and T. Fukao, "A closed-Loop Operation of Super High-Speed Reluctance Motor for Quick Response", IEEE Transactions on IA, Vol. 28, No. 3, 1992, pp. 600–606.

[3] A. Chiba, K. Chida and T. Fukao, "Principles and Characteristics of a Reluctance Motor with Windings of Magnetic Bearing", IEEJ Proc. IPEC-Tokyo 1990 (International Power Electronics Conf.), 1990, pp. 919–926.

[4] A. Chiba, M. A. Rahman and T. Fukao, "Radial Force in a Bearingless Reluctance Motor", IEEE Transactions on IA, Vol. 27, No. 2, 1991, pp. 786–790.

[5] O. Ichikawa, C. Michioka, A. Chiba and T. Fukao, "An Analysis of Radial Forces and a Rotor Position Control Method of Reluctance Type Bearingless Motors", Transactions of IEE Japan, Vol. 117-D, No. 9, 1997, pp. 1123–1131 (*in Japanese*).

[6] O. Ichikawa, C. Michioka, A. Chiba and T. Fukao, "A Decoupling Control Method of Radial Rotor Positions in Synchronous Reluctance Type Bearingless Motors", IEEJ Proc. IPEC-Yokohama 1995 (International Power Electronics Conf.), 1995, pp. 346–351.

[7] A. Chiba, M. Hanazawa, T. Fukao and M. A. Rahman, "Effects of Magnetic Saturation on Radial Forces of Bearingless Synchronous Reluctance Motors", IEEE Transactions on IAS, Vol. 32, No. 2, 1996, pp. 354–362.

[8] C. Michioka, T. Sakamoto, O. Ichikawa, A. Chiba and T. Fukao, "A Decoupling Control Method of Reluctance-Type Bearingless Motors Considering Magnetic Saturation", IEEE Transactions on IAS, Vol. 32, No. 5, 1996, pp. 1204–1210.

[9] L. Hertel and W. Hofmann, "Magnetic Couplings in a Bearingless Reluctance Machine", Proc. ICEM 2000 (International Conf. on Electrical Machines), Vol. 3, 2000, pp. 1776–1780.

13

Bearingless induction motors

Akira Chiba

In this chapter the bearingless induction motor (or induction type bearingless motor) is introduced. In a squirrel cage induction machine, short-circuit paths are constructed in the rotor iron from cast aluminium or fabricated copper in a cage-like structure. The revolving magnetic field set up by the stator winding currents induces voltages in the rotor circuit producing current. Since there is a phase shift between the rotor current and stator field, a torque is generated due to the interaction of the rotor current and the revolving magnetic field. However, in a bearingless induction motor the rotor current causes interference in the suspension force generation if the rotor circuit is not constructed correctly [1,2] i.e., a simple cage, as found in a standard cage induction motor, is not very suitable and modification is recommended. Hence, in the first section, the rotor structure and transfer characteristics are discussed. In the second section, a primitive controller block diagram is given in order to provide the basic concepts of a bearingless induction motor drive employing an indirect field-oriented controller [3,4]. Then an airgap flux-linkage-oriented controller is introduced. In the third section a direct field-oriented controller is put forward [5,6]. This detects the airgap flux linkage from flux sensors or terminal voltage and current. In the final section a comparison between the 2-pole and 4-pole motor drive is made, including the self-excitation phenomenon of the suspension winding [7].

13.1 Rotor structure and suspension force

Figure 13.1(a) shows the typical rotor lamination of a cage induction motor. The circular silicon steel sheet has a centre hole punched for the shaft, and the small holes (or slots) around the circumference are to accommodate the rotor bars for construction of the cage. The rotor slots can be open to the rotor surface or closed depending on the design. Open slots reduce leakage inductance

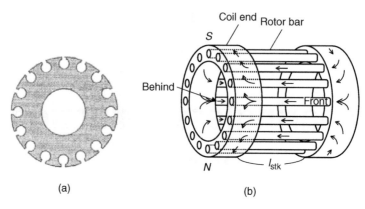

Figure 13.1 Iron core and squirrel cage windings: (a) rotor iron core; (b) squirrel cage rotor windings

whereas closed slots make casting an aluminium cage more straightforward. For high speed applications the slots are circular and closed for improved rotor robustness. Lamination thickness is typically 0.5, 0.35 mm or thinner. Thin sheets reduce the iron core loss although the manufacturing cost is increased. They form a cylindrical iron stack for the rotor core. In super-high-speed machines the rotor iron core is made from a solid cylindrical piece of iron rather than laminated sheets to enhance shaft stiffness with the sacrifice of rotor losses and performance.

Figure 13.1(b) shows the typical short-circuit conductor arrangement of the squirrel cage rotor. The bars are located in the rotor slots with an axial core length of l_{stk}. Two end-rings connect the rotor bars together. The cage, as previously mentioned, is fabricated from copper bars and rings or cast in aluminium. In super-high-speed applications, a copper compound metal is employed to improve robustness but with a conductivity reduction. In most bearingless induction motors the rotor structure is more complex than the simple cage arrangement shown in Figure 13.1(b) and, since they are still prototype machines, the rotor cage is fabricated from copper bars, inserted into the core and connected by copper end-pieces (end-ring sections), which are braised or soldered on. This is the easiest way to fabricate a prototype rotor.

Suppose that a 2-pole revolving magnetic field is applied to a squirrel cage rotor. Electromotive forces (EMFs) are induced so that a 2-pole current distribution is generated in the rotor short circuit (the cage). The arrows in Figure 13.1(b) show the direction of the current flow at an instance in time. The current in the short circuits generate a 2-pole flux wave as indicated by N and S in the figure. The 2-pole flux rotates at the slip frequency with respect to the rotor reference frame (i.e., a reference frame rotating at rotor speed) in steady-state conditions where slip frequency = slip × supply frequency. Torque is generated as a product of the rotor current and magnetic field. To enhance torque, low resistance in the rotor circuit is necessary. If a 4-pole revolving magnetic field is also applied to the squirrel cage rotor then a 4-pole current distribution flows in the rotor short circuits. Hence the squirrel cage rotor can generate torque for both the 4-pole and 2-pole revolving magnetic fields (and even higher pole numbers) and it is not pole-specific.

A standard squirrel cage rotor causes problems when used in a bearingless induction machine to generate a suspension force. Suppose that an additional 2-pole winding is used for suspension force generation. For a step change in suspension force there has to be a step change in the 2-pole stator flux. Because voltage is induced into the rotor circuit, which generates rotor current, the rotor flux does not change suddenly. Hence the interaction of stator (primary) and rotor (secondary) circuits produces a delay in the 2-pole flux which in turn delays the suspension force response. In addition to the response delay, the rotor current also makes alignment of the flux — hence the force is also difficult to orientate.

To overcome the problems associated with the squirrel cage rotor, a pole-specific short circuit can be used. Figure 13.2(a) shows a pole-specific rotor short circuit for a 2-pole induction motor. Two rotor bars are connected by two

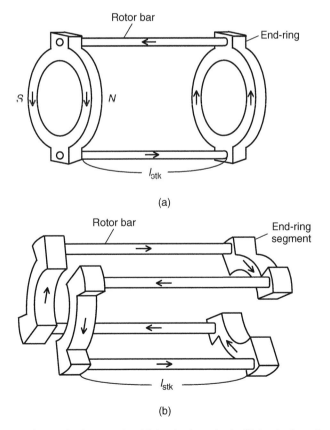

(a)

(b)

Figure 13.2   Pole-specific short circuits: (a) 2-pole short circuit; (b) 4-pole short circuit

end-rings. If a rotating 2-pole magnetic field is generated in the airgap by the stator winding currents, as indicated by N and S, EMFs are induced and current is generated in the rotor; with the flow shown by the arrows. If a 4-pole revolving magnetic field is applied, rotor current does not flow because the net rotor flux linkage between the bars is zero since the N and S poles linking to the short circuit cancel each other out, and the EMFs induced into the short circuits sum to zero. Hence the short circuit is 2-pole pole-specific. If a rotor has 16 slots then eight short circuits should be constructed so that eight end-rings are required at each end. These eight short circuits need to be isolated.

Figure 13.2(b) shows a 4-pole pole-specific short circuit. The four rotor bars are connected by four end-ring segments (rather than complete rings) – two at each end – to form one short circuit. If the current flows in the direction indicated by the arrows then a 4-pole magnetic field is generated in the airgap. If there are 16 slots in the rotor core then 4 short circuits should be constructed with a mechanical phase-shift angle of 22.5 deg. If a 4-pole revolving magnetic field is applied then current is generated in the short circuits. However, there is no current when a 2-pole revolving magnetic field is applied because the EMFs induced into the circuits sum to zero.

Now, let us go back and consider a standard cage rotor to understand why it is impractical in a bearingless induction machine application. Figure 13.3 shows equivalent circuits for a bearingless induction motor with both 4-pole and 2-pole 3-phase windings. R_s and R_r are the stator and rotor resistances, l_s and l_r are stator and rotor leakage inductances and L_m and R_m are the magnetizing inductance and iron loss resistance. The subscripts 2 and 4 indicate the 2-pole

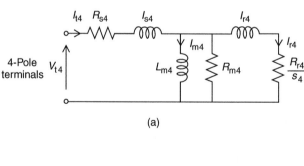

(a)

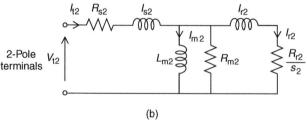

(b)

Figure 13.3 Equivalent circuits of the bearingless induction motor: (a) equivalent circuit of 4-pole winding terminals; (b) equivalent circuit of 2-pole winding terminals

and 4-pole windings. The slip is s_2 for the 2-pole and s_4 for the 4-pole winding current. These equivalent circuits are usually used for steady-state characteristics calculation in conventional induction motors driven with sinusoidal voltage and current.

Let us suppose that the 4-pole and 2-pole windings are assigned as the motor and suspension force windings respectively. When no torque is generated, s_4 is zero resulting in a high value (infinity) of the effective resistance R_{r4}/s_4 so that I_{r4} is zero (i.e., there is no rotor current) and the terminal current I_{t4} is equal to excitation current I_{m4}, assuming that iron loss is negligible. Note that the excitation current I_{m4} generates the airgap flux.

When torque is required, s_4 increases and R_{r4}/s_4 decreases so that we now get a rotor current I_{r4}. Torque is the product this rotor current and the stator component of the airgap flux. The airgap flux is mostly determined by the airgap length as well as the saturation of the rotor and stator iron material. R_{r4}/s_4 can be split into R_{r4} (rotor resistance, i.e., rotor copper loss $= I_{t4}^2 R_{r4}$) and $R_{r4} \times (1 - s_4)/s_4$ (electromechanical energy conversion $= I_{t4}^2 R_{r4} \times (1 - s_4)/s_4$). Therefore R_{r4} and s_4 are normally designed to be as small as possible at rated power for efficient operation (though there is often a compromise with the starting torque requirement, which requires a higher rotor resistance). For general-purpose 3-phase induction motors with ratings of a few kW (connected to a standard fixed-frequency supply), the rated slip s_4 is in the range of 0.02 (standard cage-rotor motors) to 0.1 (wound-rotor and specialized machines). For high speed or large motors (which are usually higher in efficiency), s_4 decreases to perhaps 0.005 or even less. If the rotor shaft rotational speed is ω_{rm} then the slip for the 4-pole winding is

$$s_4 = \frac{\omega_4/2 - \omega_{rm}}{\omega_4/2} \qquad (13.1)$$

where ω_4 is the frequency (in rad/s) of the 4-pole winding voltage and current. If a squirrel cage rotor is employed, the rotor bars and end-rings are optimized to make R_{r4} (and also R_{r2}) as small as possible. For the 2-pole winding, the slip is

$$s_2 = \frac{\omega_2 - \omega_{rm}}{\omega_2} \qquad (13.2)$$

where ω_2 is frequency of the 2-pole winding voltage and current. The 4-pole winding is assigned as the motor winding so that, at low slip, $\omega_{rm} \approx \omega_4/2$. For static suspension force generation, the current frequency of the windings should be equal so that $\omega_2 = \omega_4$, hence $s_2 = 0.5$. This is a high slip so that the rotor current is the dominant component of the terminal current I_{t2} rather than the excitation current I_{m2}. This leads to a very inefficient machine and problems with the suspension force generation.

The 4-pole and 2-pole winding fluxes are generated by the excitation currents I_{m4} and I_{m2}. I_{m4} is usually kept constant to provide excitation for torque generation so that I_{m2} is proportional to the suspension force. To regulate the excitation current, the terminal current I_{t2} is controlled by a regulator. The current transfer function I_{m2}/I_{t2} is

$$\frac{I_{m2}}{I_{t2}} = \frac{1}{1 + j\dfrac{\omega_2}{\omega_{a2}}\left\{\dfrac{R_{r2}/s_2}{R_{m2}} + \dfrac{1}{1 + j\omega_2/\omega_{b2}}\right\}} \tag{13.3}$$

where

$$\omega_{a2} = \frac{R_{r2}/s_2}{2L_{m2}}$$

and

$$\omega_{b2} = \frac{R_{r2}/s_2}{2l_{r2}}$$

This transfer function illustrates that the first pole frequency is ω_{a2} and the amplitude starts to role off at a rate of $-20\,\mathrm{dB/dec}$. The second pole frequency ω_{b2} is between 12 and 20 times higher than the first pole frequency since l_{r2}/L_{m2} is in the range of 0.05–0.08. The suspension force response is influenced by this current transfer function.

Next, let us consider the case when a pole-specific rotor is employed. If a 4-pole arrangement is employed then R_{r2} is infinite because there is no 2-pole rotor current due to the 2-pole revolving magnetic field. Hence the excitation current is proportional to the terminal current I_{t2} (neglecting iron loss). Therefore the transfer function I_{m2}/I_{t2} is unity. In practice I_{m2}/I_{t2} decreases at high frequency because of the iron loss. However, the frequency at this point is usually high enough to not affect the mechanical dynamics.

In Figure 13.4, the curves show the phase angle characteristics. The curve for the squirrel cage rotor is calculated using (13.3) while the curve for the pole-specific rotor includes the iron loss resistance. Note that the pole-specific rotor exhibits a better transfer function. The data points indicate experimental results for verification. A static suspension force is applied to the test machine and the direction of the suspension force command vector is obtained. At zero speed, the suspension force command is exactly opposite to the applied suspension force. As the speed increases, the suspension force command is automatically generated with a phase-lead angle to compensate for the phase lag of the generated suspension force. The command amplitude also increases to compensate for the decrease in gain. For the cage rotor, at a speed of about 5 rev/s, the phase lag angle is too large to maintain magnetic suspension. However, the pole-specific rotor provides a good frequency characteristic.

When the roles of the 4-pole and 2-pole windings are reversed, i.e., when the 2-pole winding is assigned as the motor winding, the transfer function of the squirrel cage improves but the pole-specific rotor is still superior.

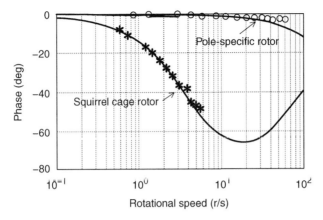

Figure 13.4 Radial force/suspension winding current transfer function phase angle

13.2 Indirect type drives

In this section the indirect type of bearingless induction motor drive is introduced. A primitive controller structure is first considered for ease of understanding. The controller structure is almost the same as that used for the primitive bearingless motor except for amplitude and phase angle compensation of the motor winding current. This controller is valid under steady-state conditions. For fast response, an airgap field-oriented controller is then introduced. It is found that the airgap flux linkage is regulated even under fast acceleration during a short transient over-current period.

Figure 13.5 shows the block diagram of a primitive controller. The controller is divided into two parts: one is a radial position controller and the other is a motor controller. The motor controller structure is slightly different from that of the primitive bearingless motor although the radial position controller has the same structure. The bearingless machine has 4-pole and 2-pole windings designated as motor and suspension force windings respectively. The mechanical synchronous speed command ω_{4m}^* is doubled to generate an electrical synchronous speed command $2\omega_{4m}^*$ and the cosine and sine functions of $2\omega_{4m}^* t$ are obtained. These functions are the airgap flux linkages and hence they are the commands used for the radial displacement derivatives of airgap flux linkages λ_{gc}' and λ_{gs}', where λ_{gc}' and λ_{gs}' are proportional to airgap flux linkages λ_{gc} and λ_{gs}. This assumes that the mutual inductance between the 4-pole and 2-pole windings is proportional to the rotor displacement. Since the amplitudes of the cosine and sine functions are unity, the amplitude of the airgap flux linkage is unity. The orientation of the airgap flux linkage is given by $2\omega_{4m}^* t$. The amplitude and direction of the airgap flux linkage is independently generated and the current amplitude I_{mp}^* and phase angle θ^* are adjusted so that the

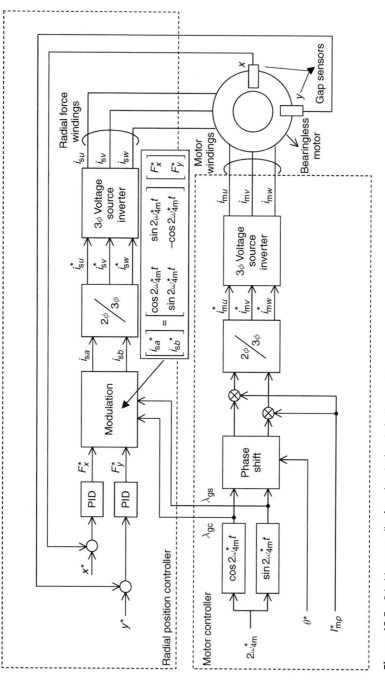

Figure 13.5 Primitive controller for a bearingless induction motor

actual airgap flux linkage follows the commands λ_{gc} and λ_{gs}. In the modulation block, suspension currents are generated from suspension force commands, and the amplitudes of flux linkage derivatives are taken to be unity in the calculation.

Current commands are generated with an amplitude of I_{mp}^* and phase-lead angle of θ^*. For example, the u-phase motor current is

$$i_{mu}^* = \sqrt{\frac{2}{3}} I_{mp}^* \cos(2\omega_{4m}^* t + \theta^*) \tag{13.4}$$

The amplitude and phase-lead angle should be adjusted so that the motor airgap flux linkage corresponds to the flux commands λ_{gc} and λ_{gs}. Since the airgap flux linkage is generated by the excitation current I_{m4} in Figure 13.3, the line current is obtained from (13.3). Substituting 4 for 2 and solving for I_{t4}/I_{m4} yields

$$\frac{I_{t4}}{I_{m4}} = 1 + j \frac{\omega_4}{\omega_{a4}} \left\{ \frac{R_{r4}/s_4}{R_{m4}} + \frac{1}{1 + j\omega_4/\omega_{b4}} \right\} \tag{13.5}$$

where

$$\omega_{a4} - \frac{R_{r4}/s_4}{2L_{m4}}$$

and

$$\omega_{b4} = \frac{R_{r4}/s_4}{2l_{r4}}$$

The amplitude and phase angle of the transfer function of I_{t4}/I_{m4} is shown in Figure 13.6. Note that correspondence of the motor airgap flux linkage with respect to the flux command is realized when the amplitude and phase-lead angle are gradually increased as a function of slip. This fact can be explained from Figure 13.3(a) as follows:

(a) When the slip is zero, the rotor current is zero so that the terminal current is equal to the excitation current I_{m4} (neglecting the iron loss current).
(b) As the slip increases, the impedance of R_{r4}/s_4 decreases and the rotor current I_{r4} is no longer zero. Since the impedance of the rotor circuit is composed of l_{r4} and R_{r4}/s_4 in series, but the mutual reactance is the inductance L_{m4} only, I_{r4} will lead with respect to I_{m4}.
(c) The line current I_{t4} is the phasor sum of the rotor and excitation currents so that I_{t4} is advanced with respect to I_{m4}. The amplitude also increases.

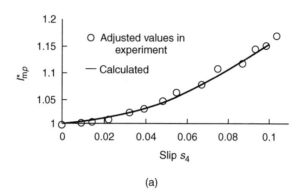

(a)

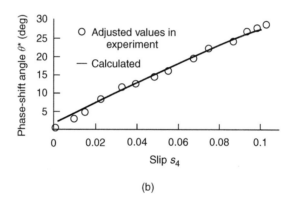

(b)

Figure 13.6 Amplitude and phase-shift angle of motor current: (a) current amplitude compensation; (b) current phase-shift angle θ compensation

To see the effectiveness of the primitive controller several tests can be carried out. The primitive controller is effective when the motor is generating torque in the following tests [3]:

(a) The suspension force command is decoupled when a static force is applied to a shaft.
(b) No interference in the perpendicular radial axes is seen in a disturbance force suppression test.
(c) The airgap flux linkage commands λ_{gc} and λ_{gs} correspond to the detected flux waveforms obtained through search coils wound around stator teeth.

These facts indicate that airgap flux linkage components in the bearingless motor correspond to the airgap flux linkage commands.

However, there is a problem if there is a sudden change in rotational speed command $2\omega^*_{4m}$ since the primitive controller is only valid under steady-state conditions. In order to realize airgap flux linkage steering during fast acceleration and deceleration in an application, field-oriented control theory is required (which is often referred to as vector control theory).

Figure 13.7 shows an example of the indirect type of a vector controller [8,9]. The controller regulates the airgap flux linkage vector even during transient conditions. The angular speed of the shaft ω_{rm} is detected by a rotary encoder. The shaft speed is multiplied by the number of pole pairs, i.e., 2 for a 4-pole motor winding. Then $2\omega_{rm}$ is compared with the rotational speed command $2\omega_{rm}^*$ and a torque current command i_{qs} is generated through the PI speed controller. An excitation current command i_{dm} is given as a constant. The slip frequency $2\omega_s$ and d-axis current are generated using the following equations

$$2\omega_s = \frac{(R_{r4} + sl_{r4})i_{qs}}{(L_{m4} + l_{r4})i_{dm} - l_{r4}i_{ds}} \tag{13.6}$$

$$i_{ds} = i_{dm} + \frac{2\omega_s \dfrac{l_{r4}}{R_{r4}} i_{qs}}{1 + s\dfrac{l_{r4}}{R_{r4}}} \tag{13.7}$$

One can see a derivative operator in the numerator of (13.6). However, in a practical system, an integration function is implemented in a later block to counteract this derivative operation. Note that the induction machine parameters, such as the rotor resistance and the mutual and leakage inductances, are required for the indirect vector controller. These parameters may vary with temperature, magnetic saturation or operation point. On-line identification of the machine parameters is possible using several methods, such as a model reference adaptive system, observers and so on. Advanced readers are encouraged to study papers on induction motor parameter identification.

The slip frequency is added to $2\omega_{rm}$ so that the instantaneous speed of the airgap flux linkage $2\omega_0$ is obtained. Integration of the angular speed, i.e.,

$$\phi_0 = \int 2\omega_0 dt \tag{13.8}$$

provides the instantaneous angular position of airgap flux linkage. The sinusoidal and cosinusoidal functions of $2\phi_0$ provide the components of the airgap flux linkage vectors λ_{gc} and λ_{gs} respectively. These components are used in the modulation block.

In the current generator, 3-phase symmetrical current commands are generated using the d- and q-axis stator currents i_{ds} and i_{qs}, as well as the airgap flux linkage position ϕ_0. For an example, the u-phase motor current is

$$i_{mu} = \sqrt{\frac{2}{3}}(i_{ds} \cos 2\phi_0 - i_{qs} \sin 2\phi_0) \tag{13.9}$$

The v- and w-phase currents have phase shifts of $2\pi/3$. The motor currents are regulated using these instantaneous currents. Hence the rotational angular position of airgap flux linkage in an actual bearingless induction motor should correspond to ϕ_0 and the amplitude of the airgap flux linkage is constant, which

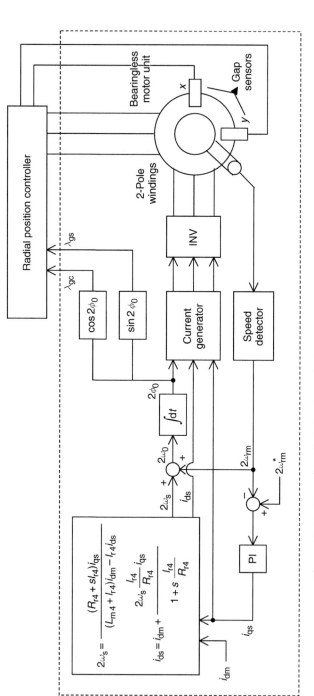

Figure 13.7 Airgap-flux-oriented vector controller

is determined by the current command i_{dm}. Even in transient conditions, such as during fast acceleration and deceleration, the control is valid.

Figure 13.8 shows the performance during a fast acceleration. A step in speed command $2\omega^*_{rm}$ from 1200 to 2700 r/min is given to the bearingless induction machine drive. The torque current command i_{qs} increases as a step function to the maximum value of 10 A in this case. The speed increases as a ramp function. The dynamic response is determined by the rotational inertia of the system. If the vector controller is not used then significant fluctuations of 100 μm are seen in the radial displacements x and y during the period from 20 ms after the acceleration start-up to the end of the acceleration. The fluctuation is delayed for 20 ms because of a delay in the change of airgap flux linkage.

In the radial displacement waveforms, there is no serious fluctuation when the airgap-oriented vector controller is employed. Only a small variation in x and y, caused by shaft unbalance and mechanical tolerance, is seen. The frequency of these small radial vibrations gradually increases as the speed increases.

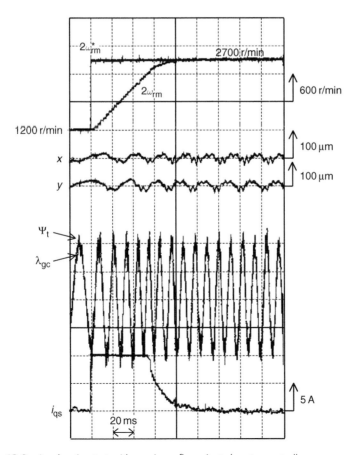

Figure 13.8 Acceleration test with an airgap-flux-oriented vector controller

The cosinusoidal waveform of airgap flux linkage λ_{gc} and detected flux waveform Ψ_t are also shown. Ψ_t is detected by a search coil wound around the stator teeth which corresponds to the position of cosinusoidal flux component. If a primitive controller is employed, a phase-shift and amplitude difference is seen in Ψ_t after 20 ms of the step in speed command. However, good correspondence is seen in this figure. This result indicates that the airgap flux linkage in the actual bearingless induction machine corresponds to the flux command [4].

13.3 Direct type drives

In the previous section an indirect vector controller was described. The indirect vector controller is cost-effective because it does not require flux sensors. However, it does require the machine parameters, which may vary with operational conditions. On the other hand a direct vector controller does not need the machine parameters because the revolving magnetic field is detected by flux sensors. Thus the direct vector controller is robust with respect to temperature variation. Some recent direct vector controllers do not need flux sensors, with the fluxes being estimated from the terminal voltage and line current, as well as the machine parameters; and a combination of indirect and direct control has also been recently proposed.

In the bearingless induction motor, instantaneous rotational position and amplitude of the airgap flux linkage vector are required for the radial position controller. These values are given by the motor vector controller described in the previous section. In this section, the airgap flux linkage is detected by flux sensors and sampling of the terminal voltage and current. There are advantages to direct detection of airgap flux linkage in bearingless induction motor drives:

(a) The motor driver can be operated independently of the radial position controller so that the motor controller does not need to provide airgap flux linkage information.
(b) Hence the motor driver can be a general-purpose inverter, operated with a constant V/f profile. These inverters are mass produced so that the cost is low.
(c) The motor windings can be supplied from a commercial power system with constant voltage and frequency.

In this section two types of direct flux detection are introduced.

13.3.1 Search coil method

Figure 13.9 shows a cross section of the stator iron core and search coil arrangement. For the 4-pole and 2-pole windings, one phase is shown. The 4-pole conductor phase-belts are located at the bottom of two adjacent stator slots and are distributed in four sections. The other stator slots will contain the v- and w-phases; the 4-pole winding is single layer, so there will only be one coil side

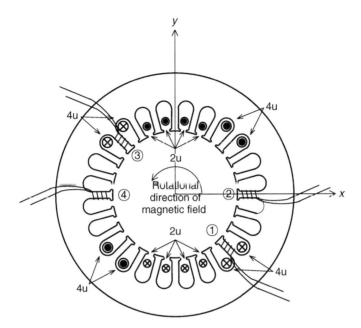

Figure 13.9 Search coil arrangement in the stator iron

per slot and phase coils will not share slots. The 2-pole conductor phase-belts are located at the top of four successive slots distributed in two bands. The individual 4-pole and 2-pole conductor arrangements are similar to single-layer windings found in conventional induction motors.

Four search coils are wound around different stator teeth and numbered 1 to 4. Let us suppose that the 4-pole windings are excited by a symmetrical 3-phase current to generate a revolving magnetic field. The 4u conductors are excited by a cosinusoidal current so that search coil 2 detects the cosinusoidal flux component and search coil 1 detects the sinusoidal component. Coils 3 and 4 are placed on opposite stator teeth with respect to search coils 1 and 2. These additional search coils are needed to cancel the influence of the 2-pole suspension force flux component as shown in the next figure. Note that Hall probes can also be installed to detect airgap fluxes instead of search coils.

Figure 13.10 shows the detected waveforms for search coil fluxes Ψ_1 and Ψ_2. The voltage induced at the search coil terminals are integrated by an RC integrator circuit. The flux command λ_{gc} in the indirect controller is also shown. No suspension force is generated in the upper trace, but 2.6 kgf is generated in the middle and the lower traces. From these figures the following observations can be made:

(a) The amplitudes of Ψ_1 and Ψ_2 are not exactly equal, with Ψ_2 being slightly larger in amplitude.
(b) The phases of λ_{gc} and Ψ_2 agree only when no suspension force is generated.
(c) λ_{gc} is in phase with the average of Ψ_2 and Ψ_4 even though a suspension force is generated.

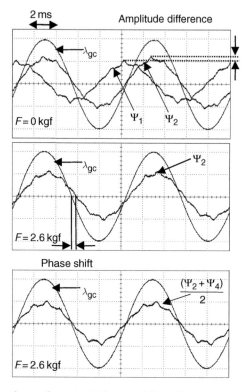

Figure 13.10 Waveforms of search coil fluxes and flux reference in the controller

When a suspension force is generated, both 4-pole and 2-pole revolving magnetic fields are generated so that the search coils detect the sum of these magnetic fields. However, the average of the output of two opposite search coils eliminates the 2-pole revolving magnetic field so that the averaged search coil voltage is independent of suspension force as noted in (c).

The amplitude difference noted in (a) is caused by space harmonics in the MMF distribution. The calculation is quite involved and MMF harmonics should be considered for the 24-slot stator; however, simply multiplying constants will provide compensation [10,11].

Figure 13.11 shows a system block diagram for direct flux detection in a bearingless induction motor drive. The sinusoidal and cosinusoidal flux components are obtained from

$$\lambda_{gs} = \frac{1}{2\sqrt{3}}(\Psi_1 + \Psi_3) \tag{13.10}$$

$$\lambda_{gc} = \frac{1}{4}(\Psi_2 + \Psi_4) \tag{13.11}$$

Note that the constants $1/(2\sqrt{3})$ and 1/4 depend on the conductor arrangement and number of stator slots. One practical method is to adjust the detector gains

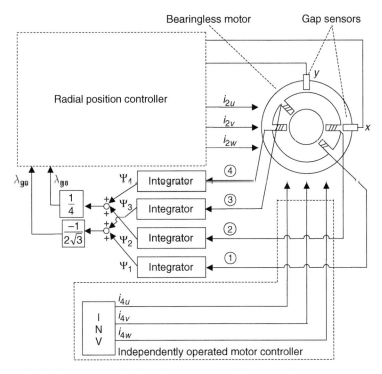

Figure 13.11 Block diagram for direct flux detection with search coils

after the integrator outputs are added. The detected flux components are used in the modulation block of the radial position controller. An open-loop inverter is connected to the motor winding terminals to drive the motor and it operates independently. In experiments, a step speed acceleration using a square-waveform inverter supply was performed and it was found that the magnetic suspension was stable during the acceleration [5].

13.3.2 Detection from line voltage and current

Figure 13.12 shows a block diagram of another method for detecting the airgap flux linkage vector. The airgap vector is calculated from the voltage and current at the motor winding terminals so that search coils are not required. The stator flux linkage vector $[\lambda_{abs}]$ is in a 2-phase coordinate system and obtained by the integration of the 2-phase voltage $[v_{abs}]$ and subtracting the stator resistance voltage drop, i.e., the product of stator resistance R_s and 2-phase current vector $[i_{abs}]$ so that

$$[\lambda_{abs}] = \int ([v_{abs}] - R_s[i_{abs}]) \mathrm{d}t \qquad (13.12)$$

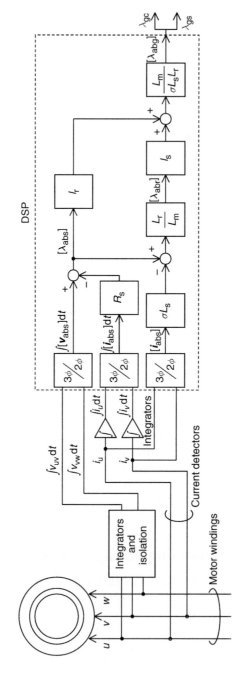

Figure 13.12 Block diagram for direct flux detection from motor terminal voltage and current

The rotor flux linkage vector $[\boldsymbol{\lambda}_{\mathrm{abr}}]$ in 2-phase coordinates is

$$[\boldsymbol{\lambda}_{\mathrm{abr}}] = \frac{L_{\mathrm{r}}}{L_{\mathrm{m}}}([\boldsymbol{\lambda}_{\mathrm{abs}}] - \sigma L_{\mathrm{s}}[\boldsymbol{i}_{\mathrm{abs}}]) \tag{13.13}$$

where L_{r} and L_{s} are the rotor and stator self-inductances. These are the sum of mutual inductance L_{m} and the respective rotor or stator leakage inductances. The leakage coefficient σ is given by

$$\sigma = 1 - \frac{L_{\mathrm{m}}^2}{(L_{\mathrm{s}} L_{\mathrm{r}})}$$

so that the airgap flux linkage vector is obtained from the stator and rotor flux linkage vectors

$$[\boldsymbol{\lambda}_{\mathrm{abg}}] = \frac{L_{\mathrm{m}}}{\sigma L_{\mathrm{s}} L_{\mathrm{r}}}(l_{\mathrm{r}}[\boldsymbol{\lambda}_{\mathrm{abs}}] + l_{\mathrm{s}}[\boldsymbol{\lambda}_{\mathrm{abr}}]) \tag{13.14}$$

where l_{s} and l_{r} are stator and rotor leakage inductances. The components of the airgap flux linkage vectors are cosinusoidal and sinusoidal.

In the figure, the line-to-line voltages are integrated and isolated and then transformed into 2-phase coordinates to calculate the integral of the voltage vector. The line currents are detected so that the current vector can be calculated. The above equations are calculated to obtain the airgap flux linkage vector. The outputs λ_{gc} and λ_{gs} are fed to the modulation block in the radial position controller. These blocks can be processed using a digital signal processor. To verify the control strategy a detection and sampling system was also constructed and tested and successful magnetic suspension, with an open-loop inverter, was confirmed [12].

Note that the sensitivity of parameter R_{s} cannot be neglected at low speed. The stator resistance varies significantly with temperature and this may cause an error in the detected airgap flux linkage vector. It should also be noted that the amplitude of airgap flux may not always be constant; for example, in some control strategies the rotor flux amplitude rather than the total airgap flux is kept constant. In this case the airgap flux linkage vector should be calculated or detected for the modulation block. How this is done is described in the following section.

13.3.3 Modulation block suspension current commands

In the previous section, which described indirect vector controllers, the amplitude of the airgap flux linkage was kept constant. However, in practical applications, the excitation level may change depending on the operational conditions so that the detected airgap amplitude may vary. If this is the case, the suspension current calculation in the modulation block can be modified [13]. In a squirrel cage induction motor, the relationship between the suspension forces and currents can be written as [6]

$$\begin{bmatrix} F_x \\ F_y \end{bmatrix} = M' \begin{bmatrix} i_{4gd} & i_{4gq} \\ i_{4gq} & -i_{4gd} \end{bmatrix} \begin{bmatrix} i_{2gd} \\ i_{2gq} \end{bmatrix} \tag{13.15}$$

where i_{4gd} and i_{4gq} are the airgap flux linkage component currents in the d- and q-axes for the 4-pole winding and i_{2gd} and i_{2gq} are the corresponding 2-pole winding currents. These currents are dc under steady-state conditions and this equation indicates that a suspension force is generated by the interaction of the 2-pole and 4-pole airgap flux linkage currents. Let us assume a 4-pole motor with a pole-specific rotor so that i_{2gd} and i_{2gq} are simply given as

$$\begin{bmatrix} i_{2gd} \\ i_{2gq} \end{bmatrix} = \begin{bmatrix} \cos 2\omega_4 t & \sin 2\omega_4 t \\ -\sin 2\omega_4 t & \cos 2\omega_4 t \end{bmatrix} \begin{bmatrix} i_{2a} \\ i_{2b} \end{bmatrix} \tag{13.16}$$

Let us also assume that the airgap flux linkage is aligned with the d-axis so that $i_{4gq} = 0$. The product of M' and i_{4gd} is defined as λ'_g. Hence the suspension forces are given by

$$\begin{bmatrix} F_x \\ F_y \end{bmatrix} = \begin{bmatrix} \lambda'_g & 0 \\ 0 & -\lambda'_g \end{bmatrix} \begin{bmatrix} \cos 2\omega_4 t & \sin 2\omega_4 t \\ -\sin 2\omega_4 t & \cos 2\omega_4 t \end{bmatrix} \begin{bmatrix} i_{2a} \\ i_{2b} \end{bmatrix} \tag{13.17}$$

$$= \lambda'_g \begin{bmatrix} \cos 2\omega_4 t & \sin 2\omega_4 t \\ \sin 2\omega_4 t & -\cos 2\omega_4 t \end{bmatrix} \begin{bmatrix} i_{2a} \\ i_{2b} \end{bmatrix}$$

The radial winding current commands are generated from the inversion of the matrix in the above equation

$$\begin{bmatrix} i^*_{2a} \\ i^*_{2b} \end{bmatrix} = \frac{1}{\lambda'_g} \begin{bmatrix} \cos 2\omega_4 t & \sin 2\omega_4 t \\ \sin 2\omega_4 t & -\cos 2\omega_4 t \end{bmatrix} \begin{bmatrix} F^*_x \\ F^*_y \end{bmatrix} \tag{13.18}$$

The equations (13.17) and (13.18) provide some important points about the control of the bearingless induction motor:

(a) In (13.17) the suspension forces are proportional to λ'_g, which is the product of M' and airgap flux linkage current i_{4gd}. M' is the displacement derivative of mutual inductance between motor and suspension windings. Let us consider λ_g without the derivative; λ_g is the flux linkage of the suspension winding due to the airgap flux and it is almost proportional to the rotor displacement. It is easily obtained from inductance measurements. λ'_g is the slope of the flux linkage characteristic of the suspension winding with respect to the rotor displacement so that λ'_g is proportional to λ_g.
(b) The cosine and sine functions originate from the rotational coordinate transformation of the current. The angular position is $2\omega_4 t$ (aligned with the d-axis with $t = 0$) so that $\lambda_g \cos 4\omega_4 t$ and $\lambda_g \sin 4\omega_4 t$ indicate perpendicular components of the airgap flux linkage.
(c) From points (a) and (b) the radial forces are given by a matrix product of the cosine and sine airgap flux components and the suspension winding currents.
(d) In a controller, the suspension currents can be calculated by detection or by feed-forward commands of the airgap flux in two perpendicular axes.

13.4 Two-pole motor drive

Since the induction motor has a cylindrical rotor, either of the 4-pole and 2-pole windings can be assigned as the motor drive winding. In this section a comparison of winding assignment is made.

A comparison of a machine with a 2-pole motor drive with 4-pole suspension winding with respect to the 4-pole motor winding arrangement is summarized below:

(a) Better transfer function of the suspension force to the currents when using a squirrel cage rotor.
(b) Reduced VA requirement at suspension force winding terminals.
(c) Self-excitation of the suspension force winding terminals with a squirrel cage rotor.
(d) Increased layers of end-ring segments in the pole-specific rotor.

Characteristic (a) is an advantage of the 2-pole motor drive. The transfer function of the suspension force to the suspension force winding current, shown in the first section of this chapter, has a phase-lag and amplitude reduction caused by the rotor short circuit. The phase-lag causes interference of the two perpendicular suspension forces and a decrease in phase margin of the radial position controller. Calculation of the transfer function of the 4-pole suspension winding should be based on an instantaneous flux representation rather than the equivalent circuit. The analysis should start from the flux linkage to current relationships, with a 12×12 matrix to determine the flux linkages and currents of the rotor and both 3-phase stator winding sets, including a consideration of the radial rotor displacement [6]. The matrix relationship is simplified by a 3-phase to 2-phase transformation which results in an 8×8 matrix and then the stored magnetic energy is calculated. Suspension forces are derived from the partial derivatives of the stored magnetic energy so that the suspension force is a product of the airgap flux currents in the 2-pole and 4-pole windings as well as a derivative of mutual inductance with respect to a radial displacement. A state-space equation with eight states, i.e., 4-pole and 2-pole rotor and stator flux linkages in two perpendicular axes, is set up and the airgap flux currents and suspension forces are calculated. The suspension force and current relationship is derived through simulation to obtain the frequency characteristics. For low-power bearingless machines, squirrel cage circuits are acceptable because of the short time constant.

Characteristic (b) is also an advantage of the 2-pole motor drive. The frequency of a 2-pole motor drive is about a half of the 4-pole motor drive. In addition, the frequency of the suspension force winding current in the 2-pole motor is halved for static suspension force generation because the suspension current frequency is equal to the motor current frequency. Since the current frequency is halved, only half the terminal voltage is required so that the VA requirement, i.e., the product of voltage and current, is reduced.

Characteristic (c) can be an advantage in some applications. In generating a static suspension force for suspending the shaft weight, ω_4 and ω_2 are the same value. In a 2-pole motor drive, ω_2 is nearly equal to the shaft speed ω_{rm} so that the slip s_4 is almost -0.5. A negative slip indicates generator operation, hence real power is generated. The output power is sufficient to supply the winding copper losses and inverter losses if the speed and suspension force are high enough so that the suspension power is provided by self-excitation [7].

An effective application can be vibration suppression at high speeds. A shaft is suspended by external bearings. As the speed increases, self-excitation occurs. Then a damping force is generated by the bearingless induction rotor which is located at the centre of the shaft length so that the shaft vibration is suppressed effectively.

Characteristic (d) is a disadvantage of the 2-pole motor drive. As shown in the first section of this chapter, a pole-specific short circuit can be constructed for better suspension force generation. The number of end-ring segments at each end is equal to the number of rotor bars. For example, 16 rotor bars require eight end-ring segment layers at each end. For the 4-pole motor drive, only 4 coil end-ring layers are necessary. Reduction of end-ring segment layers is possible if more than two rotor bars are connected to an end-ring segment. However, the radial force transfer characteristics are inferior because of the rotor short circuits between adjacent bars which produce non-pole-specific current [1].

References

[1] A. Chiba, R. Miyatake, S. Hara and T. Fukao, "Transfer Characteristics of Radial Force of Induction-Type Bearingless Motors with Four-Pole Rotor Circuits", International Symposium on Magnetic Bearings '96, Kanazawa (ISMB '96), August 1996, pp. 319–325.

[2] A. Chiba and T. Fukao, "Optimal Design of Rotor Circuits in Induction-Type Bearingless Motors", IEEE Transactions on Magnetics, Vol. 34, No. 4, July 1998, pp. 2108–2110.

[3] A. Chiba, R. Furuichi, Y. Aikawa, K. Shimada, Y. Takamoto and T. Fukao, "Stable Operation of Induction-Type Bearingless Motors Under Loaded Conditions", IEEE Transactions on Industry Applications, Vol. 33, No. 4, July/August 1997, pp. 919–924.

[4] A. Chiba, K. Yoshida and T. Fukao, "Transient Response of Revolving Magnetic Field in Induction-Type Bearingless Motors with Secondary Resistance Variations", International Symposium on Magnetic Bearings (ISMB '98), Boston, USA, August 7, 1998, pp. 461–475.

[5] K. Yasuda, T. Kuwajima, A. Chiba and T. Fukao, "A Proposed Controller for Bearingless Induction Drives with Search Coils Wound Around Stator Teeth", Proceedings of the Maglev 2000, June 7–10, 2000, Rio de Janeiro, Brazil, pp. 435–440.

[6] T. Fujishiro, R. Hanawa, Y. Sakata, A. Chiba and T. Fukao, "An Analysis of an Induction Bearingless Motor with a Squirrel Cage Rotor", Proceedings of the 8th International Symposium on Magnetic Bearing, ISMB 8, August 2002 at Mito, Japan.

[7] A. Chiba, M. Yamashita, K. Kobayashi and T. Fukao, "Principles of Self-excitation at Radial Force Winding Terminals in Bearingless Induction Motors with a Squirrel Cage Rotor", IEEE IAS 2000 Annual Meeting, Rome 00CH37129, pp. 235–240.

[8] T. Suzuki, A. Chiba, M. A. Rahman and T. Fukao, "An Air-Gap-Flux-Oriented Vector Controller for Stable Operation of Bearingless Induction Motors", IEEE Transactions on Industry Applications, Vol. 36, No. 4, July/August 2000, pp. 1069–1076.

[9] D. W. Novotny and T. A. Lipo, "Vector Control and Dynamics of AC Drives", Oxford Science Publications, 1996.

[10] K. Kiryu, A. Chiba and T. Fukao, "A Radial Force Estimation with Search Coil Fluxes in a Bearingless Induction Motor Driven by Multi Inverters", European Power Electronics Conference (EPE 2001) CDROM, August 2001, Graz, Austria, 11 pages.

[11] K. Kiryu, A. Chiba and T. Fukao, "A Radial Force Detection and Feedback Effects on Bearingless Motors", IEEE Industry Applications Conference, September 30, 2001 at Chicago IEEE01CH37248C 0-7803-7114-3/01, 6 pages.

[12] K. Yasuda, T. Kuwajima, A. Chiba and T. Fukao, "A Feedback Controller With Airgap Flux Estimation in Induction Type Bearingless Motors", 2000 National Convention Record IEE Japan 5–123, March 24, 2000 at Tokyo Institute of Technology, pp. 2090–2091 (*in Japanese*).

[13] T. Kuwajima, K. Yasuda, A. Chiba and T. Fukao, "Decoupling Control of the Radial Forces with Gap Flux Estimation", IEEJ Technical Meetings on Static Power Converters, SPC-00-44, June 30, 2000 at Mie University, pp. 41–46 (*in Japanese*).

14

Homopolar, hybrid and consequent-pole bearingless motors

Osamu Ichikawa

Bearingless motors can be classified into several groups. The major group produces suspension force by the interaction of n-pole and $(n \pm 2)$-pole fluxes (or p-pole-pair and $p \pm 1$-pole-pair). However, homopolar and hybrid bearingless motors belong to another group where the suspension force is generated by the interaction of a homopolar flux and a 2-pole flux, which is similar to the homopolar type of radial magnetic bearing, although the theory of operation can be unified by considering the homopolar airgap flux as a flux wave with zero pole number.

In this chapter, the structures and characteristics of homopolar, hybrid and consequent-pole bearingless motors are described.

14.1 Structures and principles

14.1.1 Homopolar magnetic bearing

Figure 14.1 shows the structure of a homopolar-type radial magnetic bearing. Permanent magnets or field windings are normally placed between the stator cores to produce a magnetizing flux. This flux flows along the rotor axis and through both pairs of the cores. Therefore the surface of each rotor core is magnetized with a unidirectional flux or a single pole (which can also be considered as a zero pole since it is constant). Radial forces for rotor suspension can be generated with an additional suspension flux superimposed onto the magnetizing flux. Figure 14.2 shows the cross section of an N-pole (north-pole) rotor and stator with the main paths of the magnetizing and suspension fluxes illustrated.

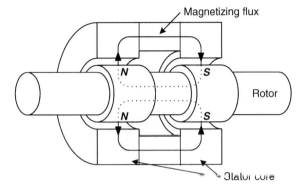

Figure 14.1 Structure of a homopolar magnetic bearing

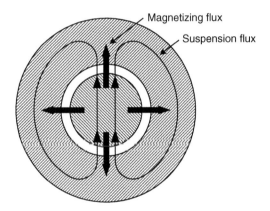

Figure 14.2 Cross section of a homopolar magnetic bearing

Radial arrows show the magnetizing flux; the other core combination would have these flowing in the opposite direction. The 2-pole suspension windings are wound separately on each stator core and the flux is always perpendicular to the rotor axis. In Figure 14.2, the flux is increased in the upper section of the airgap and decreased in the lower section of the airgap. The unbalance of this flux distribution produces a radial force in the upward direction on the rotor. The magnitude of the radial force is adjusted by varying either the magnetizing flux or the suspension flux. A horizontal radial force can be produced with a 2-pole suspension flux that is perpendicular to the 2-pole flux in Figure 14.2. Therefore, an adjustable radial force on the rotor can be generated in all directions on each rotor core so that the rotor radial position has four degrees of freedom; these can be controlled in the same way as the homopolar magnetic bearing shown in Figure 14.1. Since the rotor is symmetrical and the zero-pole flux is uniform, the radial force is produced independently of the rotor rotational position. The advantages of homopolar magnetic bearings are:

1. No bias current is required if permanent magnet magnetization is employed.
2. Low current requirement for magnetic suspension because the suspension flux does not flow through the permanent magnets so that the effective airgap length is relatively small.

Both the homopolar bearingless motor and the hybrid bearingless motor, as described below, have homopolar structures. The magnetizing flux of a hybrid motor flows along the rotor axis, while the mechanism for rotor suspension is the same as that of the homopolar magnetic bearing. The structures and mechanisms for torque and suspension force production are discussed in the following sections.

14.1.2 Homopolar bearingless motors

Figure 14.3 shows the structure and magnetizing flux path of a homopolar bearingless motor [1–3]. There are two rotor cores and two stator cores in the motor. Each stator core has both motor and suspension windings. The 8-pole motor winding is wound in both cores together while the 2-pole suspension windings are wound separately in each core. Each rotor core has four salient poles but no windings or permanent magnets (similar to the switched reluctance type machine). A field winding is wound between the stator cores. This winding

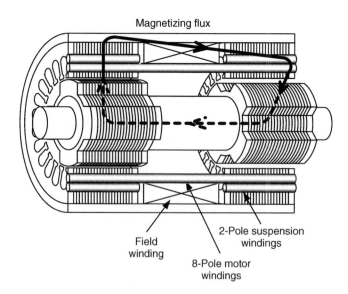

Figure 14.3 Homopolar bearingless motor

can be replaced with permanent magnets between the stator or rotor cores. If this is the case then it becomes a hybrid bearingless motor.

The main flux (as generated by the field current and a *d*-axis motor current) passes through the stator cores and circulates between the cores via the casing and shaft as shown in Figure 14.3. The salient poles on the left-hand rotor core are north poles whereas they are south poles on the right-hand rotor core. Looking into the rotor in the axial direction from the left, we can see that the south poles and north poles are offset by one rotor pole pitch as shown in Figure 14.4. The arrangement of the magnetized poles is similar to that of an 8-pole motor so that torque is generated in a similar manner to that of an 8-pole synchronous machine. The magnetizing flux can be generated by both the field current and *d*-axis motor current. With variation of the ratio of the field current and *d*-axis motor current, the power factor and the terminal voltage of the machine are adjustable.

The mechanism of producing radial force is virtually the same as for the homopolar magnetic bearing, as described in the previous section. By super-imposing a 2-pole suspension flux on the magnetizing flux a suspension force in any direction can be generated. Although the airgap flux is concentrated in the salient poles of the rotor cores, a constant suspension current generates a constant radial force, which is independent of the rotor rotational position if the motor has an appropriate number of poles. One of the remarkable characteristics is that the rotational position of the revolving magnetic field is not required for the control of the radial force so that it is possible to suspend the rotor with a simple and robust controller.

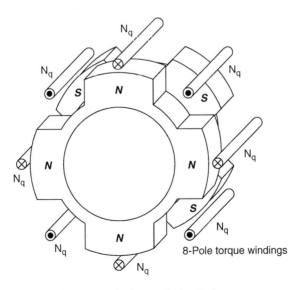

Figure 14.4 Axial view of the rotor of a homopolar bearingless motor

14.1.3 Hybrid bearingless motors

There are several different arrangements of the hybrid bearingless motor based on the permanent magnet locations. Some typical structures are shown in Figures 14.5, 14.6 and 14.7. The structure shown in Figure 14.5 has permanent magnets between the stator cores to develop the magnetizing flux. The flux path of the magnetizing flux is the same as that of the homopolar magnetic bearing. Each stator core has a 2-pole suspension winding to produce radial force so that four radial degrees of freedom of rotor position can be magnetically regulated.

Several further variations and modifications are possible. For instance, one of the rotor cores can be cylindrical and act as a magnetic bearing only. Also the permanent magnets can be on the rotor surface. As an example, [4] describes a 6-pole permanent magnet motor. Figure 14.6 shows a magnetizing permanent

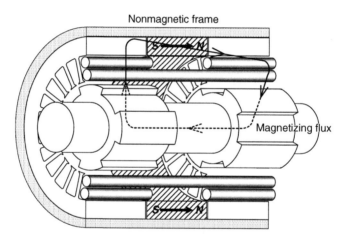

Figure 14.5 Hybrid bearingless motor with stator permanent magnets

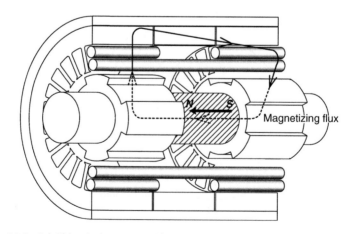

Figure 14.6 Hybrid bearingless motor with rotor permanent magnet

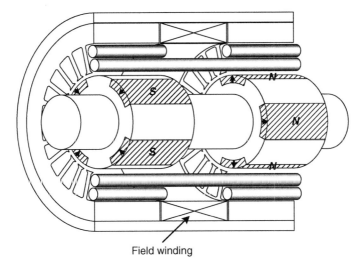

Field winding

Figure 14.7 Hybrid bearingless motor with permanent magnets between rotor poles

magnet on the rotor where the permanent magnet is oriented in the axial direction and located around the shaft. Another modification uses four permanent magnets attached to the rotor surface [5]. The permanent magnets can be embedded between rotor poles as shown in Figure 14.7. This structure is similar to the consequent-pole bearingless motor.

14.1.4 Consequent-pole bearingless motors

The rotor and stator of the consequent-pole bearingless motor do not work in tandem and there is no homopolar flux, although the rotor has similar topology to the homopolar and hybrid machines (which have one magnet per rotor pole-pair); the production of the radial force is slightly different. However, the radial force production of consequent-pole machines is also different from the more usual type of permanent magnet bearingless motor (where there are two magnets per rotor pole-pair). Therefore this type of machine lies between the homopolar and multi-polar types of machine – it exhibits characteristics from both, in terms of torque and suspension force production.

Figure 14.8 shows a cross-sectional view of a consequent-pole bearingless motor [6]. All the flux flows perpendicular to the rotor axis. The thick arrows indicate the polarities of the permanent magnets and the dotted lines show the 8-pole flux paths of the permanent magnets. The polarities are different from those of the conventional 4-pole inset permanent magnet motors (which have the same number of magnets). The surfaces of the permanent magnets are magnetized as north poles, and the surfaces of the iron poles are consequently magnetized as south poles. The stator has both an 8-pole motor winding and a 2-pole suspension winding (whereas a more standard p-pole-pair bearingless permanent magnet machine would have a p-pole-pair motor winding and a $p+1$

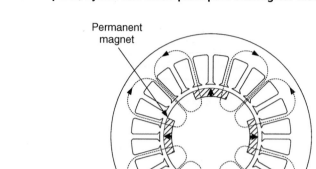

Permanent
magnet

Figure 14.8   Cross section of a consequent-pole bearingless motor

or $p - 1$ pole-pair suspension winding). Torque is produced by the interaction of the 8-pole permanent magnet flux with the 8-pole motor winding currents. 2-pole suspension flux flows mainly through the salient iron poles on the rotor because the permanent magnets have high magnetic reluctance. Superposition of the permanent magnet and suspension current fluxes under the salient rotor poles produces an unbalanced airgap flux distribution which results in a radial force. Since the permanent magnet flux can be thought of as a homopolar flux under the iron rotor poles (where the suspension winding current generates a 2-pole flux) we can see that we have met the criterion for suspension, where we have airgap fields with pole-pairs differing by 1 (2 and homopolar – since 1 pole is not possible because there is no start or finish of the pole, we may also consider homopolar flux as zero-pole or dc). Therefore we can say that the radial force production is similar to that of the homopolar bearingless motor, where a constant radial force is generated with constant suspension winding current at any rotor angular position.

However, in terms of actual airgap fields (and not just flux under the iron poles) a more rigorous mathematical analysis for the consequent-pole machine considers the permeance modulation of the consequent-pole rotor on the 2-pole suspension winding MMF. For a p-pole-pair consequent-pole machine (in this case 8-pole), the rotor will modulate the 2-pole suspension winding MMF to produce $p \pm 1$ pole-pair flux waves (i.e., 6-pole and 10-pole) which produces suspension. Hence, in terms of airgap flux waves, suspension production is different from the homopolar and hybrid machines, which have 2-pole and zero-pole airgap flux waves. Although, as already stated, in a consequent-pole machine, the homopolar action is really quasi-homopolar since it is thought to exist under the iron salient poles only. In the next section, a mathematical derivation is put forward which suggests a 6-pole consequent-pole rotor is not suitable due to a component of the radial force that is a function of rotor rotational position. Using the airgap permeance argument, we can see that a

6-pole consequent-pole rotor will modulate the 2-pole suspension winding MMF, producing 4-pole and 8-pole airgap components. Interaction of the 4-pole flux wave with the two-pole flux wave will produce this rotational-position-dependent radial force. This does not occur in the 8/2-pole winding combination. Hence an 8-pole consequent-pole rotor does not produce this rotational-position component of suspension force.

14.2 Number of poles

14.2.1 Features

Most bearingless motors generate radial force by adding a revolving suspension flux onto the revolving motor flux. In order to generate radial force in a certain direction, the direction of the suspension flux must be varied in accordance with the direction of the motor flux. On the other hand, homopolar and hybrid bearingless motors generate radial force by superimposing a 2-pole suspension flux onto the homopolar flux. Hence the advantage of these types of bearingless motor (and the consequent-pole machine) is that the radial force in a certain direction can be generated with a fixed suspension current for any rotor rotational position and load condition. In the following sections, the radial force is shown to be independent of the rotor angular position and motor torque current.

14.2.2 Flux distributions

Due to the homopolar structure, the flux distributions in the homopolar and hybrid bearingless motors and also the quasi-homopolar distribution of the consequent-pole bearingless motor are quite different from other types of bearingless motor. In this section, the airgap flux density distribution of a homopolar bearingless motor is analysed in order to derive the characteristics of the suspension force. The conditions and assumptions for the analysis are as follows:

1. The magnetizing curve is linear.
2. Only field current I_f, q-axis motor current i_{mq} and suspension current i_{sa} are assumed and the MMFs of the motor and suspension currents are distributed sinusoidally.
3. The motor winding is m-pole and the suspension winding is 2-pole. Each rotor core has m-poles. The pole arc is λ mechanical radians where $0 < \lambda \leq 2\pi/m$. The permeance between the poles is zero.

The airgap flux density can be calculated using the following equation:

$$B_g(\phi_s) = \frac{1}{lr}P(\phi_s)A(\phi_s) \qquad (14.1)$$

where l is the axial stack length of the laminated rotor and stator cores, r is the mean radius of the airgap under the salient poles, $P(\phi_s)$ is the permeance of the

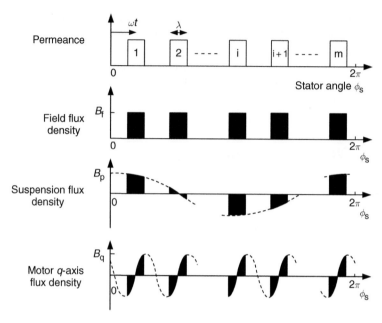

Figure 14.9   Airgap flux densities of a homopolar bearingless motor

airgap as a function of stator angle ϕ_s, and $A(\phi_s)$ is the MMF as a function of ϕ_s and the rotor angular position with respect to time is ωt. The sign of the flux density indicates the direction of flux across the airgap. Figure 14.9 shows airgap flux density distribution due to the field current, suspension current and q-axis motor current. The sum of these flux densities is

$$B_g(\phi_s) = B_f + B_p \cos \phi_s + B_q \sin m(\phi_s - \omega t) \tag{14.2}$$

for rotor poles, and

$$B_g(\phi_s) = 0 \tag{14.3}$$

for the inter-pole areas.

14.2.3 Suspension force

The radial forces in the x and y directions are calculated from the square of the sum of the airgap flux density $B_g(\phi_s)$:

$$F_x = \int_0^{2\pi} \frac{1}{2\mu_0} B_g(\phi_s)^2 lr \cos \phi_s d\phi_s \tag{14.4}$$

$$F_y = \int_0^{2\pi} \frac{1}{2\mu_0} B_g(\phi_s)^2 lr \sin \phi_s d\phi_s \tag{14.5}$$

If $m = 2$, i.e., each rotor core has 2 salient poles and the motor winding has 4 poles, the suspension forces can be obtained from:

$$F_x = \frac{lr}{2\mu_0}\left(2B_f B_p(\lambda + \sin\lambda\cos 2\omega t) + \frac{B_p B_q}{2}(\sin 2\lambda - 2\lambda)\sin 2\omega t\right) \quad (14.6)$$

$$F_y = \frac{lr}{2\mu_0}\left(2B_f B_p \sin\lambda\sin 2\omega t - \frac{B_p B_q}{2}(\sin 2\lambda - 2\lambda)\cos 2\omega t\right) \quad (14.7)$$

These suspension forces are functions of rotational angle ωt and q-axis motor flux B_q.

If $m = 3$, i.e., each rotor core has 3 salient poles and the motor winding has 6 poles, the suspension forces can be obtained from:

$$F_x = \frac{lr}{2\mu_0}\left(3B_f B_p\lambda + \frac{B_p^2}{2}\sin\frac{3\lambda}{2}\cos 3\omega t\right) \quad (14.8)$$

$$F_y = \frac{lr}{2\mu_0}\left(\frac{B_p^2}{2}\sin\frac{3\lambda}{2}\sin 3\omega t\right) \quad (14.9)$$

These suspension forces are generated independent of the torque current. However, the radial forces vary with the rotor rotational angle ωt.

If $m = 4$, i.e., each rotor core has 4 poles and the motor windings are 8-pole, the suspension forces can be obtained from:

$$F_x = \frac{lr}{2\mu_0}mB_f B_p\lambda \quad (14.10)$$

$$F_y = 0 \quad (14.11)$$

It is obvious from (14.10) and (14.11) that the radial force is independent of the rotor rotational angle ωt and the q-axis motor flux B_q. Thus, Figures 14.3 to 14.8 are drawn for the case of an 8-pole motor. For higher pole numbers, such as 10, 12, 14, and 16, the radial force is also independent of rotor position and q-axis motor current.

If permanent magnets are mounted on the surface of the rotor shaft or if the rotor core does not have salient iron poles [4–5] then radial forces of $m = 3$ can be derived by substituting $\lambda = 2\pi/3$ into (14.8) and (14.9), which yields the same equations as for the case of $m = 4$. It is obvious that the control system for a homopolar, hybrid or consequent-pole bearingless motor is more straightforward compared to that of the other bearingless motors if the number of poles is more than eight for salient pole rotors or more than six for surface-mount (non-salient) permanent magnet rotors. Generally, the radial forces of these bearingless motors can be written as:

$$\begin{bmatrix} F_x \\ F_y \end{bmatrix} = [M_f' I_f + M_d' i_{md}]\begin{bmatrix} i_{sa} \\ i_{sb} \end{bmatrix} \quad (14.12)$$

where M_f' and M_d' are constants related to the differential of the mutual inductance with respect to radial displacement.

14.3 Drive systems

Figure 14.10 shows a block diagram of the control system of a homopolar bearingless motor which has an 8-pole motor winding and a 2-pole suspension winding. Although four axis radial positions can be actively controlled by the system, only the left-half position control system is shown.

A field current I_f is supplied by an adjustable dc power supply. The field winding and the power supply can be removed if permanent magnets are used for the excitation. By regulating the field current, the terminal voltage and the power factor of the motor or generator terminals can be controlled in order to realize high-speed and high-efficiency operation.

Motor currents are controlled on the d and q axes and supplied with a 3-phase inverter. The rotational angle is required in the transformation from a rotational d-q coordinate into a fixed stator a-b coordinate system where

$$\begin{bmatrix} i_{ma}^* \\ i_{mb}^* \end{bmatrix} = \begin{bmatrix} \cos 4\omega t & -\sin 4\omega t \\ \sin 4\omega t & \cos 4\omega t \end{bmatrix} \begin{bmatrix} i_{md}^* \\ i_{mq}^* \end{bmatrix} \tag{14.13}$$

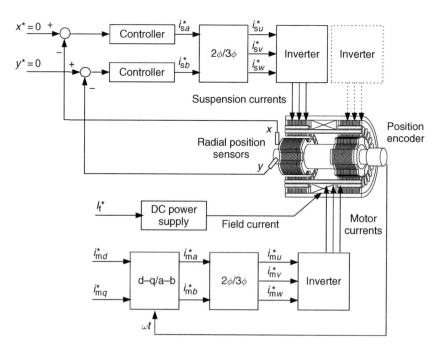

Figure 14.10 Block diagram for the control system of a homopolar bearingless motor

The 2-phase motor current commands are transformed into a 3-phase set using

$$\begin{bmatrix} i_{mu}^* \\ i_{mv}^* \\ i_{mw}^* \end{bmatrix} = \sqrt{\frac{2}{3}} \begin{bmatrix} 1 & 0 \\ -\frac{1}{2} & \frac{\sqrt{3}}{2} \\ -\frac{1}{2} & -\frac{\sqrt{3}}{2} \end{bmatrix} \begin{bmatrix} i_{ma}^* \\ i_{mb}^* \end{bmatrix} \tag{14.14}$$

The radial position of the rotor is detected by contactless gap sensors and the signals from these are used to generate the radial force commands. Suspension current commands i_{sa}^* and i_{sb}^* are directly obtained from the radial force commands in the x- and y-axes and the 2-phase suspension current commands are similarly transformed using (14.14). A 3-phase inverter injects currents into the suspension winding; note that the suspension control system is independent of the motor control system and there is no cross-coupling caused by the main torque-producing current.

Figure 14.11 shows the experimental results from a homopolar bearingless motor with the control system shown in Figure 14.10. An external static radial force was applied to the rotor and the bearingless motor produced 120 N in the x direction in order to suspend the rotor. The radial position and radial-force suspension currents were measured in this experiment. The field current and the d-axis motor current were kept constant, where $I_f = 1.5$ A and $I_{md} = 0$ A. Figure 14.11 shows that the suspension currents were almost constant, i.e., $i_{sa}^* = 10$ A and $i_{sb}^* = 0$ A, even though the rotor speed was approximately 290 r/min.

Example 14.1 The test machine in Figure 14.10 has the following dimensions: rotor diameter $= 72.4$ mm, stator diameter $= 135$ mm and the laminated axial length of each rotor core $= 56$ mm. The machine also has a radial force constant $M_f' = 8$ N/A^2. Calculate the suspension currents required to

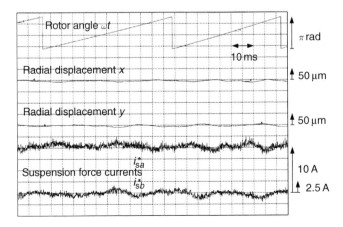

Figure 14.11 Current waveforms of a homopolar bearingless motor

produce a radial force $F_x = 120\,\text{N}$, when the motor is revolving at a speed of $3000\,\text{r/min}$ and the motor currents are $I_f = 1.5\,\text{A}$, $i_{md} = 0\,\text{A}$ and $i_{mq} = 3\,\text{A}$.

Answer

From (14.12), it is clear that the suspension currents are independent of rotor rotational angle and q-axis motor current. Therefore the suspension currents are $i_{sa} = 10\,\text{A}$ and $i_{sb} = 0\,\text{A}$ for any rotational angle.

References

[1] C. Michioka, Y. Toyoshima, O. Ichikawa, T. Fukao and A. Chiba, "Radial Force of Homopolar-Type Bearingless Motors in no Load Condition", IEEJ Proc. of Meeting of Rotating Machinery, RM-96-24, 1996, pp. 91–100 (*in Japanese*).

[2] O. Ichikawa, A. Chiba and T. Fukao, "Principles and Structures of Homopolar-Type Bearingless Motors", IEEJ Proc. IPEC-Tokyo 2000 (International Power Electronics Conf.), Vol. 1, 2000, pp. 401–406.

[3] O. Ichikawa, A. Chiba and T. Fukao, "Inherently Decoupled Magnetic Suspension in Homopolar-Type Bearingless Motors", IEEE Trans. on IA, Vol. 37, No. 6, November/December 2001, pp. 1668–1674.

[4] Y. Okada, K. Shinohara, S. Ueno and T. Ohishi, "Hybrid AMB Type Self Bearing Motor", Proc. 6th ISMB (International Symp. on Magnetic Bearings), 1998, pp. 497–506.

[5] H. Kanebako and Y. Okada, "Development of Hybrid Type Self-Bearing Motor Without Extra Bias Permanent Magnets", Proc. 7th ISMB (International Symp. on Magnetic Bearings), 2000, pp. 347–352.

[6] T. Takenaga, Y. Kubota, A. Chiba and T. Fukao, "A Principle and a Design of a Consequent-Pole Bearingless Motor", Proc. 8th ISMB (International Symp. on Magnetic Bearings), 2002, pp. 259–264.

15

Switched reluctance bearingless motors

Masatsugu Takemoto

In this chapter the bearingless switched reluctance motor is introduced. Conventional switched reluctance motors with mechanical bearings have generated significant interest among researchers and industrial engineers. For particular environments and applications, the switched reluctance motor can have superior performance because of its inherent features such as being fail-safe, robustness, low cost and possible operation at high temperature or in intense temperature variation [1–7]. Industrial drives which utilize the switched reluctance machine have been available for some time.

The switched reluctance motor also offers the possibility of operation as a bearingless motor. Torque is generated by magnetic attraction between rotor and stator poles. In this process a significant amount of attractive radial force is generated because the switched reluctance motor has salient poles and a short airgap length between these poles in order to effectively produce a reluctance torque. It is quite possible to take an advantage of the inherent high radial force that is produced for rotor shaft magnetic suspension. Therefore bearingless switched reluctance motors with differential stator winding configurations have been developed [8–14]. These motors have motor windings and suspension windings wound in the same stator in order to produce a suspension force that can realize rotor shaft suspension. Hence these motors are expected to be suitable for maintenance-free drives used in special circumstances such as high temperature, wide temperature variation and high acceleration operation.

15.1 Configuration of stator windings and principles of suspension force generation

Figure 15.1 shows only the A-phase stator winding of a 3-phase system with additional differential windings. The motor winding N_{ma} consists of four coils connected in series while the suspension windings N_{sa1} and N_{sa2} consist of two coils each. These coils are separately wound around the top of opposite stator poles. The B-phase winding configuration is located at 30 mechanical degrees to the A phase, and the C-phase winding configuration is located at 60 mechanical degrees to the A phase. The axes $a1$ and $a2$ of a perpendicular coordinate system can be defined using the A-phase winding as a reference, as shown in Figure 15.1. In this case, the axes x and y are aligned with $a1$ and $a2$ respectively. Similarly, the axes of coordinates $b1$, $b2$, $c1$ and $c2$ can be defined for the B-phase and C-phase windings. Since the number of stator poles is 12, the step angle is 15 mechanical degrees. The rotor angular position ϕ is defined as $\phi = 0$ when aligned with the A phase. The rotor angular position ϕ in Figure 15.1 is $-\pi/18$ ($-10\,\mathrm{deg}$).

Figure 15.2 shows the principle of suspension force generation for the bearingless switched reluctance motor. The thick solid lines show the symmetrical 4-pole flux produced by the 4-pole motor winding current i_{ma}. The broken lines show the symmetrical 2-pole flux produced by the 2-pole suspension winding current i_{sa1}. In this situation, the flux density in air-gap section 1 is increased because the direction of the 2-pole flux is the same as that of the 4-pole flux. In contrast, the flux density in airgap section 2 is decreased since the direction of the 2-pole

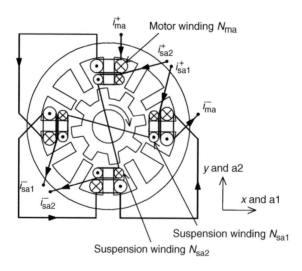

Figure 15.1 A-phase winding configuration

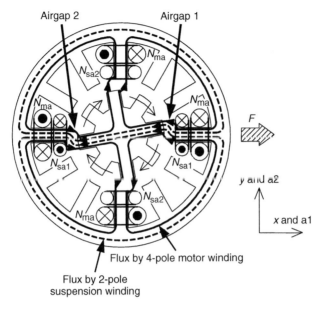

Figure 15.2 Principle of suspension force generation

flux is opposite to that of the 4-pole flux. Therefore, superposition of the airgap magnetic flux waves results in a suspension force F acting along the x-axis. A negative suspension force along the x-axis can be generated with a negative current i_{sa1}. A suspension force on the y-axis can be produced using a 2-pole suspension winding current i_{sa2}. Hence a suspension force can be generated in any desired direction by a vector sum of these forces. It should be noted that the 4-pole flux produced by the motor winding currents is essential as the bias magnetic flux for suspension force generation. So far only one main and suspension winding has been considered; however, this principle can be applied to the B- and C-phases so that a suspension force can be continuously generated by these three phases in turn for every 15 deg, i.e., from the start of the switching overlap up to the aligned position (turn-off).

If rotor and stator poles are aligned, a high suspension force is generated. As the rotor rotates away from the aligned position the suspension force becomes less because of the increase in magnetic reluctance causing a decrease in airgap flux (for constant phase current). The relationships between suspension force and stator winding currents are found to be dependent on the rotor angular position; so for successful magnetic suspension, the force must be controlled at all rotor angular positions. Therefore it is essential to derive an accurate theoretical algorithm of the suspension force for successful control because these formulae are required to maintain constant loop gain in the controller when using a negative feedback loop.

15.2 Derivation of inductances

Figure 15.3 shows the magnetic equivalent circuit for the A phase only. Voltage sources represent the MMFs of the 4-pole motor windings and the 2-pole suspension windings and, in addition, the magnetic permeances of airgap are represented by electrical conductances. The definitions of the variables in Figure 15.3 are as follows: N_m is the number of turns of the motor winding, N_s is the number of turns of the suspension winding, $P_{a1}-P_{a4}$ are the permeances of the airgap under the A-phase winding poles and $\psi_{a1}-\psi_{a4}$ are the magnetic fluxes of each pole. Only the A-phase magnetic equivalent circuit is used in the simple calculation in this section. This is possible because the switched reluctance motor has little mutual inductance between phases.

If each of the permeances (shown as P_{a1} to P_{a4}) can be simply assumed, the self inductances and the mutual inductances of each winding of the A phase can be calculated from Figure 15.3. The self inductances and the mutual inductances can be written as [9]:

$$L_{ma} = \frac{4N_m^2(P_{a1}+P_{a3})(P_{a2}+P_{a4})}{P} \tag{15.1}$$

$$L_{sa1} = \frac{N_s^2[P_{a1}(P_{a2}+2P_{a3}+P_{a4})+P_{a3}(2P_{a1}+P_{a2}+P_{a4})]}{P} \tag{15.2}$$

$$L_{sa2} = \frac{N_s^2[P_{a2}(P_{a1}+P_{a3}+2P_{a4})+P_{a4}(P_{a1}+2P_{a2}+P_{a3})]}{P} \tag{15.3}$$

$$M_{ma,sa1} = \frac{2N_mN_s(P_{a1}-P_{a3})(P_{a2}+P_{a4})}{P} \tag{15.4}$$

$$M_{ma,sa2} = -\frac{2N_mN_s(P_{a1}+P_{a3})(P_{a2}-P_{a4})}{P} \tag{15.5}$$

$$M_{sa1,sa2} = -\frac{N_s^2(P_{a1}-P_{a3})(P_{a2}-P_{a4})}{P} \tag{15.6}$$

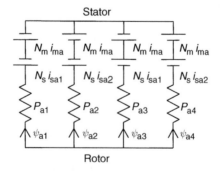

Figure 15.3 Magnetic equivalent circuit of phase A

where:

L_{ma} is the self inductance of N_{ma},
L_{sa1}, L_{sa2} are the self inductances of N_{sa1} and N_{sa2},
$M_{ma,sa1}$ is the mutual inductance between N_{ma} and N_{sa1},
$M_{ma,sa2}$ is the mutual inductance between N_{ma} and N_{sa2},
$M_{sa1,sa2}$ is mutual inductance between N_{sa1} and N_{sa2} and
P is the sum of the permeances P_{a1} to P_{a4}.

15.3 Assumption and calculation of permeances

The gap permeance P_{a1} of the airgap section 1 can be divided into three permeance components P_1–P_3 as shown in Figure 15.4; this is an enlargement of Figure 15.2. P_1 can be derived with straight magnetic paths while P_2 and P_3 can be derived with the assumed (as illustrated) magnetic paths. The variables in Figure 15.4 can be defined as follows: r is the radius of the rotor pole, g is the airgap length, t_s is a position on the circular surface (outer radius) of the rotor pole, ϕ_s is an angle of t_s in polar coordinates, dt_s is the derivative of t_s, and dP_2 and dP_3 are the permeances of the assumed infinitesimal-width magnetic paths. The following assumptions are considered in calculating the permeances:

1. Magnetic saturation is neglected. A calculation for suspension force and torque under magnetic saturation is given in [10] for advanced readers.
2. The square of rotor displacement is small compared with the square of airgap length.
3. Flux paths which do not link the rotor are neglected.

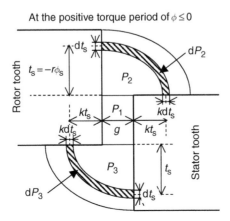

At the positive torque period of $\phi \leq 0$

Figure 15.4 Fringing magnetic path assumed with elliptical line

4. Flux paths between stator poles and rotor inter-polar area are neglected.
5. Only fringing fluxes at the aligned position can be neglected because the airgap length is very short.

Figure 15.5 shows an enlargement of the magnetic paths at the pole edge obtained from a two-dimensional finite element analysis. The shape of the fringing magnetic paths is elliptical so that the fringing magnetic paths of the permeances P_2 and P_3 can be assumed to follow elliptical curves using a variable k. If the semi-minor axis of the elliptical curve is t_s, the semi-major axis can be expressed as $g + kt_s$.

It is important to evaluate the value of constant k to determine the shape of the ellipse. k can be calculated from each magnetic path in Figure 15.5. Figure 15.6 shows the relationship between k and t_s as derived from results of the finite element analysis. This analysis is carried out with airgap lengths of 100, 220 and 300 μm in order to observe the dependence of the fringing fluxes on the airgap length.

It can be seen that the relationship between k and t_s is dependent on the airgap length g. Therefore a general expression between k and t_s in terms of g

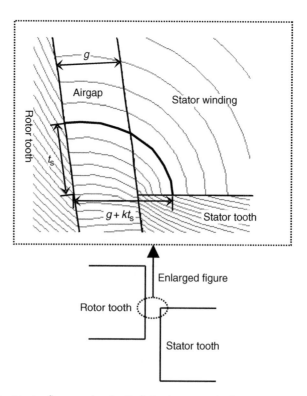

Figure 15.5 Fringing fluxes analysed with finite-element method

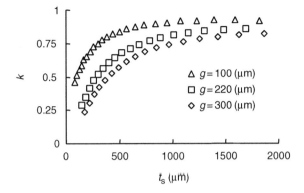

Figure 15.6 Relationship between k and t_s

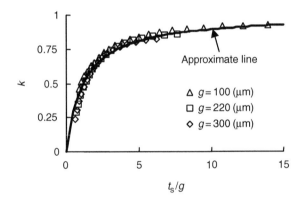

Figure 15.7 Relationship between k and t_s/g

should be derived. Figure 15.7 shows the relationship between k and t_s/g, and k is found to be a function of only t_s/g.

From Figure 15.7, the relationship between k and t_s/g can be approximated to

$$k = \frac{t/g}{a+t/g} = -\frac{r\phi_s}{ag - r\phi_s} \qquad (15.7)$$

where a is a constant calculated, using the method of the least squares, to be 1.18. The average cross section area S of dP_2 and dP_3 can be expressed as:

$$S = \frac{h(dt_s + kdt_s)}{2} = -\frac{hr(l+k)}{2}d\phi_s \qquad (15.8)$$

where h is the iron stack length. Assuming that l is the magnetic path length of the permeances dP_2 and dP_3, l can be approximated to [10]:

$$l = g - \frac{\pi}{2} r\phi_s \qquad (15.9)$$

Consequently, the permeances dP_2 and dP_3 can be derived from (15.7), (15.8) and (15.9) so that

$$dP_2 = dP_3 = -\frac{\mu_0 hr(ag - 2r\phi_s)}{(ag - r\phi_s)(2g - \pi r\phi_s)} d\phi_s \qquad (15.10)$$

where μ_0 is permeability of the air. The suspension force is proportional to the derivative of the gap permeance with respect to the rotor displacement. It is necessary to take into account the rotor displacement in the expressions for the airgap length g as used in the expression for permeance P_{a1}; hence we can write

$$g = g_0 - x \qquad (15.11)$$

where g_0 is the average airgap length, x is the rotor displacement on the x-axis. Therefore P_2 and P_3 can be derived from (15.10) and (15.11) using the equation

$$P_2 = P_3 = \int_0^\phi -\frac{\mu_0 hr[a(g_0 - x) - 2r\phi_s]}{[a(g_0 - x) - r\phi_s][2(g_0 - x) - \pi r\phi_s]} d\phi_s \qquad (15.12)$$

The permeance P_1 of the airgap between the rotor and stator poles can be simply derived as

$$P_1 = \int_0^{\frac{\pi}{12} + \phi} \frac{\mu_0 hr}{g_0 - x} d\phi_s \qquad (15.13)$$

The theoretical formula of the gap permeance P_{a1} during the positive torque period of $\phi \le 0$ can be derived from (15.12) and (15.13) so that:

$$P_{a1} = \frac{\mu_0 hr (\pi + 12\phi)(g_0 + x)}{12 g_0^2}$$

$$+ \frac{2\mu_0 h}{\pi(\pi a - 2)} \left[\pi a \ln \left(\frac{a g_0^2 - r\phi(g_0 + x)}{a g_0^2} \right) \right. \qquad (15.14)$$

$$\left. + (\pi a - 4) \ln \left(\frac{2 g_0^2 - \pi r\phi(g_0 + x)}{2 g_0^2} \right) \right]$$

provided that x^2 can be neglected. This should be the case because g_0 is large compared to x. Similarly, the theoretical formulae of the gap permeances P_{a2} to P_{a4} can also be derived.

15.4 Theoretical formulae of suspension force and torque

The self and mutual inductances can be derived by substituting gap permeances P_{a1} to P_{a4}, calculated from the above method, into (15.1) to (15.6). It is possible to construct a 3×3 inductance matrix $[L_3]$ from the derived self and mutual inductances. The inductance matrix $[L_3]$ of phase A can then be written as

$$[L_3] - \begin{bmatrix} L_{\text{ma}} & M_{\text{ma,sa1}} & M_{\text{ma,sa2}} \\ M_{\text{ma,sa1}} & L_{\text{sa1}} & M_{\text{sa1,sa2}} \\ M_{\text{ma,sa2}} & M_{\text{sa1,sa2}} & L_{\text{sa2}} \end{bmatrix} \tag{15.15}$$

Next, the stored magnetic energy W_a in phase A can be expressed using (15.15) so that

$$W_a = \frac{1}{2}[i_{\text{ma}} \ i_{\text{sa1}} \ i_{\text{sa2}}][L_3]\begin{bmatrix} i_{\text{ma}} \\ i_{\text{sa1}} \\ i_{\text{sa2}} \end{bmatrix} \tag{15.16}$$

The suspension forces F_x and F_y produced by the current in phase A can be obtained from the derivatives (with respect to rotor displacements x and y) of the stored magnetic energy W_a. This leads to expressions for the suspension forces, as derived from (15.16):

$$F_x = 4 K_f(\phi) N_m i_{\text{ma}} N_s i_{\text{sa1}} \tag{15.17}$$

$$F_y = 4 K_f(\phi) N_m i_{\text{ma}} N_s i_{\text{sa2}} \tag{15.18}$$

The proportional coefficient $K_f(\phi)$ is a function of the rotor angular position ϕ and the dimensions of the motor. There are two different expressions for this depending on the torque. It becomes $K_{fp}(\phi)$ when there is positive torque ($\phi \leq 0$), where

$$K_{fp}(\phi) = \frac{1}{2}\left[\frac{\mu_0 hr(\pi + 12\phi)}{12g_0^2} - \frac{2\mu_0 hr\phi(ag_0 - 2r\phi)}{g_0(ag_0 - r\phi)(2g_0 - \pi r\phi)} \right] \tag{15.19}$$

This is correct when the rotor radial position is under stable control at the centre of the stator bore, i.e., x and y are zero. On the other hand, it is equal to the coefficient $K_{fn}(\phi)$ when there is negative torque ($\phi > 0$), where

$$K_{fn}(\phi) = \frac{1}{2}\left[\frac{\mu_0 hr(\pi - 12\phi)}{12g_0^2} + \frac{2\mu_0 hr\phi(ag_0 + 2r\phi)}{g_0(ag_0 + r\phi)(2g_0 + \pi r\phi)} \right] \tag{15.20}$$

The first term in the square brackets of (15.19) and (15.20) is due to the straight magnetic paths across the airgap between the rotor and stator poles so that the first term is dominant. This decreases linearly as the rotor rotates away from the aligned position. At $\phi = \pm 15\,\mathrm{deg}$, i.e., at the start of overlap position, the first term becomes zero. However, a suspension force is generated by the second term which is due to the fringing magnetic paths (which are assumed as elliptical lines as shown earlier). Therefore the second term plays an important role during the overlap.

Similarly, the suspension forces due to phases B and C can be derived. The suspension forces along axes $a1$, $a2$, $b1$, $b2$, $c1$ and $c2$ can be obtained:

1. During the conduction period of phase $A(-15\,\mathrm{deg} \le \phi < 15\,\mathrm{deg})$

$$
\begin{bmatrix} F_{a1} \\ F_{a2} \end{bmatrix} = \begin{bmatrix} F_x \\ F_y \end{bmatrix} = 4\,K_{\mathrm{f}}(\phi)\,N_{\mathrm{m}}\,i_{\mathrm{ma}}\,N_{\mathrm{s}} \begin{bmatrix} i_{\mathrm{sa1}} \\ i_{\mathrm{sa2}} \end{bmatrix} \tag{15.21}
$$

2. During the conduction period of phase C $(0\,\mathrm{deg} \le \phi < 30\,\mathrm{deg})$

$$
\begin{bmatrix} F_{c1} \\ F_{c2} \end{bmatrix} = \begin{bmatrix} \cos 240° & -\sin 240° \\ \sin 240° & \cos 240° \end{bmatrix} \begin{bmatrix} F_x \\ F_y \end{bmatrix} = 4\,K_{\mathrm{f}}(\phi - \pi/12)\,N_{\mathrm{m}}\,i_{\mathrm{mc}}\,N_{\mathrm{s}} \begin{bmatrix} i_{\mathrm{sc1}} \\ i_{\mathrm{sc2}} \end{bmatrix} \tag{15.22}
$$

3. During the conduction period of phase B $(15\,\mathrm{deg} \le \phi < 45\,\mathrm{deg})$

$$
\begin{bmatrix} F_{b1} \\ F_{b2} \end{bmatrix} = \begin{bmatrix} \cos 120° & -\sin 120° \\ \sin 120° & \cos 120° \end{bmatrix} \begin{bmatrix} F_x \\ F_y \end{bmatrix} = 4\,K_{\mathrm{f}}(\phi - \pi/6)\,N_{\mathrm{m}}\,i_{\mathrm{mb}}\,N_{\mathrm{s}} \begin{bmatrix} i_{\mathrm{sb1}} \\ i_{\mathrm{sb2}} \end{bmatrix} \tag{15.23}
$$

A suspension force can be continuously generated by the switching of phases A, B and C using (15.21) to (15.23). As a result successful magnetic suspension can be realized for all rotor angular positions.

The torque T_a due to phase A can be written as

$$
T_a = G_{\mathrm{t}}(\phi)\,\left(4N_{\mathrm{m}}^2\,i_{\mathrm{ma}}^2 + 2N_{\mathrm{s}}^2\,i_{\mathrm{sa1}}^2 + 2N_{\mathrm{s}}^2\,i_{\mathrm{sa2}}^2\right) \tag{15.24}
$$

The second and third terms in the parenthesis are due to the suspension winding currents. The proportional coefficient $G_{\mathrm{t}}(\phi)$ is again a function of the rotor angular position ϕ and the dimensions of the motor. During the positive torque period $(\phi \le 0)$ it becomes

$$
G_{\mathrm{tp}}(\phi) = \frac{1}{2}\left[\frac{\mu_0 h r}{g_0} - \frac{2\mu_0 h r(a g_0 - 2r\phi)}{(a g_0 - r\phi)(2g_0 - \pi r\phi)}\right] \tag{15.25}
$$

(under stable control at the central position) and during the negative torque period $(\phi > 0)$ it is modified to:

$$
G_{\mathrm{tn}}(\phi) = \frac{1}{2}\left[-\frac{\mu_0 h r}{g_0} + \frac{2\mu_0 h r(a g_0 + 2r\phi)}{(a g_0 + r\phi)(2g_0 + \pi r\phi)}\right] \tag{15.26}
$$

The torques due to phases B and C can be similarly derived.

15.5 A drive system

In this section a drive system which controls the average torque and instantaneous suspension force to produce stable operation is discussed [11]. This is done by means of controlling the advanced angle of the conduction period of the stator winding currents under any torque condition from no load to full load.

15.5.1 Basic concept of a drive system

The phase-advance angle θ_m is the angle from the mid-point of the square-wave current conduction period to the aligned position of rotor and stator poles as shown in Figure 15.8. If θ_m is fixed at $\pi/24$ (7.5 deg) for example, the conduction period of phase A is fixed from -15 to 0 deg and positive torque is produced. We can use the magnitude of the current to vary the torque (where θ_m is constant), in which case the amplitude of motor winding currents i_m has to be zero at zero torque. However, total stable operation cannot be realized since the suspension force requires a motor winding current, i.e., $i_m \neq 0$, even if there is a large suspension winding current. Therefore, it is particularly important for stable operation to bias the magnetic fluxes sufficiently with a non-zero i_m over the whole torque range. Generation of the bias magnetic fluxes can be realized by controlling θ_m (rather than controlling the torque solely via the current amplitude).

Figure 15.8 shows the idealized current and instantaneous torque waveforms for the drive system. Here i_{sl} is the limit of the suspension winding current. The motor winding current i_m has a square waveform for system simplification. Both the motor winding and the suspension winding currents are simultaneously excited even during the negative torque period by reducing θ_m from the maximum value of $\pi/24$ (7.5 deg). Hence the bias magnetic flux required for generating suspension force is always produced by i_m. At no load ($T_{avg} = 0$), θ_m becomes zero and the positive torque is equal to the negative torque. The desired average torque T_{avg} and instantaneous suspension force F can be produced by determining the correct values for θ_m and i_m. In the case of a motoring action, θ_m can be in the range indicated by Figure 15.8 where

$$0 \leq \theta_m \leq \frac{\pi}{24}(7.5 \text{ deg}) \tag{15.27}$$

If rotor and stator poles are aligned, the suspension force is high. As the rotor moves away from the aligned position the suspension force becomes less because of the increasing magnetic reluctance. Accordingly, the suspension winding current has a maximum value $i_{s.on}$ at the turn-on angle θ_{on} as shown in Figure 15.8. This is because θ_{on} corresponds to the most unaligned position during the conduction period. It is necessary for $i_{s.on}$ to be below the limit value i_{sl} when determining the advanced angle θ_m. It was reported in [12] that the optimal conduction period is 15 deg, i.e., one phase excitation, so that the average

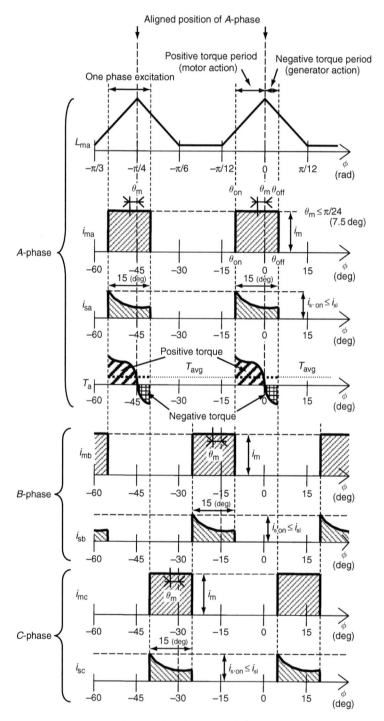

Figure 15.8 Idealized current and instantaneous torque waveforms

torque is regulated with square-wave currents in motor windings. It is necessary to reduce θ_m until the suspension winding current $i_{s.on}$ is below the limit value i_{sl}. At the same time the motor winding current i_m should be increased in order to generate the desired average torque T_{avg} (which will also increase the suspension force).

15.5.2 Drive system structure

The desired values of θ_m and i_m, which can simultaneously produce the desired T_{avg} and F, can be calculated by means of deriving the relationship between T_{avg} and F with respect to θ_m and i_m from (15.17), (15.18) and (15.24). It is necessary to include the condition that θ_m is a maximum value for the following reasons:

1. It minimizes the negative instantaneous torque.
2. The maximum positive torque necessary for the desired T_{avg} can be reduced.
3. Hence it reduces the torque-ripple.

However, it should be remembered that the suspension winding current $i_{s.on}$ at the turn-on angle θ_{on} should be below the limit value i_{sl}. Based on these considerations, the motor winding current i_m is determined from average torque and suspension force commands [11].

Figure 15.9 shows the structure of a suitable drive system. The calculator for controlling θ_m^* and i_m^* is added to the system with a negative feedback loop. The calculator consists of a look-up table with outputs for θ_m^* and i_m^* for corresponding inputs of T_{avg}^* and F^*. The table is written in order to reduce the sampling time of the digital signal processor. The controller of the suspension winding currents uses (15.21), (15.22) and (15.23). It can be seen that the controller of motor main winding currents is separate from the digital signal processor. Therefore it is possible to realize a high-speed drive without a time delay in the control cycle of motor winding currents.

15.5.3 Experimental results

A test machine was built as shown in Figure 15.9. The bottom of the rotor shaft is held by a spherical roller bearing and it is connected to the load machine through a torque meter. The radial rotor position is controlled in two perpendicular axes x and y. A weight hangs via a pulley which pulls the shaft along the negative x-axis to apply a suspension force $F = 73.5\,\text{N}$. The dimensions of the test motor are shown in Table 15.1. Figure 15.10 shows the waveforms of rotational speed ω, rotational speed command ω^*, rotor displacements x and y, average torque command T_{avg}^*, suspension force command F^*, advanced angle command θ_m^*, and finally motor winding current command i_m^*, for a step change in speed reference of the drive system from 2000 to 3000 r/min. Figure 15.11 shows the waveforms

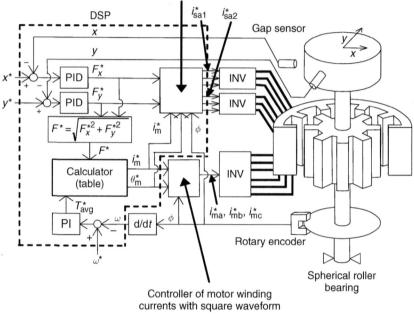

Figure 15.9 Drive system

Table 15.1 Dimensions of test motor

Number of turns of motor main winding, N_m (0.8 mm ϕ, 3 parallels, 1.51 mm^2)	14 turns
Number of turns of suspension winding, N_s (0.8 mm ϕ, 2 parallels, 1.01 mm^2)	11 turns
Arc angle of rotor and stator teeth	15 deg
Stack length, h	50 mm
Outside diameter of stator core	100 mm
Inside diameter of stator pole	50 mm
Radius of rotor pole, r	24.78 mm
Average airgap length, g_0	0.22 mm

from the Z pulse in the rotary encoder (giving angular position), the A-phase main winding current i_{ma} and the A-phase suspension winding current i_{sa1} when $T^*_{avg} = 0.5$ and 0.0 Nm. During full-torque acceleration, θ^*_m and i^*_m are increased in order to produce maximum torque. Before and after the acceleration when the torque is low, the suspension-force bias flux is still generated by the i_m at $\theta^*_m = 0$ and the operation of the bearingless switched reluctance motor is still stable despite the sudden step acceleration. Hence the drive system can realize stable operation by means of controlling θ_m and i_m under any torque condition from no load to full load.

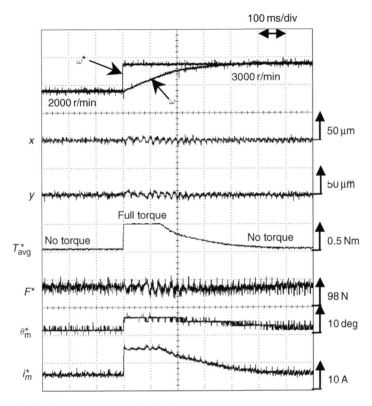

Figure 15.10 Step acceleration with the drive system

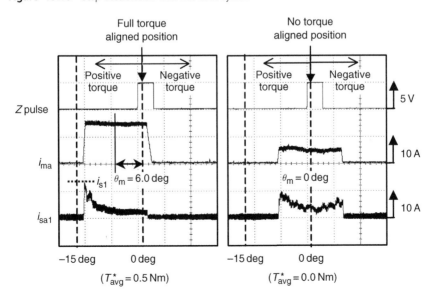

Figure 15.11 Current waveforms

15.6 A feed-forward compensator for vibration reduction considering magnetic attraction force

Regardless of whether the machine is a conventional or a bearingless switched reluctance motor, rotor eccentricity due to mechanical error causes large radial forces due to the magnetic attraction force. This is a major cause of vibration and subsequent noise emission in switched reluctance motors. In conventional switched reluctance motors, compensation of the magnetic attraction force is not possible (hence they can be noisy). On the other hand, the bearingless switched reluctance motor does generate a controlled radial force. Therefore, it is possible to actively compensate for the radial magnetic force due to rotor eccentricity using the suspension force. The time delay in the compensation is minimized by a feed-forward compensator in the rotor radial positioning. Hence in this section a feed-forward compensator for vibration reduction is discussed.

15.6.1 Relationship between magnetic centre and magnetic attraction force

Rotor displacement (eccentricity) from the magnetic centre of the stator causes non-uniformity in the magnetic flux distribution with a concentration of flux where the airgap is narrowest. Therefore the magnetic attraction force acts in the direction of rotor displacement since the flux density in the airgap is higher at this point. To assess the effects of eccentricity it is necessary to measure accurately the radial position of the magnetic centre. Figure 15.12 shows the measured radial position as a function of rotational direction [13] of the 3-phase machine given in Table 15.1. The rotor centre draws a locus in the shape of a circle with a radius of about 13 μm. The displacement of the rotor from the stator centre gives the direction and magnitude of the required suspension force to bring the rotor back to the centre. Hence the rotor centre and the magnetic centre can be described. The origin for the rotor radial position controller is set at the central point of the locus of the magnetic centres as shown in Figure 15.12.

Figure 15.13 shows the waveforms of rotational speed ω, rotational speed command ω^*, rotor displacements x and y, Z pulse, average torque command T_{avg}^*, and motor winding current command i_m^*, for a step change in speed reference from 1500 to 2500 r/min. It is seen that stable operation can be obtained at zero mean torque. However, the rotor shaft vibrates during the acceleration, i.e., during the high torque period. A magnetic attraction force (unbalanced magnetic pull) is always produced since the rotor is never at the stator centre as

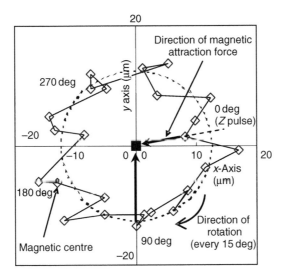

Figure 15.12 Locus of magnetic centres

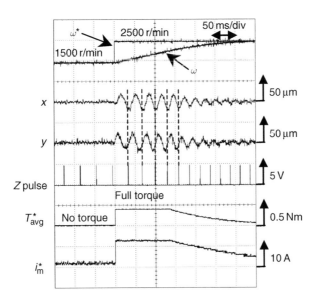

Figure 15.13 Step acceleration with controller neglecting magnetic attraction force

shown in Figure 15.12. The vibration of the rotor is higher during this period because the magnetic attraction force is in proportion to the square of airgap flux density, which is significantly increased during acceleration as can be observed by the increase in motor winding current.

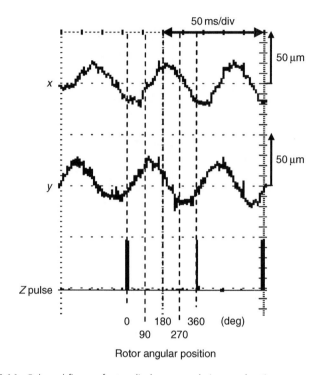

Figure 15.14 Enlarged figure of rotor displacements during acceleration

Figure 15.14 shows an enlargement of the waveforms during acceleration. The vibration of rotor shaft corresponds to the direction of magnetic attraction force caused by the circular movement of rotor centre. For example, at a rotor angular position of $\phi = 0$ deg, the rotor radial position is situated along the negative x-direction from the stator centre and the magnetic attraction force acts in the same direction. As a result, the rotor is pulled further along the negative x-direction. Similarly, at a rotor angular position of $\phi = 90$ deg, the rotor is pulled along the positive y-direction. This confirms that the magnetic attraction force caused by the circular movement of rotor centre is one cause of vibration.

15.6.2 A feed-forward compensator

In order to suppress the radial vibration it is necessary to compensate the magnetic attraction force by cancelling the magnetic forces caused by the circular movement of rotor centre around the stator centre. For that purpose, it is important to derive accurate theoretical formulae for the magnetic attraction force.

The locus of rotor radial position in Figure 15.12 is described by:

$$x_e = \gamma_e \cos \phi \tag{15.28}$$

$$y_e = -\gamma_e \sin \phi \tag{15.29}$$

where x_e and y_e are the radial positions of the rotor in a two-axes reference frame and γ_e is the radius of the locus which was measured as 13×10^{-6} m. The magnetic attraction force due to phase Λ when considering the locus of the rotor position, as expressed in (15.28) and (15.29), can be written as

$$F_{umx} = 4 K_{um}(\theta_e) N_m^2 (x - \gamma_e \cos \phi) i_{ma}^2 \tag{15.30}$$

$$F_{umy} = 4 K_{um}(\theta_e) N_m^2 (y + \gamma_e \sin \phi) i_{ma}^2 \tag{15.31}$$

$$F_{usx} = K_{us}(\theta_e) N_s^2 (x - \gamma_e \cos \phi) i_{sa1}^2 \tag{15.32}$$

$$F_{usy} = K_{us}(\theta_e) N_s^2 (y + \gamma_e \sin \phi) i_{sa2}^2 \tag{15.33}$$

where, F_{umx}, F_{umy}, F_{usx} and F_{usy} are the magnetic attraction forces produced by i_{ma}, i_{sa1} and i_{sa2} in the x and y directions. $K_{um}(\theta_e)$ and $K_{us}(\theta_e)$ are coefficients of the magnetic attraction force where

$$K_{um}(\theta_e) = \frac{1}{2} \frac{\mu_0 h r (\pi - 12|\theta_e|)}{12 g_0^3} \tag{15.34}$$

$$K_{us}(\theta_e) = \frac{1}{2} \frac{\mu_0 h r (\pi - 12|\theta_e|)(\pi + 12|\theta_e|)}{6 \pi g_0^3} \tag{15.35}$$

and θ_e is the rotor angular position from the aligned position of the excited phase. The magnetic attraction forces for the B and C phases can be similarly derived. The magnetic attraction forces are proportional to the rotor displacements x and y and the square of stator winding currents. In addition, if rotor and stator poles are aligned, the magnetic attraction force is the maximum. As the rotor rotates from the aligned position, the magnetic attraction force becomes less (as previously stated).

Figure 15.15 shows the rotor radial position controller with the addition of a feed-forward compensator based on (15.30–15.33). (The feed-forward compensator is surrounded by the broken line.) If the locus of the rotor centre is not a circle, it is still possible to compensate the magnetic attraction force without a time delay by means of accurately measuring the locus.

Figure 15.16 shows the experimental waveforms for a step change in speed reference with the feed-forward compensator. The conditions are the same as Figure 15.13. The operation of the motor is now stable in spite of the sudden step change. Therefore it is found that the feed-forward compensator is extremely effective in reducing the radial position variation caused by the circular movement of the unbalanced magnetic pull.

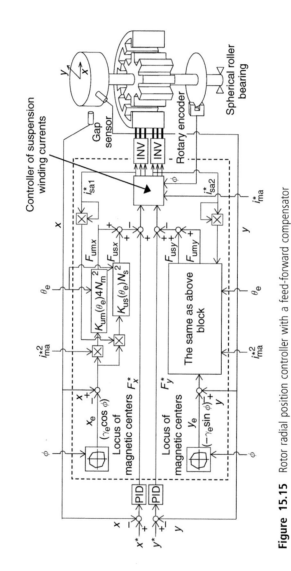

Figure 15.15 Rotor radial position controller with a feed-forward compensator

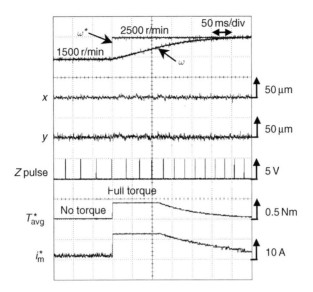

Figure 15.16 Step acceleration with the feed-forward compensator

References

[1] C. M. Stephens, "Fault Detection and Management System for Fault Tolerant Switched Reluctance Motor Drives", in Conf. Record of IEEE-IAS Annual Meeting, 1989, pp. 574–578.

[2] C. A. Ferreira, S. R. Jones, W. S. Heglund and W. D. Jones, "Detailed Design of a 30-kW Switched Reluctance Starter/Generator System for a Gas Turbine Engine Application", IEEE Trans. on IA, Vol. 31, May/June 1995, pp. 553–561.

[3] A. V. Radun, C. A. Ferreira and E. Richter, "Two-Channel Switched Reluctance Starter/Generator Results", IEEE Trans. on IA, Vol. 34, September/October 1998, pp. 1026–1034.

[4] R. Krishnan, R. Arumugan and J. F. Lindsay, "Design Procedure for Switched-Reluctance Motors", IEEE Trans. on IA, Vol. 24, May/June 1988, pp. 456–461.

[5] T. J. E. Miller, "Faults and Unbalance Forces in the Switched Reluctance Machine", IEEE Trans. on IA, Vol. 31, March/April 1995, pp. 319–328.

[6] T. Sawata, P. C. Kjaer, C. Cossar, T. J. E. Miller and Y. Hayashi, "Fault-Tolerant Operation of Single-Phase SR Generators", IEEE Trans. on IA, Vol. 35, July/August 1999, pp. 774–781.

[7] K. M. Rahman, B. Fahimi, G. Suresh, A. V. Rajarathnam and M. Ehsani, "Advantages of Switched Reluctance Motor Applications to EV and HEV: Design and Control Issues", in Conf. Record IEEE-IAS Annual Meeting, 1998, pp. 327–334.

[8] K. Shimada, M. Takemoto, A. Chiba and T. Fukao, "Radial Forces in Switched Reluctance Type Bearingless Motors" (*in Japanese*), in Proc. 9th Symp. Electromagnetics and Dynamics (The 9th SEAD), 1997, pp. 547–552.

[9] M. Takemoto, K. Shimada, A. Chiba and T. Fukao, "A Design and Characteristics of Switched Reluctance Type Bearingless Motors", in Proc. 4th Int. Symp. Magnetic Suspension Technology, Vol. NASA/CP-1998-207654, May 1998, pp. 49–63.

[10] M. Takemoto, A. Chiba, H. Akagi and T. Fukao, "Radial Force and Torque of a Bearingless Switched Reluctance Motor Operating in a Region of Magnetic Saturation" in Conf. Record IEEE-IAS Annual Meeting, 2002, pp. 35–42.

[11] M. Takemoto, A. Chiba and T. Fukao, "A Method of Determining the Advanced Angle of Square-Wave Currents in a Bearingless Switched Reluctance Motor", IEEE Trans. on IA., Vol. 37, November/December 2001, pp. 1702–1709.

[12] M. Takemoto, A. Chiba and T. Fukao, "A New Control Method of Bearingless Switched Reluctance Motors using Square-Wave Currents", in Proc. 2000 IEEE Power Engineering Society Winter Meeting, CD-ROM, January 2000.

[13] M. Takemoto, A. Chiba and T. Fukao, "A Feed-Forward Compensator for Vibration Reduction Considering Magnetic Attraction Force in Bearingless Switched Reluctance Motors", in Proc. 7th Int. Symp. Magnetic Bearings, Zurich, Switzerland, August 2000, pp. 395–400.

[14] M. Takemoto, H. Suzuki, A. Chiba, T. Fukao and M. A. Rahman, "Improved Analysis of a Bearingless Switched Reluctance Motor", IEEE Trans. on IA, Vol. 37, January/February 2001, pp. 26–34.

16

Winding arrangement variations

Akira Chiba

Table 1.4 lists the winding variations for the bearingless machines described in this book – a brief description was also provided in the introduction chapter. In this chapter, further electromagnetic arrangements with integrated magnetic bearing and motor functions are described. A modification is put forward for the original radial magnetic bearing together with a split winding structure for the original motor. Also a p-pole and $(p \pm 2)$-pole winding bearingless motor is described.

16.1 Modified radial magnetic bearings [1,2]

Figure 16.1 shows the cross section of a radial magnetic bearing. There are four windings consisting of two adjacent series-connected coils as shown. The winding current is the sum of the bias current and the feedback current (for radial positioning). Normally the bias current is dc. If the bias current has sinusoidal and co-sinusoidal variations in the x- and y-oriented windings then a revolving magnetic field is generated. The rotor can have conductors in order to act as an induction motor or it can be made of hysteresis or permanent magnet materials; with the interaction of rotor structure and the revolving magnetic field producing a rotational torque. The concept is quite simple and may be suitable for low-power motors, i.e., rather than being considered as a bearingless motor it can be considered as a magnetic bearing with limited torque-generating capability.

However, for high power motors, there are several problems that have to be addressed:

1. DC bias field optimization is required to reduce the braking torque in induction motors.
2. Elimination of induced parasitic rotor current.

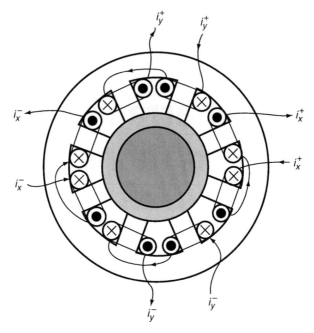

Figure 16.1 Revolving magnetic field in radial magnetic bearing

3. The reduction of space harmonics.
4. Decoupling of the revolving magnetic field and radial forces.

If the rotor has saliency, as in a case of a switched reluctance machine, more torque can be generated; however, radial force generation is affected by rotor rotational position. If a permanent magnet rotor is inserted then it generates torque in the same way as a permanent magnet motor where there should be current on the rotor q-axis, so the radial force is again dependent on rotor rotational position. Hence in both cases the radial force regulation requires angular rotor position feedback (unlike magnetic bearings).

With a minor structural modification it is possible to obtain an improved bearingless motor design. Figure 16.2 shows the cross section of a $y+$ electromagnet with small teeth on the rotor and stator surface, similar to a stepping motor arrangement. Rotational torque is generated by magnetic saliency between the rotor and the stator teeth. Conductors carry the motor currents i_{ma} and i_{mb}; in the figure, if i_{ma} flows then the generated airgap flux produces an anti-clockwise reluctance torque. The principle is similar to the hybrid stepping motor. The suspension conductors carry a current i_s. In this arrangement radial force generation is not influenced by rotor rotation because the rotor and stator teeth pitches are not equal. The suspension current generates a dc excitation flux Ψ_s, which also provides considerable torque. This bearingless motor is suitable for low-speed motor applications.

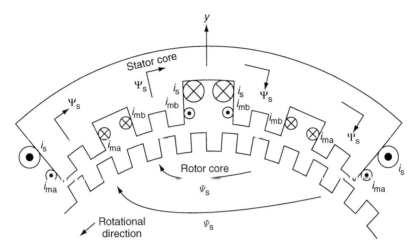

Figure 16.2 Tooth on radial magnetic bearing surface

16.2 Modified motors [3–5]

Figure 16.3 shows an example of a split motor winding structure. At the mid-point of the original motor windings the circuit is split and terminals u |, and u are inserted so that the original neutral connections are separated. If the original windings are the u, v and w phases then $u+$, $u-$, $v+$, $v-$, $w+$ and $w-$ windings are generated. This allows the currents i_u^+ and i_u^- in windings $u+$ and $u-$ to be independently regulated, and the difference in current values results in a radial force. For example, a y-axis radial force is generated by a u phase current. The new neutral point N has six winding connections and the total number of inverter sections is six, requiring a more complex control strategy. The winding modification is easy to produce and it allows a 4-pole winding layout to also generate a 2-pole MMF wave so that there is no need for the insertion of an additional suspension winding. The rotor can be an induction, reluctance or permanent magnet type.

In a permanent magnet motor with a concentrated short-pitch winding, the phase coils are wound around the stator poles and the winding currents can be independently regulated to adjust the radial attractive force between the stator pole and rotor surface. With both this arrangement and with the split winding arrangement, to generate static radial force, the winding current has to be unbalanced. The number of phases can be more than three; to reduce the number of terminal connections a neutral point can also be connected internally.

These split winding configurations are suitable for low-power motors. The induced winding voltage increases as the speed increases and the instantaneous

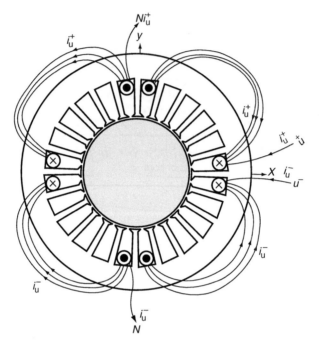

Figure 16.3 Split winding conductor connection

current is regulated by the difference between the induced voltage and inverter output voltage. Hence, for stable magnetic suspension, increased inverter voltage is required in order to regulate instantaneous current and obtain a fast response. The current drivers for the inverters provide both torque and radial force current components simultaneously.

Figure 16.4 shows a single-phase motor with a 4-pole permanent magnet rotor. The motor has a single-phase winding carrying a current i_m. This sort of winding is popular in low power and low cost applications. The 2-phase suspension winding carries suspension currents i_{sx} and i_{sy} and these are responsible for the radial force generation in the directions x and y. In this situation the voltage and current requirements for the suspension winding are rather low due to the differential winding structure.

It is possible to modify a coreless motor to produce a bearingless drive [6–8]. Coreless motors are usually low-power motors without stator teeth or slots (in order to minimize cogging torque and torque-ripple) with the windings located in the airgap. The principle of force generation is based on Fleming's law rather than the magnetic attractive force (Maxwell stress in the normal direction) and the force can be calculated from the force on the current-carrying conductors in the airgap. To achieve a reasonable force the conductor fill factor and current density need to be improved.

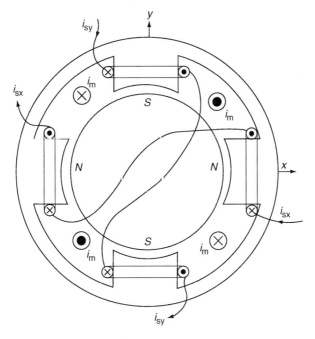

Figure 16.4 Single-phase motor winding

16.3 *p*-Pole and (*p*±2)-pole windings [9–11]

A *p*-pole and ($P\pm 2$)-pole winding set consists, for example, of a 4-pole rotor and 4-pole motor winding with a 6-pole suspension winding, or a 6-pole rotor and 6-pole motor winding with a 4 pole suspension winding or even a 6-pole rotor and 6-pole motor winding with an 8-pole suspension winding. These winding sets, in theory, are capable of generating a radial force by the interaction of MMF and flux waves.

Figure 16.5 shows the cross section of a permanent magnet bearingless motor. Let us define the number of pole-pairs on the rotor as M and stator suspension winding pole-pairs as N. The figure shows the case when $M = 2$ and $N = 3$ and the rotational position is $\omega_{rm} t$ and stationary angular coordinate is ϕ_s. The airgap flux density B is a sum of permanent magnet flux density B_{pm} and suspension current flux density B_s while the current in the main M-pole-pair motor winding is zero. Hence the airgap flux density can be written as

$$B = B_{pm} \cos(M\phi_s - \omega_{rm}t) + B_s \cos(N\phi_s - \omega_{rm}t - \theta_s) \tag{16.1}$$

where θ_s is the phase angle of the suspension winding current. Note that the airgap flux density distribution is approximated by a sinusoidal function. In addition, the magnetic circuit is assumed to be linear.

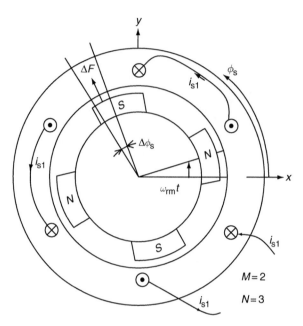

Figure 16.5   4-Pole permanent magnet bearingless motor with 6-pole suspension winding arrangement

Let us suppose that ΔF is the radial stress force in the airgap area for a small angle $\Delta \phi_s$. The airgap area is given by

$$\Delta S = Rl\Delta\phi_s \tag{16.2}$$

where R and l are airgap radius and axial length respectively. Then ΔF is:

$$\Delta F = \frac{B^2}{2\mu_0}\Delta S \tag{16.3}$$

so that the x-axis radial force is obtained from the integration of ΔF around the airgap; substituting (16.2) into the equation produces

$$F_x = \frac{Rl}{2\mu_0}\int_0^{2\pi} B^2 \cos\phi_s d\phi_s \tag{16.4}$$

Substituting flux density B, as given in (16.1), into the integration part of (16.4) yields

$$\begin{aligned}
B^2 \cos\phi_s &= B_{pm}^2 \cos^2(M\phi_s - \omega_{rm}t)\cos\phi_s \\
&\quad + B_s^2 \cos^2(N\phi_s - \omega_{rm}t + \theta_s)\cos\phi_s \\
&\quad + 2B_{pm}B_s\cos(M\phi_s - \omega_{rm}t)\cos(N\phi_s - \omega_{rm}t - \theta_s)\cos\phi_s
\end{aligned} \tag{16.5}$$

Note that M and N are greater than one so that integration of the first and the second terms results in zero. Only the third term is important, which can be rewritten as

$$B_{pm} B_s \{ \cos[(M+N)\phi_s - 2\omega_{rm}t - \theta_s]\cos\phi_s + \cos[(M-N)\phi_s + \theta_s]\cos\phi_s \} \tag{16.6}$$

When integrated, the first term becomes zero, but the second term may not be zero if $M - N = \pm 1$. When $M - N = 1$, executing the integration produces

$$\int_0^{2\pi} B^2 \cos\phi_s \, d\phi_s = \frac{B_{pm} B_s}{2} \cos\theta_s \tag{16.7}$$

so that the radial force is

$$F_x = \frac{\pi R l}{2\mu_0} B_{pm} B_s \cos\theta_s \tag{16.8}$$

In the similar way, the y-axis radial force can also be derived. These can be written in a matrix form

$$\begin{bmatrix} F_x \\ F_y \end{bmatrix} = \frac{\pi R l B_{pm} B_s}{2\mu_0} \begin{bmatrix} \cos\phi_s \\ -\sin\theta_s \end{bmatrix} \tag{16.9}$$

For the case of $M - N = -1$, the radial forces are

$$\begin{bmatrix} F_x \\ F_y \end{bmatrix} = \frac{\pi R l B_{pm} B_s}{2\mu_0} \begin{bmatrix} \cos\theta_s \\ \sin\theta_s \end{bmatrix} \tag{16.10}$$

From these equations, the following can be observed:

1. The radial force amplitude is proportional to the product of the flux densities B_{pm} and B_s, the airgap radius and the axial length.
2. The direction of the radial force is regulated by the phase angle of the suspension current.
3. A condition where $M - N = \pm 1$ indicates that the difference in pole numbers in motor and suspension windings is two, i.e., p-pole and $(p\pm 2)$-pole combination; for example, in a 4-pole motor the suspension winding should have six or two poles.

In practice high pole numbers can only be applied to low-speed motors and the flux amplitude should be less at high frequency to avoid excessive iron loss.

We can confirm radial force equation as derived in section 6.6. For a 4-pole motor and a 2-pole suspension winding, the suspension winding currents are

$$i_{2a} = I_s \cos(\omega t + \theta_s) \tag{16.11}$$

$$i_{2b} = I_s \sin(\omega t + \theta_s) \tag{16.12}$$

to satisfy (16.1). In addition, the angular frequency of the current should be twice the rotor rotational angular speed, where $\omega = 2\omega_{rm}$. We then obtain the equation set

$$\begin{bmatrix} F_x \\ F_y \end{bmatrix} = M'I_4I_s \begin{bmatrix} \cos\theta_s \\ -\sin\theta_s \end{bmatrix} \tag{16.13}$$

showing a correspondence between (16.9) and (16.13).

Example 16.1 Suppose that the stator teeth flux density due to the permanent magnets and suspension current in a bearingless PM motor are 0.8 T and 0.25 T respectively. Let us also suppose that the tooth width is about half the tooth pitch. Find the radial force using (16.9).

Answer
In the airgap, the flux density due to the permanent magnet excitation is $0.8/2 = 0.4$ T. When the suspension winding is excited the airgap flux density is either increased or decreased by 0.125 T, i.e., in the airgap $B_{pm} = 0.4$ T and $B_s = 0.125$ T. Substituting these values into (16.9) produces

$$F_x = 3 \times 10^4 \, Dl$$

where D is a rotor diameter. If D and l are given in cm, then the force is $F = 0.3 \times D \times l$ kgf. This corresponds to the primitive bearingless motor in section 8.3.

References

[1] K. Sakai and I. Morino, "Self-levitated Motor", Japan patent H3-2540.
[2] T. Higuchi, "Magnetically Floating Actuator Having Angular Positioning Function", United States Patent No. 4 683 391, March 12, 1985.
[3] P. Meinke and G. Flachenecker, "Electromagnetic Drive Assembly for Rotary Bodies using a Magnetically Mounted Rotor", United States Patent No. 3 988 658, 29 July 1974.
[4] J. A. Santisteban, A. O. Salazar and R. M. Stephan, "Characteristics of a Bearingless Motor with Split Windings", IPEC, Japan, April 2000, pp. 367–370.
[5] W. Amrhein, S. Silber and K. Nenninger, "Levitation Force in Bearingless Permanent Magnet Motor", IEEE Trans. on Mag., Vol. 35, No. 5, September 1999, pp. 4052–4055.
[6] L. S. Stephens and H.-M. Chin, "Robust Stability of the Lorentz-type Self Bearing Servomotor", ISMB-8, August 2002 at Mito, pp. 27–33.
[7] T. Tokumoto, D. Timms, H. Kanebako, K-I Matsuda and Y. Okada, "Development of Lorentz Force type Self-Bearing Motor", ISMB-8 2002 August at Mito, pp. 59–64.

[8] S. Aoyagi, A. Chiba and T. Fukao, "An Analysis of Radial Forces in a Coreless Motor", IEEJ paper in Technical Meeting on Linear Drives LD-01-98 at Kyoto, June 6, 2001, pp. 1–6.

[9] Y. Okada, K. Dejima and T. Ohishi, "Analysis and Comparison of PM Synchronous Motor Induction Motor Type Magnetic Bearings", IEEE Trans. on IA, Vol. 31, No. 5, September/October 1995, pp. 1047–1053.

[10] T. Ohishi, Y. Okada and S. Miyamoto, "Levitation control of IPM Type Rotating Motor", ISMB, Kanazawa, Japan, 1996, pp. 327–332.

[11] J. Bichsel, "The Bearingless Electrical Machine", NASA Conf. Publ, USA, 1992, pp. 561–573.

17

Mechanical structure and position regulation

Osamu Ichikawa

There are two main mechanical arrangements for rotational electric machines – radial gap machines and axial gap machines. Although most of the proposed bearingless motors are radial gap machines it is possible to construct axial flux types of this machine. In this chapter, different mechanical structures for the bearingless motor are introduced and the axial-gap type of bearingless motor is described in detail. This actively regulates the thrust and inclination movements of the rotor. One aspect highlighted is that radial vibration is shown to be suppressed with a small rotor incline at the critical speeds.

17.1 Mechanical structure

17.1.1 Radial gap bearingless motor

Figure 17.1 shows the typical structure of a radial gap motor. All the airgap flux components flow in the radial direction. One of the advantages of the radial gap motor is that the normal forces exerted on the rotor surface cancel each other out if the rotor is positioned at the stator centre.

Some modifications of the rotor and stator structures are possible as described in Chapter 1. For example, two tandem radial gap bearingless motor units can regulate four radial degrees of freedom. Radial gap machines also produce a small axial force along the z-axis due to the rotor tending to centre itself on the axial centre of the stator – this force is due mainly to the main motor flux. If the rotor is allowed to have a small amount of axial movement then this axial force may be sufficient for the z-axis suspension. However, if precise z-axis positioning is needed then an additional magnetic thrust bearing is required.

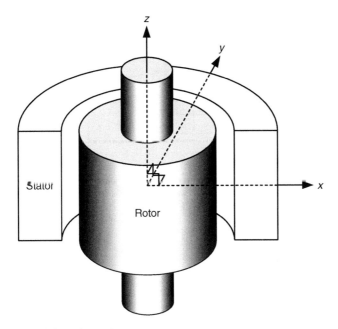

Figure 17.1 Radial gap bearingless motor

17.1.2 Axial gap bearingless motor

Figure 17.2 shows the typical structure of an axial-gap motor. All the flux in the airgap flows in the axial or z direction. Most of the flux is produced to generate the motor torque and this also generates a large axial force between the rotor and stator which is not cancelled out as in the radial-flux machine. However, by arranging two axial gap machines as shown in Figure 17.2, it is possible to cancel the axial forces. The axial position of the rotor can be maintained with the motor current. Inclination torques (i.e., torques that do not act around the normal z-axis of the motor) around the x- and y-axes (in the rotor disk plane) can be generated by an unbalanced airgap flux distribution caused by injecting suspension currents. These torques are used to keep the rotor disk position at the horizontal and axial centre (or any angle of incline that is required). Therefore axial and rotational positioning is realized. There also exists a small radial force, similar to the axial force in a radial-gap bearingless motor, which keeps the rotor at the radial centre.

Another possible structure is shown in Figure 17.3 where the radial and axial magnetic suspensions are combined. Radial positioning of the rotor can be controlled by the superposition of the motor and suspension fluxes while the axial position of the rotor can be controlled by the motor flux. Therefore a magnetic thrust bearing is not necessary.

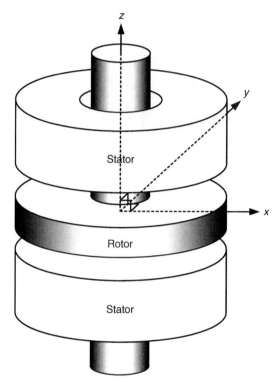

Figure 17.2 Axial gap bearingless motor

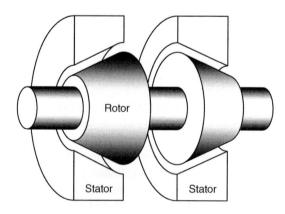

Figure 17.3 Corn type bearingless motor

17.2 Axial gap bearingless motors

17.2.1 Structures

Figure 17.4 shows an example of an axial-gap bearingless motor. A rotor disk between the two stators is magnetically suspended without mechanical contact. If a wheel is attached, either outside or inside the rotor, then a compact pump can be realized. Permanent magnets in a 4-pole arrangement are mounted on the upper surface of the disk rotor producing, in conjunction with the upper stator, a permanent magnet motor. Four salient poles are located on the lower surface so that a synchronous reluctance motor structure is produced in conjunction with the lower stator. The stator of the synchronous reluctance motor has both motor windings and suspension windings located in slots.

Figure 17.5 shows the main flux paths of the synchronous reluctance motor. All the airgap flux is assumed to flow in the axial direction producing a large axial force. By superimposing the $(n \pm 2)$-pole suspension flux on the n-pole motor flux, the airgap flux density becomes unbalanced and an inclination torque is produced.

The axial position and the incline of the rotor have to be regulated for successful magnetic suspension. The axial position of the rotor can be maintained by the reluctance motor currents while the incline of the rotor around x- and y-axes can be maintained by the synchronous reluctance motor suspension currents. The radial position is passively stable due to the centring force caused by flux in both motors.

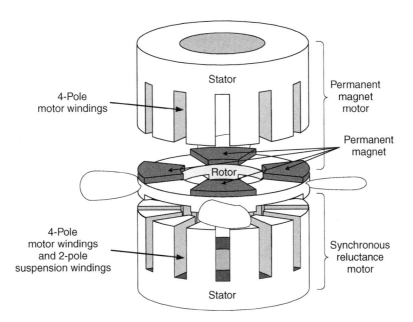

Figure 17.4 Test machine for an axial gap bearingless motor

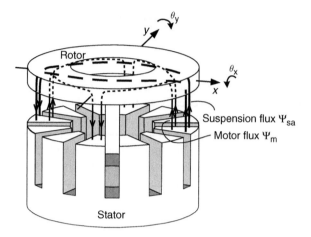

Figure 17.5 Main flux paths of synchronous reluctance motor part

17.2.2 Axial position regulation

The permanent magnets on the rotor surface produce an axial force in the upward direction. On the other hand, the winding currents of the synchronous reluctance motor produce an adjustable axial force in the opposite (downward) direction, and the axial position of the rotor can be controlled with the synchronous motor currents. Figure 17.6 shows the axial forces generated by the d- and q-axis reluctance motor currents i_{rd} and i_{rq}. These axial forces were measured in a test machine with an airgap length of 0.48 mm and a stator outer diameter of 120 mm

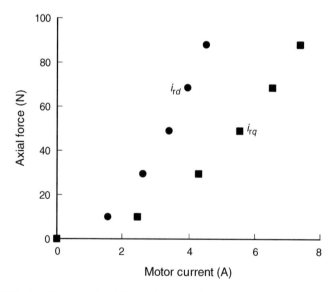

Figure 17.6 Axial forces produced by synchronous reluctance motor currents

for the synchronous reluctance motor. Both the d- and q-axis currents produce axial force [1] which can be written as

$$F_{rz} = k_{zd}i_{rd}^2 + k_{zq}i_{rq}^2 \qquad (17.1)$$

The q-axis reluctance motor current i_{rq} is injected to produce a reluctance torque where

$$T_r = k_t i_{rd} i_{rq} \qquad (17.2)$$

Therefore, the motor current commands i_{rd}^* and i_{rq}^* should be determined using (17.1) and (17.2). For zero torque, i_{rd} is maximum and $i_{rq} = 0$. If torque is required then the d-axis reluctance motor current should be decreased as the q-axis reluctance motor current is increased in order to produce constant axial force. From (17.1) and (17.2), the torque can be written in terms of the axial force and q-axis current [2] where

$$|T_r| = \frac{k_t}{\sqrt{k_{zd}}} \sqrt{-\left(\sqrt{k_{zq}i_{rq}^2} - \frac{F_{rz}}{2\sqrt{k_{zq}}}\right)^2 + \frac{F_{rz}^2}{4k_{zq}}} \qquad (17.3)$$

which has a maximum value of

$$\frac{k_t}{2\sqrt{k_{zd}k_{zq}}}F_{rz} \quad \text{at} \quad i_{rq}^2 = \frac{F_{rz}}{2k_{zq}}$$

It is obvious that the q-axis reluctance motor current should be in the range

$$\sqrt{\frac{F_{rz}}{2k_{zq}}} \leq i_{rq} \leq \sqrt{\frac{F_{rz}}{2k_{zq}}} \qquad (17.4)$$

Figure 17.7 shows the measured torque when a constant axial force of 62.5 N is required to maintain the axial position of the rotor. This illustrates the increase of the q-axis current and the decrease of the d-axis current as the torque increases to the maximum point. The torque response of the synchronous reluctance motor should be determined by taking this restriction into account. Note that the permanent magnet motor can produce significant torque without this restriction since axial force is generated without any motor winding current and only q-axis current is required for torque production.

17.2.3 Inclination control

If an $(n \pm 2)$-pole suspension flux is superimposed on the n-pole motor flux then the flux distribution in the airgap is unbalanced and an inclination torque is generated on the rotor. This inclination torque enables the control of rotor

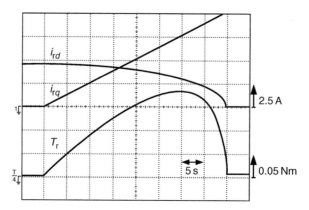

Figure 17.7 Reluctance torque at a constant axial force

angular positions around x- and y-axes. For example, the 2-pole suspension flux Ψ_{sa} and the 4-pole motor flux Ψ_m are superimposed as shown in Figure 17.5. The flux density is increased in the right-hand side airgap and decreased in the left-hand side airgap. This unbalance of flux distribution produces a clockwise inclination torque around the y-axis. Hence, the inclination torques around x- and y-axes can be written in a similar fashion to the radial forces in a radial-gap synchronous reluctance bearingless motor as described in Chapter 12:

$$
\begin{bmatrix} T_x \\ T_y \end{bmatrix} = \begin{bmatrix} 0 & -1 \\ 1 & 0 \end{bmatrix} \begin{bmatrix} M_d' i_{md} & -M_q' i_{mq} \\ M_q' i_{mq} & M_d' i_{md} \end{bmatrix} \begin{bmatrix} \cos 2\phi & \sin 2\phi \\ \sin 2\phi & -\cos 2\phi \end{bmatrix} \begin{bmatrix} i_{sa} \\ i_{sb} \end{bmatrix} \tag{17.5}
$$

Figure 17.8 shows a block diagram of a rotor inclination control system. The position feedback for the incline of the rotor is obtained from the difference in measurements from a pair of axial-position sensors. Torque commands T_{1x}^* and

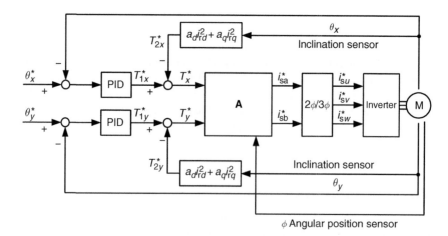

Figure 17.8 Block diagram of an inclination control system

T^*_{1y} are calculated from the inclination angular positions around each radial axis which allows the cancellation of the displacement-originated torques T^*_{2x} and T^*_{2y}. The torque commands T^*_x and T^*_y are transformed into suspension current commands in block A using (17.5) so that:

$$
\begin{bmatrix} i^*_{sa} \\ i^*_{sb} \end{bmatrix} = \frac{1}{(M'_d i_{md})^2 + (M'_q i_{mq})^2} \begin{bmatrix} \cos 2\phi & \sin 2\phi \\ \sin 2\phi & -\cos 2\phi \end{bmatrix} \begin{bmatrix} -M'_q i_{mq} & M'_d i_{md} \\ -M'_d i_{md} & -M'_q i_{mq} \end{bmatrix} \begin{bmatrix} T^*_x \\ T^*_y \end{bmatrix}
$$
(17.6)

17.2.4 Radial damping control

Radial positioning of the rotor is passively stable without an additional magnetic bearing due to the radial centring force. However, continuous vibrations can occur at a certain rotational speeds because of low damping. It is possible to improve the damping by using rotational torque control of the synchronous reluctance motor and also introducing a small rotor incline [3].

Figure 17.9 shows a mechanism for radial force production which provides improved damping. The rotor is slightly inclined while generating reluctance torque. The flux distribution is unbalanced because of the unsymmetrical airgap length under the salient poles. A flux decrease results in a torque decrease under salient pole 1 and a flux increase results in a torque increase under salient pole 3. While the sum of the torque produced across all the poles is unchanged, the torque unbalance under poles 1 and 3 produces radial force along the negative y-axis. A radial force can be produced in any desired radial direction by adjusting the rotor incline angle. However, the torque unbalance is opposite for the

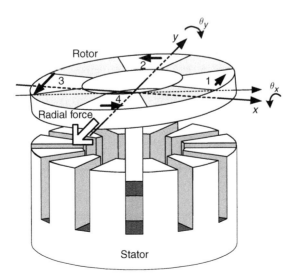

Figure 17.9 Mechanism of a radial force production with intentionally inclined disk rotor

permanent magnet motor side but the large airgap, due to the permanent magnets, results in a negligible torque unbalance compared to the synchronous reluctance motor.

Figure 17.10 shows measured radial forces for the test machine. These radial forces were generated from both the permanent magnet and synchronous reluctance motors. These confirm that a small radial force is produced by a slight incline of the rotor. The magnitude of the radial force is approximately proportional to both the incline angle and the torque current of synchronous reluctance motor so that:

$$\begin{bmatrix} F_x \\ F_y \end{bmatrix} = k_{xq} I_{rq} \begin{bmatrix} \theta_x \\ \theta_y \end{bmatrix} \tag{17.7}$$

The radial force shown in Figure 17.10 is too small to suspend the 5 kg rotor. However, it is high enough to improve the radial damping. Figure 17.11 shows a block diagram of the radial damping control system. The radial position of the rotor is detected with radial gap sensors. Radial force commands F_x^* and F_y^* are produced by the damping regulators, which include derivative controllers and low-pass filters. The incline angle commands are supplied to the inclination control system, as shown in Figure 17.8, producing the radial forces. The bearingless motor produces a radial force to centre the rotor; this force has a force-displacement factor k_x, and this can cause a continuous radial vibration at a natural frequency. By controlling the radial damping forces F_x and F_y these vibrations can be suppressed.

Figure 17.12 shows the experimental results from the radial damping system. During this experiment, the q-axis reluctance motor current was constant at 5 A. Without the damping system, serious radial vibration is seen during the initial acceleration. However, when the damping system is activated the radial

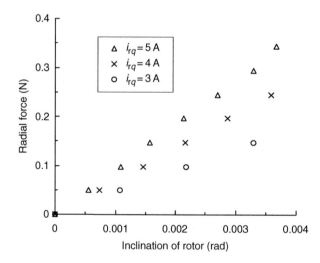

Figure 17.10 Measured radial forces

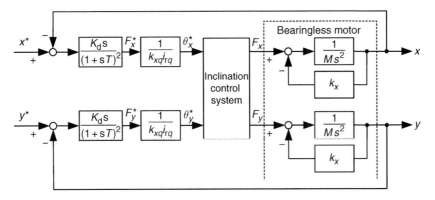

Figure 17.11   Block diagram of a radial damping control system

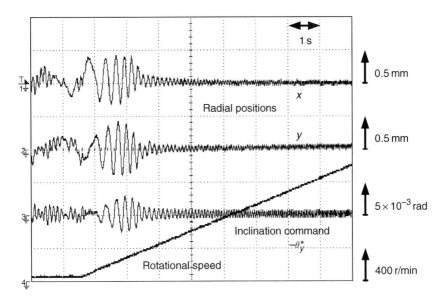

Figure 17.12   Radial positions with a radial damping control

vibrations at rotational speed are initially amplified but quickly decay, illustrating the radial damping effect.

17.2.5 Nonlinear characteristics

The control systems of an axial-gap bearingless motor are described above and are based on approximate and simplified models of the bearingless motor. In an actual system, there are motor parameter variations and control system interactions which depend on the operating conditions. In order to operate the motor with acceptable performance some modifications may be required, such as

1. Advanced robust controller design which accounts for parameter variation and gives disturbance suppression [4].
2. Identification and cancellation of parameter variations using an applicable method [5].

Example 17.1 In the test machine shown in Figure 17.4, the excitation flux of the permanent magnet motor produces an axial force of 120 N and the rotor weight is 5.0 kg. Suppose that the synchronous reluctance motor has constants of $k_{zd} = 3.0\,\text{N/A}^2$, $k_{zq} = 1.0\,\text{N/A}^2$ and $k_t = 9.0 \times 10^{-3}\,\text{N/A}^2$. Determine the maximum torque produced by the synchronous reluctance motor in the steady-state condition.

Answer
If the synchronous reluctance motor produces an axial force F_{rz} along the negative z-axis, the following force equation should be satisfied:

$$120 = F_{rz} + 5.0 \times 9.8 \tag{17.8}$$

Therefore $F_{rz} = 71\,\text{N}$. Substituting this force and the motor constants into the maximum torque condition, derived from (17.3), we obtain the maximum torque as:

$$\frac{k_t}{2\sqrt{k_{zd}k_{zq}}}F_{rz} = 0.18\,\text{Nm} \tag{17.9}$$

References

[1] J.Q. Li, T. Satoh, O. Ichikawa, C. Michioka, A. Chiba and T. Fukao, "The Axial Force of the Shaftless Axial Gap Bearingless Motor", 1995 National Convention Record I.E.E. Japan, Vol. 5, 1996, pp. 5–166 (*in Japanese*).
[2] O. Ichikawa, I. Tomita, C. Michioka, A. Chiba and T. Fukao, "An Operating Method of the Shaftless Axial-gap Bearingless Motors Considering the Axial Force and the Torque", The 9th Symposium on Electromagnetics and Dynamics, 1997, pp. 525–528 (*in Japanese*).
[3] O. Ichikawa, A. Chiba and T. Fukao, "Two-axis Vibration Control of Axial Gap Bearingless Motors Using Unbalanced Rotational Torque", Proc. of the 2000 Japan Industry Applications Society Conf., Vol. 1, 2000, pp. 265–266 (*in Japanese*).
[4] O. Ichikawa, A. Chiba and T. Fukao, "Construction Method of a Position Controller of Axial Gap Bearingless Motors", IEEJ Proc. of Meeting of Rotating Machinery, RM-98-141, 1998, pp. 127–132 (*in Japanese*).
[5] F. Saito, O. Ichikawa, K. Takahashi and T. Fukao, "Hierarchical Levitation and Rotation Control System of Axial-gap Bearingless Motors", IEEJ Proc. of Meeting of Semiconductor Power Converter, SPC-99-11, 1999, pp. 61–66 (*in Japanese*).

18

Displacement sensors and sensorless operation

Akira Chiba

In active magnetic suspension, displacement sensors are necessary to detect the radial (and sometimes axial) movement of the suspended object. Reducing the cost of the sensor is important in developing a bearingless drive since these can be expensive. Hence practical problems and solutions concerning sensor development are described in this chapter. The requirements for a displacement sensor can be summarized as follows:

(a) *Wide frequency response* – The frequency response bandwidth should be high enough so that the phase margin of the magnetic suspension loops does not need to be decreased. For example, if the cross-over frequency of a magnetic suspension loop is 100 Hz then the sensor response should be more than 5 kHz.

(b) *Low noise* – If there are high frequency noise components then these will be amplified by the derivative operation of a PID controller so that there will be significant noise in the suspension force references. If this occurs then the suspension current commands also include significant noise leading to current regulator saturation for a short time. This saturation results in a serious delay in the magnetic suspension loops which may cause shaft touch-downs in the stator bore.

(c) *Low interference noise* – In some displacement sensors, interference noise is generated if two or more sensors are installed. This is usually the case in order to detect *x*- and *y*-axis displacements. The sensor interference can take the form of a signal with a voltage of a few mV and a frequency of a few kHz. This can be a significant problem as described in (b) above.

In addition to the above requirements, low temperature drift, good linearity, compactness, as well as reliability are required.

In the first section of this chapter, the principles and problems of a displacement sensor with a magnetic core and winding are discussed. In the second section some methods to improve sensor performance are described, and a performance comparison of the inductive and eddy current types of sensor is put forward. In the third section the problems associated with a rotating target is highlighted. In the final section the sensorless displacement operation of a bearingless motor is introduced.

18.1 Principles of displacement sensor

Displacement sensors detect the linear position during the movement of an object without a mechanical contact. To the author's best knowledge these devices take the form of three basic types: capacitive, laser and electromagnetic. In capacitive sensors, the airgap length is detected using a variation in capacitance. Therefore good isolation between the sensor and the shaft is necessary. In addition, the air must be clean and oil and other particles should not be present because this will affect the dielectric. In laser sensors, displacement is detected by reflected laser light so that a uniform target surface is required to prevent noise. These sensors can be used, in some cases, in a bearingless motor application; however, electromagnetic sensors are still considered better for general-purpose applications. Therefore electromagnetic sensors are described in the following sections since they are the usual type of device used in bearingless drives.

Figure 18.1 shows the structure and principle of operation of an electro-magnetic displacement sensor. An E-shaped magnetic core has a winding with two terminals. A target (rotor shaft) is drawn as a rectangular solid having airgap length of l_g. The input impedance at the terminals varies with the airgap length. If the input terminals are excited by a high frequency voltage then the coil impedance will be dominated by the inductance (which is the variable part of the impedance) and it is obtained by detecting the terminal voltage and current.

There are two types of electromagnetic sensors. One is inductive while the other is an eddy current type. In inductive sensors, the target is made of a ferromagnetic material of high permeability such as laminated silicon steel, ferrite and carbon steel. The excitation voltage frequency is in the range of 20–100 kHz and the inductance varies as a function of airgap length (approximately inversely). If the airgap length is small then there is high impedance.

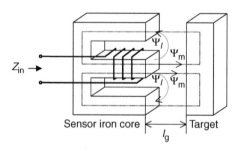

Figure 18.1 Principle of displacement sensor

In eddy current sensors, the target is made of a conductive material of low resistance such as copper, non-magnetic stainless steel, aluminium, carbon steel and other metallic material. The excitation voltage frequency is high, for an example 2 MHz, so that eddy currents are generated in the target. If the target is close to the sensor core then eddy currents are induced into the target which reduces the flux (almost as a short-circuit transformer) and produces a low input impedance. As the target moves away, the coupling decreases which increases the input impedance (which is opposite to the inductive type).

Figure 18.2 shows current variation with respect to the airgap length for different target materials; the iron core of the sensor is ferrite which has good permeability at 100 kHz and it takes the form of a cylindrical pot core with an outer diameter of 3.3 mm. The target materials are copper, aluminium, carbon steel and ferrite. The carbon steel is referred to as iron in the following sections. The terminal voltage is constant. The following characteristics can be observed:

(a) The winding current increases with airgap length when the iron and ferrite targets are used because these materials have high permeability so that the inductance decreases with increasing airgap length.
(b) On the other hand the winding current decreases when the aluminium and copper targets are used because these materials have low resistance and allow eddy current flow.
(c) For the iron and ferrite targets the current is significant even when the airgap length is zero. This is because the magnetic reluctance in the iron core is finite so that it is not dominated by the reluctance of the airgap alone.

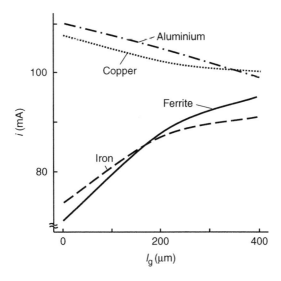

Figure 18.2 Sensor current and airgap length with several target materials at 400 kHz with a ferrite core

(d) For all materials the current variation with respect to the airgap length is only 10 to 20% of the maximum line current. Hence the sensitivity of winding current to airgap length variation is not very good.

(e) In all cases the current variation becomes saturated with a large airgap length because the input impedance becomes dominated by inductance due to the leakage fluxes Ψ_l.

Next let us look at the case when laminated silicon steel is employed as the excitation core [1]. Figure 18.3 shows the frequency characteristics for the input impedance of a bearingless motor winding with a laminated silicon steel core. This is the standard material in electrical machines (as opposed to small

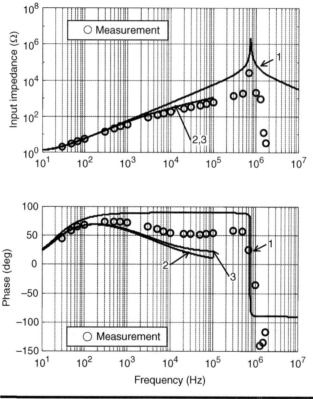

Case	Permeability (μ_r)	Iron loss resistance (R_i)
1	Infinite	Infinite
2	Infinite	Frequency dependent
3	Frequency dependent	Frequency dependent

Figure 18.3 Frequency dependence of input impedance for a sensor winding in silicon steel with influence of permeability decrease and iron loss

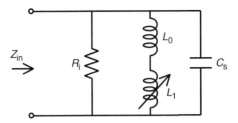

Figure 18.4 Equivalent circuit

ferrite actuator cores) because manufacturing is straightforward; however, its high frequency characteristics are not as good as for ferrite material. Three different curves are drawn using the measured points and assumed equivalent circuits. Curve 1 is obtained by assuming that the iron permeability is infinite and there are no iron losses although there is stray capacitance. The stray capacitance exists between the turns of the winding so that the equivalent circuit is drawn with a parallel capacitance across the winding inductance. Hence a resonance is seen at 700 kHz. Curve 2 is obtained by considering iron loss in the silicon steel core so that an iron loss resistance is placed in parallel with the inductance. The resistance value is a function of the excitation frequency, and the impedance decreases at high frequency. Also the phase angle is less than 90 deg (since it is an RL circuit), which more closely matches the measured input impedance. Curve 3 is similar to curve 2 except there is now a decrease in permeability of the core due to high frequency, therefore producing a better correspondence with the measured results.

From this figure the following observations about silicon steel displacement sensors can be made:

(a) The excitation frequency should be less than the resonant frequency.
(b) Core loss is not negligible in silicon steel at a frequency of more than 1 kHz.
(c) Permeability is also decreased at high frequency.

Figure 18.4 shows the equivalent circuit of a sensor winding. The inductance L_0 is constant while the inductance L_1 is the airgap-length-dependent-inductance. C_s is the stray capacitance and R_i is the iron loss resistance. Hence this circuit includes all the characteristics exhibited in curves 1–3 in Figure 18.3.

18.2 Improvements in sensitivity

As shown in the previous section, the current variation with respect to the airgap length is less than 20% of the winding current. The airgap length is detected as a function of current so that sensitivity improvement is necessary. In this section, two methods are introduced. One is a resonant circuit and the other is a differential circuit.

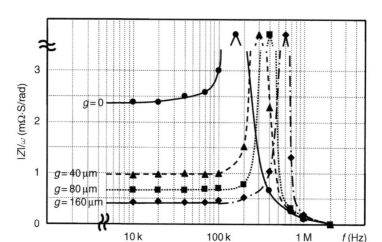

Figure 18.5 Impedance variation in a resonant circuit

18.2.1 Resonant circuit

Small variations in the sensor inductance can be amplified with a resonant circuit. This can be simply constructed with an external capacitance connected in parallel to the sensor winding terminals.

Figure 18.5 shows an example of the input impedance variation for several airgap lengths. A parallel capacitance is connected. When an airgap length is zero, the inductance value is high so that the resonant frequency, i.e., $f = 1/(2\pi\sqrt{LC})$, is low. At this resonant frequency, the input impedance is maximum, and as the airgap length increases so does the resonant frequency.

One way to take an advantage of the resonance is to set the excitation frequency to around 200 kHz giving an improved input impedance variation. This is detected by voltage or current amplitude variation (when using a current- or voltage-sourced supply). The amplitude can be transformed into a dc voltage in several different ways. The complete circuit can be realized using operational amplifiers, multiplier devices and rms-to-dc converter ICs. Another way is to track the resonant frequency with an oscillating circuit where the oscillating frequency indicates the airgap length.

18.2.2 Differential circuit

Figure 18.6(a,b) shows the magnetic and electric circuit arrangement for a differential type of displacement sensor. There are two E-shaped cores. Core 1 faces a target with a variable airgap l_g while core 2 faces a fixed target. Two isolated windings (1 and 2) are wound on the cores. Winding 1 carries the excitation current while winding 2 detects the core flux in each core. Figure 18.6(b) shows the circuit connection. The two coils of winding 1 are connected in series and excited by a voltage source v_0. The two coils of winding 2 are also connected

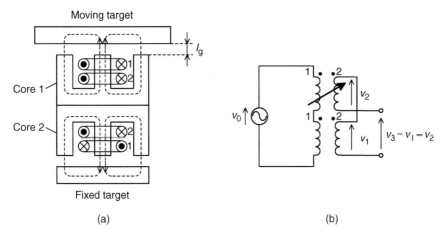

Figure 18.6 Differential sensor structure and principles: (a) iron core and winding arrangement; (b) circuit connection

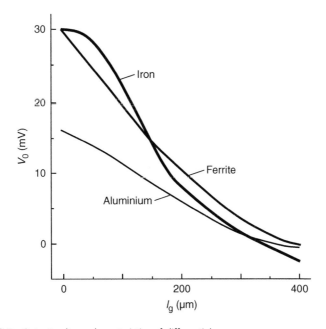

Figure 18.7 Output voltage characteristics of differential sensors

in series but one has reversed polarity so that the induced voltages v_1 and v_2 in winding 2 are subtracted at the output terminals so that $v_3 = v_1 - v_2$. If the airgap length l_g is equal to the airgap of the fixed target then the output voltage is zero. Hence the output voltage varies as a function of the airgap length of the moving target and can be set to zero when the fixed target airgap is equal to l_g

for the case of a centred rotor. The ac output voltage can be transformed into dc as described above.

Figure 18.7 shows the output voltage characteristics for a differential sensor with several different target materials. Note that the airgap length of the fixed target is set to about 400 μm. The sensor core is made from ferrite material with an excitation frequency of 400 kHz. A comparison with Figure 18.2 shows that the output voltage has a wider variation with respect to the airgap length. However, linearity still remains a problem but a microprocessor can be used to produce a linear characteristic via a look-up table or some other technique.

18.3 Inductive and eddy current sensors

One of the problems of a displacement sensor is associated with the surface quality of the rotating target ring. Since the displacement sensors detect the airgap length using the rotating shaft, the shaft surface may not be uniform and small cracks and scratches may be present. This is particularly relevant to eddy current sensors where a small mechanical crack may prevent the induced eddy currents, so errors will occur.

Figure 18.8 shows the sensor output waveforms for sensors along two perpendicular axes. The shaft is not rotating but reciprocally vibrating. The waveforms of x_E and y_E are the output voltages from eddy current sensors while x_I and y_I are the output voltages from inductive sensors. These sensors are adjacent to each other and measuring the same vibration, and the shaft is vibrating in a sinusoidal manner at 100 Hz by magnetic forces applied in the x- and y-directions. It can be seen that the output voltages of both the eddy current and the inductive sensors correspond.

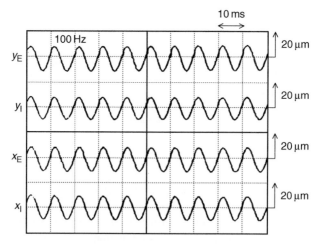

Figure 18.8 Output waveforms of inductive and eddy current sensors with a reciprocating motion

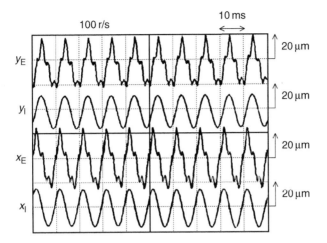

Figure 18.9 Displacement sensor output waveforms with surface imperfectness

Figure 18.9 shows the same output waveforms with the rotor rotating at 100 r/s. A mechanical unbalance causes a rotating vibration as seen in the output voltages of the inductive sensors. A similar vibration is seen in the eddy current sensor outputs; however, there are also high frequency components in the waveforms. This noise is caused by surface imperfections in the rotating target ring. The inductive sensors are not as sensitive as the eddy current sensors for the following reasons:

(a) The excitation frequency of an inductive sensor is low, so the sensor magnetic pole area is large. If the pole cross section is large then small target cracks have a negligible effect. However, small poles are required in eddy current sensors to minimize stray capacitance.
(b) The inductive sensor can be made from laminated silicon steel so that several magnetic poles can be situated around the circumference (like a motor stator). In this case the differential voltages detected in opposite magnetic poles, as well as neighbouring poles, can be processed. The signal processing eliminates the second-, third- and higher-order harmonics caused by deformation and imperfections in the target ring.

When comparing the figures, the advantages of the inductive type of sensor are obvious. However, there are disadvantages with inductive sensors as listed below:

(a) There are few, if any, inductive sensors available on the market.
(b) The shaft target ring has to be made from laminated silicon steel.
(c) The excitation frequency is low, so the filtering of the carrier frequency component may cause a delay in the sensor response.

In eddy current sensors, further improvements of the output voltage are possible:

(a) Two eddy current sensors can be placed on one axis and the output voltages are subtracted to give differential operation. The second harmonic and even harmonics are reduced and temperature drift is decreased.
(b) The target material can be replaced with non-magnetic stainless steel to avoid magnetic imperfections. Also copper can be used in some applications. In these cases, adjustment for the nonlinear characteristics should be made in the sensor amplifier.
(c) The sensor head diameter should be small so that the target ring does not interfere with the two-axis movement. For example, the sensor head diameter should be 5 mm or less for a target ring diameter of 50 mm. On the other hand, the sensor head diameter should be large compared to the airgap length for better sensitivity and linearity. For example, a head diameter of 5 mm should be used for 1 mm airgap length or less.
(d) The excitation frequency of the x-, y- and z-axis sensors should be set far enough apart to avoid interference with each other. The difference in excitation frequencies should be greater than the frequency range of the sensor. For example, for a sensor response range of 20 kHz, when the x-axis sensor is excited at 2 MHz the frequencies of the other sensors should be less than 1.96 MHz or higher than 2.04 MHz to provide enough frequency separation.

18.4 Sensorless bearingless motor

If the shaft displacement can be estimated or calculated from the terminal voltage and current then the shaft structure can be simplified and shortened because the sensor target rings are not needed. In addition, assembly, installation and resonant vibration problems associated with flexible shafts can be avoided. Many different ways to remove the position sensors in magnetic bearings have been investigated and these can be split up into the following categories:

(a) High frequency sinusoidal current injection [2].
(b) PWM carrier injection [3–6].
(c) Flux detection.
(d) Observer-based calculation [7].

These methods can also be applied to the bearingless machine.

Method (a) is similar, in principle, to the inductive displacement sensor. High frequency sinusoidal voltages and currents are injected and the impedances or induced voltages are detected to estimate the shaft displacement. The injected frequency is sufficiently high so that it does not interfere with the motor and suspension currents. However, the precision may be inferior to that of the inductive sensor because the motor and suspension fluxes are dominant and may interfere with the injected high frequency component. In order to inject the carrier signal, an analogue power amplifier, suitable for low power applications, can be used. However, for fractional horse-power applications, a PWM inverter should be used instead – in which case, the PWM carrier frequency has to be high enough to inject the sinusoidal component.

Method (b) is similar to (a); however, the PWM carrier signals are used for the displacement estimation. Modern PWM inverters are low cost and compact and also have high efficiency and low losses.

Method (c) uses flux detectors similar to Hall probes. If the rotor shaft is displaced from the stator centre then the flux distribution becomes unsymmetrical. The rotor shaft position is then estimated from the flux unbalance.

Method (d) estimates the rotor radial position from the terminal voltage and current using a mathematical electromagnetic representation. The dynamic radial position is obtained by an observer although the absolute position, i.e., the radial position at low frequency, is missing. Inductance and resistance parameters have to be obtained over wide frequency and current ranges to implement the strategy.

A combination of the above position-sensing techniques is possible.

In this section, method (a) is briefly described [8] where a high-frequency current is injected into the motor windings and detected in the suspension windings. Let us suppose that i_{ma} and i_{mb} are the instantaneous 2-phase high-frequency currents in the motor windings. Let us also suppose that the suspension current can be neglected because it is at a substantially lower frequency. Therefore the flux linkages λ_{sa} and λ_{sb} of the suspension windings can be written as

$$\begin{bmatrix} \lambda_{sa} \\ \lambda_{sb} \end{bmatrix} = \begin{bmatrix} M'x & M'y \\ M'y & -M'x \end{bmatrix} \begin{bmatrix} i_{ma} \\ i_{mb} \end{bmatrix} \tag{18.1}$$

where M' is constant (M' is the derivative of mutual inductance with respect to the rotor radial position) and x and y are the radial positions of the rotor in two perpendicular radial axes. This equation indicates that the mutual coupling between the motor winding and the suspension winding is proportional to the radial displacement of the rotor.

Let us only look at the high frequency current component, at an angular frequency of ω_h, which is generated in the ma-phase of the motor winding. Neglecting the low frequency torque and suspension current, the motor current simplifies to:

$$\begin{bmatrix} i_{ma} \\ i_{mb} \end{bmatrix} = \sqrt{2}I_4 \begin{bmatrix} \cos \omega_h t \\ 0 \end{bmatrix} \tag{18.2}$$

where I_4 is the rms value of the injected high frequency current.

Substituting (18.2) into (18.1) and executing the derivative of the flux linkage, the voltages v_{sa} and v_{sb}, at the suspension winding terminals, are obtained:

$$
\begin{bmatrix} v_{sa} \\ v_{sb} \end{bmatrix} = -\frac{d}{dt} \begin{bmatrix} \lambda_{sa} \\ \lambda_{sb} \end{bmatrix}
$$

$$
= -\sqrt{2} M' I_4 \frac{d}{dt} \left\{ \begin{bmatrix} x & y \\ y & -x \end{bmatrix} \begin{bmatrix} \cos \omega_h t \\ 0 \end{bmatrix} \right\} \tag{18.3}
$$

$$
= -\sqrt{2} M' I_4 \begin{bmatrix} \dot{x} & -x\omega_h \\ \dot{y} & -y\omega_h \end{bmatrix} \begin{bmatrix} \cos \omega_h t \\ \sin \omega_h t \end{bmatrix}
$$

The derivatives of the radial positions are normally small compared to the product of the radial position and ω_h so that the induced voltages at the suspension winding terminals can be simplified to

$$
\begin{bmatrix} v_{sa} \\ v_{sb} \end{bmatrix} = \sqrt{2} \omega_h M' I_4 \sin \omega_h t \begin{bmatrix} x \\ y \end{bmatrix} \tag{18.4}
$$

This equation indicates that high frequency voltages are induced into the suspension windings and these have amplitudes that are proportional to the rotor radial displacement. Therefore the signal processing of these induced voltages will provide the rotor displacement information.

Figure 18.10 shows the relationship between the induced voltage and the rotor radial displacement x. It can be seen that the induced voltage is almost proportional to the displacement. Also note that the differential voltage principle, as described in the previous section, is an inherent characteristic in this case because the 4-pole and 2-pole windings are employed and the mutual inductance between the two is zero at the centre position. There is also a good range of voltage variation as a function of the radial displacement. The measurements were carried out over a range of frequencies. At 20 kHz the voltage gain (2.56 mV/μm)

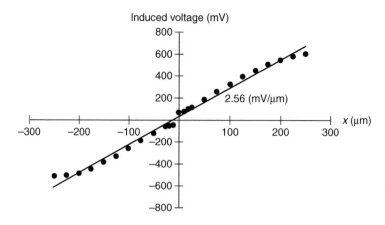

Figure 18.10　Induced voltage vs rotor radial displacement

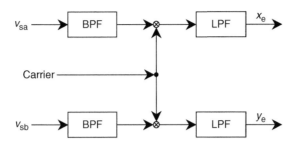

Figure 18.11 Block diagram of a sensing circuit

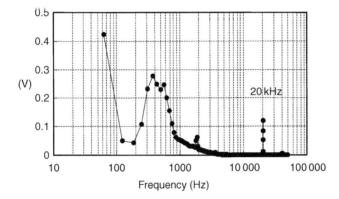

Figure 18.12 Measured frequency spectrum of suspension voltage with 20 kHz current injection

is about one-tenth of that at 50 Hz when a conventional iron core of laminated silicon steel is utilized. The reduction of permeability and the iron loss are the causes of the reduced voltage gain at a high frequency.

Figure 18.11 shows a block diagram of the displacement sensing circuit. The terminal voltages v_{sa} and v_{sb} are detected and fed to a band-pass-filter (BPF) circuit to obtain the carrier frequency component and eliminating other frequency components (such as the main suspension winding frequency and the motor revolving field components); these signals are then multiplied by a carrier frequency for demodulation. The low-pass-filter (LPF) eliminates the high frequency component (which has twice the carrier frequency) to obtain dc values so that the estimated displacements x_e and y_e are proportional to the rotor displacement.

Figure 18.12 shows a spectrum of the suspension winding terminal voltage. It can be noted that the carrier frequency is set to 20 kHz, and the main suspension voltage components are below several kHz.

Figure 18.13 shows the operating waveforms when the radial position command x^* is intentionally varied in a sinusoidal manner. The BPF output signal amplitude is almost proportional to the radial displacement. The estimated displacement x_e and the detected displacement x (as detected by an eddy current sensor) are shown and good correlation is seen.

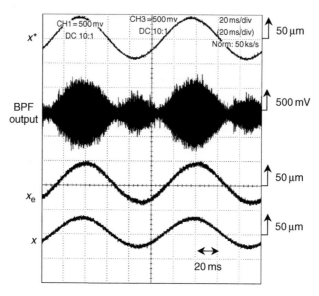

Figure 18.13 Waveforms with x* variation

The estimated radial position was used in the radial position feedback loop and stable operation was obtained. However, careful gain adjustment in the radial position regulator and a decrease in the current limit of the radial force winding were required because the frequency response of the estimated radial position was inferior to that from the eddy current sensor. Despite the relatively low stiffness of the radial suspension, a low cost and compact bearingless motor drive can be realized.

References

[1] T. Kuwajima, T. Nobe, K. Ebara, A. Chiba and T. Fukao, "An Estimation of the Rotor Displacements of Bearingless Motors Based on a High Frequency Equivalent Circuit", IEEE PEDS conference Bali, October 2001, IEEE Catalogue No. 01TH8594C, ISBN 0-7803-7234-4, pp. 725–731.

[2] C. Choi and K. Park, "A Sensorless Magnetic Levitation Systems Using a LC Resonant Circuit with PPF Control", Proc. MOVIC '98, Vol. 2, Zurich, Switzerland, August, 1998, pp. 739–744.

[3] T. Mizuno, H. Namiki and K. Araki, "Self-sensing Operations of Frequency-Feedback Magnetic Bearings", Proc. Fifth International Symposium on Magnetic Bearings, Kanazawa Univ., 1996, pp. 119–123.

[4] K. Matsuda, Y. Okada and J. Tani, "Self-sensing Magnetic Bearing Using the Principle of Differential Transformer", Proc. Fifth International Symposium on Magnetic Bearings, Kanazawa Univ., 1996, pp. 119–123.

[5] P. Tsao, S. R. Sanders and G. Risk, "A Self-sensing Homoplar Magnetic Bearing: Analysis and Experimental Results", IEEE IAS 34th Annual Meeting, 1999, pp. 2560–2565.

[6] T. Yoshida, K. Ohniwa and Y. Katsuyama, "A Position-Sensorless Magnetic Levitation Control Method", Proc. IEEJ JIASC 2001, Vol. 271, 2001, pp. 1385–1390 (*in Japanese*).

[7] T. Mizuno, K. Araki and H. Bleuler, "Stability Analysis of Self-sensing Magnetic Bearing Controllers", IEEE Trans. on Control Systems Technology, Vol. 4, No. 5, 1996, pp. 572–576.

[8] K. Muronoi, N. Andoh, A. Chiba and T. Fukao, "Characteristics of Induction type Self-sensing Bearingless Motors", Proc. IPEC-Tokyo 2000, pp. 2109–2114.

19

Controllers and power electronics

Masatsugu Takemoto

In this chapter, a digital controller and a sample program for controlling a bearingless motor are introduced. This is a general-purpose structure that can be modified and tailored to a specific bearingless drive application.

19.1 Structure of digital controllers

Currently, digital control using digital signal processors (DSP) is usually used in a bearingless drive to realize the system. This is because:

1. The drive inverters generate a lot of electromagnetic noise due to switching. However, the interference caused by this noise in the control loop of the magnetic suspension can be reduced by using digital control. In particular, digital control can protect the delicate derivative controller, which is essential for magnetic suspension.
2. Digital control allows a higher degree of flexibility and universality compared with analogue control.
3. When using analogue control, it is difficult to introduce complex control methods such as an observer, a nonlinear calculation (such as magnetic saturation) or any calculation using complex branching methods or extensive memory. However, digital control can deal with these complex control methods.

Figure 19.1 shows the typical structure of a digital control system in a bearingless drive using hysteresis current regulators. An example of a complete digital control system is illustrated in Figure 19.2.

In the motor speed controller the rotor angular position ϕ is detected by the rotary encoder. The rotational speed ω is generated from the encoder counter and speed detector. This is then compared to the rotational speed command ω^*, and the q-axis current command i_q^* (in rotating angular coordinate) is generated by the proportional-integral (PI) controller which is applied to the speed error. In the

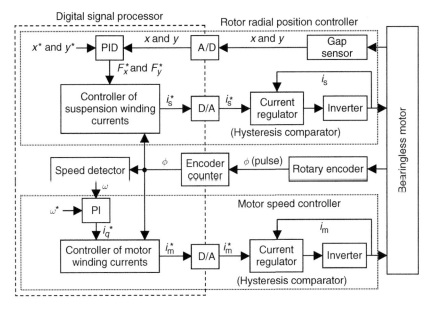

Figure 19.1   Typical structure of a digital control system in a bearingless drive using hysteresis comparators

Figure 19.2   The photograph of the digital control systems

motor winding current controller, the 3-phase motor winding current command i_m^* is calculated by transforming i_q^* from the rotating angular coordinate to the stationary coordinate and transforming from 2-phase to 3-phase. Finally, the motor winding currents are controlled by the current regulator using hysteresis comparators and the 3-phase inverter.

In the rotor radial position controller, the rotor displacements x and y (on the x and y axes) are detected by the gap sensors through an A/D converter and these displacements are compared to the displacement commands x^* and y^*. The suspension force commands F_x^* and F_y^* are generated by the proportional-integral-derivative (PID) controllers. The outputs are then fed into the suspension winding current controller where F_x^* and F_y^* are modulated to become the 2-phase suspension winding current commands by means of sinusoidal functions which are synchronized to the rotor angular position ϕ. These 2-phase commands are then transformed into 3-phase commands. The controller for the suspension winding currents has been discussed in each chapter as the various bearingless motors were introduced. The 3-phase suspension winding current command i_s^* is fed forward to the current regulator via the D/A converter where the suspension winding currents are generated by the hysteresis comparators in the regulator and the 3-phase inverter. As an example, Figure 19.2 shows a digital controller complete with current regulators, power supply and analogue interface circuits fabricated on layered printed boards.

An alternative current regulation scheme for the regulator can be realized in rotating coordinates instead of by the hysteresis method. The detected currents are transformed into the rotating reference frame coordinates and then compared with rotating current references. The current errors are amplified by PI controllers to generate voltage references, and gate pulses are generated by comparison of the voltage references with triangular carrier waveforms.

The current regulator, consisting of the hysteresis current regulators and inverter drive, was explained in detail in Chapter 5. Figure 19.3 shows the main circuit of a 3-phase inverter suitable for use in a bearingless drive. This is made up of six power-switching devices, such as IGBTs, MOSFETs, transistors, etc., each encapsulated with an anti-parallel diode.

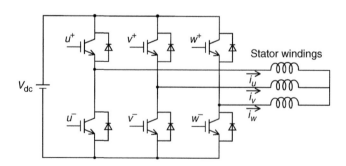

Figure 19.3 3-Phase inverter

19.2 Discrete-time systems of PID controllers with the z-transform

It is necessary to transform the PID controllers from continuous-time operation to discrete-time operation in order to construct the rotor radial position controller using a digital controller as shown in Figure 19.1. For this purpose, the z-transform is used. In this section, the discrete-time operation of the PID controller in the z-domain, as well as a sample program for realizing the controller, are introduced.

19.2.1 A PID controller in the z-domain

Figure 19.4(a) shows the block diagram of an ideal PID controller in the s-domain. K_p, K_i and K_d are the gain coefficients of the proportional, integral and derivative components of the controller respectively, while $s(t)$ and $x(t)$ are the input and output signals in the continuous-time domain. The transfer function of the PID controller in Figure 19.4(a) can be expressed as

$$h(s) = K_p + \frac{K_i}{s} + K_d s \tag{19.1}$$

The transfer function of the PID controller in the z-domain can be derived by applying the z transform to (19.1) so that

$$H(z) = K_p + K_i \frac{T_s}{2} \frac{1+z^{-1}}{1-z^{-1}} + K_d \frac{2}{T_s} \frac{1-z^{-1}}{1+z^{-1}} \tag{19.2}$$

where T_s is the sampling period. Figure 19.4(b) shows a block diagram of the PID controller in the z-domain and $S(z)$ and $X(z)$ are the input and output signals respectively in the discrete-time domain.

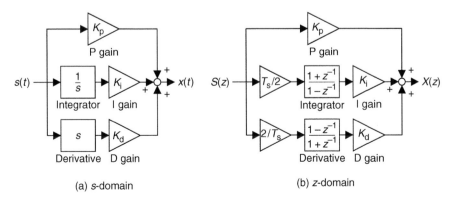

(a) s-domain   (b) z-domain

Figure 19.4 Proportional-integral-derivative controllers

19.2.2 A sample program of a PID controller

The transfer function of the integral controller can be re-written from (19.2) as

$$H_i(z) = \frac{X_i(z)}{S_i(z)} = \frac{1}{1-z^{-1}} \times (1+z^{-1}) \times K_i \times \frac{T_s}{2} \qquad (19.3)$$

where $S_i(z)$ and $X_i(z)$ are the input and output signals of the integral controller. To understand the basic operation of the integral controller in the z-domain, the controller is illustrated in a block diagram form using (19.3) as shown in Figure 19.5(a). This block diagram can be transformed into the block diagram shown in Figure 19.5(b) where $W_i(0)$ and $W_i(1)$ are medium factors of the integrator.

From Figure 19.5 the following expressions can be derived

$$W_i(0) = S_i(z) + W_i(1) \qquad (19.4)$$

$$X_i(z) = \frac{K_i T_s}{2} [W_i(0) + W_i(1)] \qquad (19.5)$$

$$W_i(1) = z^{-1} W_i(0) \qquad (19.6)$$

An outline of the programming required for the integral controller in the discrete-time domain is listed below and it is derived from (19.4), (19.5), and (19.6).

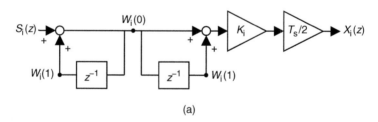

(a)

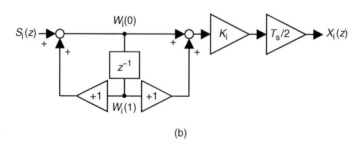

(b)

Figure 19.5 Block diagrams of the integral controller: (a) the block diagram of the integral controller based on (19.3) (b) the general block diagram of the integral controller

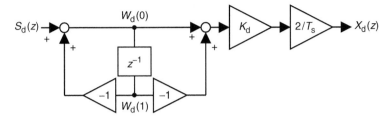

Figure 19.6 The block diagram of the derivative controller

Similarly, the derivative controller in the z-domain is illustrated in a block diagram form in Figure 19.6. The variable definitions in Figure 19.6 are as follows: $W_d(0)$ and $W_d(1)$ are medium factors of the derivative, $S_d(z)$ is the input signal and $X_d(z)$ is the output signal.

The following expressions can be obtained from Figure 19.6:

$$W_d(0) = S_d(z) - W_d(1) \tag{19.7}$$

$$X_d(z) = \frac{2K_d}{T_s}[W_d(0) - W_d(1)] \tag{19.8}$$

$$W_d(1) = z^{-1}W_d(0) \tag{19.9}$$

Again, an outline program for the derivative controller in the discrete-time domain, as obtained from (19.7), (19.8), and (19.9), is listed below.

A sample program of the PID controller in C-language is:

```
/***** Sample Program ********************/
double S, Sp, Si, Sd;        // Input Signals
double X, Xp, Xi, Xd;        // Output Signals
double Ts;                   // Sampling period [s]
double Kp, Ki, Kd;           // Gains
double Wi[2], Wd[2];         // Medium factors


/***** Proportional Controller *****/
Sp = S;
Xp = Kp * Sp;


/***** Integral Controller ********/
Si = S;
Wi[0] = Si + Wi[1];
Xi = Ki * Ts * 0.5 * (Wi[0] + Wi[1]);
Wi[1] = Wi[0];
```

```
/***** Derivative Controller *******/
Sd = S;
Wd[0] = Sd - Wd[1];
Xd = 2.0 * Kd / Ts * (Wd[0] - Wd[1]);
Wd[1] = Wd[0];

/***** Sum of These Controllers ****/
X = Xp + Xi + Xd;
/*****************************************/
```

The other components of the digital controller program are straightforward because the modulation, coordinate transformation, 2-phase to 3-phase conversion and nonlinear parameter variations are simple calculations using the inputs at the start of the sample period.

20

Design procedure and examples

Akira Chiba

In this chapter, one example of the bearingless induction motor and one example of the bearingless permanent magnet motor are introduced. In the first section, the sizes, dimensions, winding structure, inductance functions and performance figures of a bearingless induction machine with pole-specific rotor circuits are put forward and the machine is described in detail. In the following section, a bearingless permanent magnet machine, with a consequent-pole rotor, is introduced. The consequent-pole rotor machine is the most straightforward bearingless motor to initially study because rotor angular position feedback is not required for the magnetic suspension system. Also the output torque and radial force are high.

20.1 An induction type bearingless motor [1–4]

Figure 20.1 shows two cross sections of a bearingless induction motor with geometrical parameters defined for the rotor and stator iron cores. The stator has 24 slots with an outer radius of R_{styko} and an inner radius of R_{styki}. The rotor has 16 circular holes centred round the radius of R_{bar}. The outer and inner radii of the rotor iron are R and R_{s} respectively. In the enlarged figures, several dimensions, as indicated by T_{ws}, T_{gs}, T_{gd}, S_{o}, $S_{\text{lotOpenRotor}}$, R_{cIron} and airgap g, are defined. Table 20.1 summarizes these parameters.

Figure 20.2 shows an enlarged figure of the stator conductor structure. Each stator slot contains a suspension and a motor winding coil-side arranged in two layers. The outer layer coil-side (in the slot bottom), as indicated by circles, is for the 4-pole motor winding; while the inner layer coil-side is for the 2-pole suspension winding. In the figure there are 12 and 3 series-connected coil-turns representing the motor and suspension winding coil-sides. Let us define the number of series-connected coil turns as N_{4slot} and N_{2slot}. However, in reality, one series turn is made of several parallel wires and the numbers of parallel

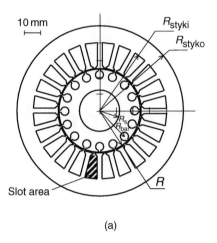

(a)

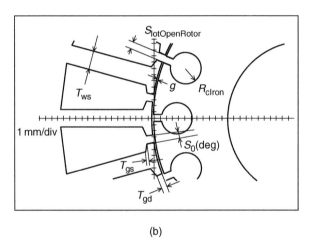

(b)

Figure 20.1   Iron core design: (a) iron core design; (b) enlarged view

wires (strands-in-hand) for each coil are defined as P_{4slot} and P_{2slot}. Let us also define the wire diameters as D_4 and D_2 and the total number of conductors in one slot as $N_{4slot}P_{4slot} + N_{2slot}P_{2slot}$. Table 20.2 summarizes winding design of the test machine. Note that N_{4slot} and N_{2slot} are 30 and 6 for the prototype machine and both the 2-pole and 4-pole wires have a diameter of 0.6 mm. The number of strands-in-hand is two for both windings, and the conductors are connected as shown previously in section 6.9. The area of one stator slot is 78.3 mm² while the total cross-sectional area of all the conductors for one slot is 20.36 mm²; this is obtained from

$$\pi \left(\frac{D_2}{2}\right)^2 N_{2slot}P_{2slot} + \pi \left(\frac{D_4}{2}\right)^2 N_{4slot}P_{4slot}$$

Table 20.1   Iron design

Shaft radius	R_s	12.5 mm
Rotor bar centre radius	R_{bar}	21 mm
Rotor iron outer radius	R	24.6 mm
Rotor slot radius	R_{cIron}	2.5 mm
Rotor slot open width	$S_{lotOpenRotor}$	1 mm
Airgap length	g	0.4 mm
Stator inner radius		25 mm
Stator yoke inner radius	R_{styki}	40 mm
Stator yoke outer radius	R_{styko}	50 mm
Stator teeth width	T_{ws}	3 mm
Stator slot open angle	S_o	3 deg
Stator teeth head thickness	T_{gd}	1 mm
Stator teeth neck	T_{gs}	0.5 mm
Stack length of iron lamination	L_s	50 mm

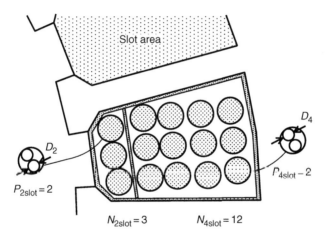

Figure 20.2   Series and parallel wire

Table 20.2   Winding design

Stator design

2-Pole series turns per slot	N_{2slot}	6 turn
2-Pole parallel wires	P_{2slot}	2
4-Pole series turns per slot	N_{4slot}	30 turn
4-Pole parallel wires	P_{4slot}	2
2-Pole wire diameter	D_2	0.6 mm
4-Pole wire diameter	D_4	0.6 mm
One slot area of stator		78.3 mm²
Conductor area per slot		20.36 mm²
Conductor slot fill factor		0.26
2-Pole conductor ratio		0.17

Rotor design

Outer radius of coil end conductor	R_{oend}	24.11 mm
Inner radius of coil end conductor	R_{iend}	12.75 mm
Rotor bar clearance	$B_{arClear}$	0.02 mm
Coil end conductor thickness	t_e	3.2 mm
Coil end insulator thickness	t_i	3 mm

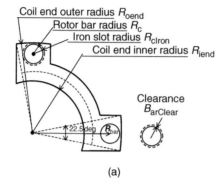

(a)

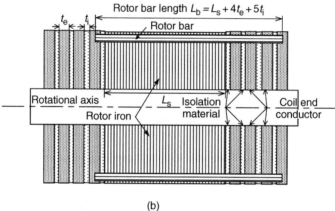

(b)

Figure 20.3   Rotor structure: (a) rotor coil end conductor; (b) cross sectional view of rotor

which gives a slot fill factor of 0.26. This value can be improved up to 0.3 or 0.35 with tight winding. The cross-sectional area ratio of the 2-pole coil-side with respect to the total conductor area is 0.17 which means that 17% more slot area is required when compared to the original induction motor in order to accommodate the suspension winding.

Figure 20.3(a) shows the coil end conductor structure (end-ring section) for the pole-specific rotor short circuit. The end-ring section outer radius R_{oend} is almost equal to the rotor radius. The rotor bar radius R_c is slightly smaller than the rotor iron slot radius. The end-ring section inner radius R_{iend} is designed so that it has a suitable width; the end-ring cross section should have the same area as the rotor bars. Four end-ring sections are required for one short circuit (which link four bars) in the 4-pole-specific rotor. In total, for a 16-slot rotor, 16 conductors are needed, which are arranged in four short circuits.

Figure 20.3(b) shows a cross-sectional view of the rotor in the axial direction. The shaft is horizontal across the centre of the section and it is surrounded by laminated silicon steel. At both the left and right ends of the rotor are the four layers of the end-ring sections, as illustrated. Two rotor bars are drawn to illustrate the connection of one rotor short circuit. The thickness of the end-ring sections is t_e and the thickness of the isolation material is t_i. The isolation layer thickness is small since the induced voltage in the rotor circuits is low. In some cases, the isolation layer is not really needed because of the metal contact resistance – the oxidized surface of the end-ring sections providing sufficient isolation. In this test machine, the insulation ring is also used to mechanically fix the end-ring sections to the rotor. Table 20.2 lists the parameters for the rotor design.

Table 20.3 summarizes the measured inductances of the test machine. The inductance measurements were carried out before the rotor short circuits were formed so that only the laminated silicon iron was employed as the rotor. Some inductance values were measured across the whole current range to assess the influence of magnetic saturation. The flux density in a stator tooth was measured using search coils with an excitation current of 3.5 A so that the tooth flux density was about 1.2 T. The excitation current was supplied at mains frequency and the inductance measurements were carried out with different rotor radial displacements. The measured inductance values were transformed into 2-phase coordinates using a 3-phase-to-2-phase transformation matrix, and the self inductances and mutual inductance derivatives M' were obtained.

The table also shows the test results of motor and suspension characteristics at a speed of 6000 r/min. At the motor terminals, the line-to-line voltage was 92 V while the input power was 840 W at a slip of 0.1 and a line current of 7 A. For the magnetic suspension, two bearingless motors generated a magnetic force of 3.3 kgf to match the shaft weight. To generate a force of 3.3 kgf, a current of 0.4 A and a voltage of 0.8 V were required at a frequency of 200 Hz, i.e., twice the rotational frequency. These values were measured by an FFT analyser to

Table 20.3 Measured values

4-Pole self-inductance in 3-phase	$L_{4(3ph)}$	9.2~8 mH	(Magnetic saturation)
2-Pole self-inductance in 3-phase	$L_{2(3ph)}$	1.4~1.5 mH	(Magnetic saturation)
4-Pole mutual-inductance in 3-phase	$M_{44(3ph)}$	−4.5 mH	
2-Pole mutual-inductance in 3-phase	$M_{22(3ph)}$	−0.75 mH	
4-Pole self-inductance in 2-phase	$L_{4(2ph)}$	15.3 mH	
2-Pole self-inductance in 2-phase	$L_{2(2ph)}$	2.42 mH	
4-Pole and 2-pole mutual-inductance derivative	M'	6.4 H/m	

Motor
Rotational speed	6000 r/min
Line-to-line voltage	92 V
Line current	7 A
Slip	0.1
Input power	840 W

Suspension
Shaft weight	3.3 kg
Weight current	0.4 A
Weight voltage	0.8 V
Peak current	2 A
Peak voltage	20 V

detect the frequency component. The peak current was 2 A and the voltage was 20 V. This voltage and current were required to control the vibrations caused by mechanical unbalance and misalignment. About 10 kgf of radial force can be generated by one bearingless unit within the current ratings of suspension windings.

20.2 A permanent magnet type bearingless motor [5,6]

In this section, a consequent-pole bearingless motor is described as an example of a permanent magnet type of bearingless motor. As described in Chapter 14, the consequent-pole rotor does not require rotor angular position feedback in the suspension control which makes it easier to construct.

Figure 20.4 shows the rotor and stator structures. The rotor has four permanent magnets in a laminated silicon steel core. These are identical and pre-magnetized. In this case, the magnet north poles are all on the rotor surface. They are manufactured from NeFeBr and are 5-mm thick. In the stator core there are two sets of 3-phase conductors, arranged in slots. One set is the 8-pole motor winding, N_{8u}, N_{8v} and N_{8w} and the other set is the 2-pole suspension winding, N_{2u}, N_{2v} and N_{2w}. The 3-phase suspension winding has been found to minimize the radial force variation with respect to rotor angular position. Since there are two 3-phase sets there are six wires connecting the bearingless unit to the inverters.

Table 20.4 summarizes the machine and winding design. The rotor radius is 24.5 mm with an axial length of 50 mm. The product $D \times L$ (diameter and the axial length) is 24.5 cm^2 and the airgap length is 0.5 mm. In the winding design, the conductor slot fill factor is a rather low value of 0.27 for easy winding installation and the cross-sectional area of suspension conductors occupies 17% of the total conductor area (similar to the bearingless induction motor above).

Figure 20.5 shows a system block diagram of the test machine. In the suspension controller, the radial positioning is detected by displacement sensors and compared with the references. The PID controllers generate radial force commands and, using these commands, the current commands I_x^* and I_y^* are generated. Instantaneous 3-phase winding current commands are generated by a 2-phase-to-3-phase transformation, and a current-regulated PWM voltage-source inverter provides the 3-phase currents for the suspension windings.

In the motor controller, the rotor angular position is detected by a rotary encoder which also provides speed feedback. The speed is compared with the speed command so that a motor current command is generated. From the d- and q-axis current commands, 2-phase currents are generated for the 8-pole motor operation. The 3-phase current commands are fed forward to a current-regulated PWM voltage-source inverter to provide the motor currents. It is possible to have sensorless angular position operation since the rotary encoder is needed only for the motor drive.

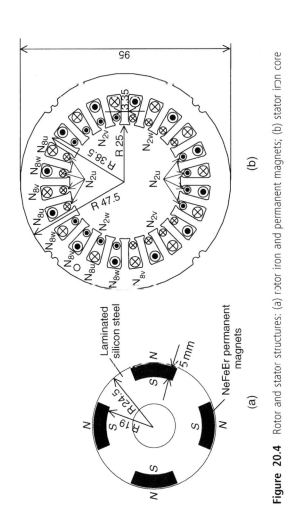

Figure 20.4 Rotor and stator structures: (a) rotor iron and permanent magnets; (b) stator iron core

Table 20.4   Machine design

Iron and PM design		
Rotor iron outer radius	R	24.5 mm
Permanent magnet thickness	L_m	5.0 mm
Permanent magnet arc		45 deg
PM remanent flux density	B_r	1.28 T
Airgap length	g	0.5 mm
Stator inner radius		25 mm
Stator yoke inner radius	R_{styki}	38.5 mm
Stator outer radius	R_{styko}	47.5 mm
Stator teeth width	T_{ws}	3.35 mm
Stack length of iron lamination	L_s	50 mm
Winding design		
2-Pole series turns per slot	N_{2slot}	10 turn
2-Pole parallel wires	P_{2slot}	1
8-Pole series turns per slot	N_{8slot}	25 turn
8-Pole parallel wires	P_{8slot}	2
2-Pole wire diameter	D_2	0.6 mm
8-Pole wire diameter	D_8	0.6 mm
One stator slot area		63.4 mm^2
Conductor slot fill factor		0.27
2-Pole conductor ratio		0.17

In constructing a test machine, special attention should be paid to the permanent magnets. Inserting them into the rotor core is not straightforward if they are pre-magnetized rare-earth magnets since there is always a strong attractive force between the rotor core and the permanent magnets; and sintered material is quite brittle. Also, inserting the rotor into the stator core can be tricky, again because of the attractive force between the rotor and stator which can be quite significant. Once inserted, the rotor should be fixed using touch-down bearings so that the shaft is movable within a radius of only about 0.1 mm from the centre position.

The drive inverter should be carefully designed so that there is fast-acting circuit protection for the occasional transient over-current. The current needs to be limited to prevent the permanent magnets from being irreversibly demagnetized (though sintered rare-earth magnets are often very difficult to magnetize and demagnetize – hence the fabrication of the motor with fully magnetized magnets).

Table 20.5 summarizes the measured test data. The voltage, current and power were measured by a digital power meter at the rated load point. The shaft torque is reasonable for the machine dimensions although the efficiency is rather low; however, it can be improved by increasing the slot fill factor and by PM thickness optimization. At the suspension winding terminals, the voltage, current and power were measured at two operating points – one is the zero torque point while the other is the full-load torque point. The radial force was 128 and 96 N, while the radial force density DL is 0.53 and 0.4 kgf/cm^2 and the VA ratio and power ratio between the suspension and motor windings are only 0.08 and 0.03 at zero and full torque, respectively.

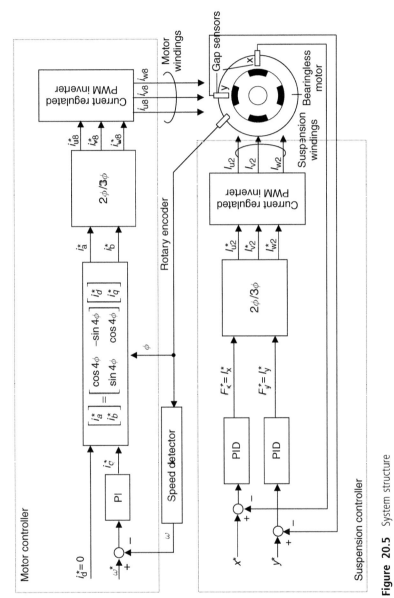

Figure 20.5 System structure

Table 20.5 Test results

Motor

Rotational speed	4000 r/min
Line-to-line voltage	143 V
Line current	4.8 A
Input power	926 W
Shaft torque	1.89 Nm
Efficiency	85.6%
Total power factor	0.721

Suspension

At no torque ($I_q = 0$ A)	
Radial force	128 N
Line-to-line voltage	25.0 V
Line current	2.47 A
Input power	30.3 W
At full torque ($I_q = 8$ A)	
Radial force	96 N
Line-to-line voltage	25.9 V
Line current	2.07 A
Input power	30.3 W

References

[1] Y. Takamoto, A. Chiba and T. Fukao, "Test Results on a Prototype Bearingless Induction Motor with Five-Axis Magnetic Suspension", Proceedings of 1995 International Power Electronics Conference (IPEC – Yokohama '95), Vol. 1, April 7, 1995, pp. 334–339, Pacifico Yokohama.

[2] A. Chiba, R. Furuichi, Y. Aikawa, K. Shimada, Y. Takamoto and T. Fukao, "Stable Operation of Induction-Type Bearingless Motors Under Loaded Conditions", IEEE Transaction on IA, Vol. 33, No. 4, July/August, 1997, pp. 919–924.

[3] A. Chiba, K. Yoshida and T. Fukao, "Transient Response of Revolving Magnetic Field in Induction type Bearingless Motors with Secondary Resistance Variations", International Symposium on Magnetic Bearings (ISMB '98), August 7, 1998, pp. 461–475, Boston, USA.

[4] T. Suzuki, A. Chiba, M. A. Rahman and T. Fukao, "An Air-Gap-Flux-Oriented Vector Controller for Stable Operation of Bearingless Induction Motors", IEEE Transaction on IA, Vol. 36, No. 4, July/August 2000, pp. 1069–1076.

[5] T. Takenaga, Y. Kubota, A. Chiba and T. Fukao, "A Principle and a Design of a Consequent-Pole Bearingless Motor", ISMB-8, August, 2000, pp. 259–264, at Mito.

[6] Y. Kubota, T. Takenaga, A. Chiba and T. Fukao, "Consequent-Pole Type Bearingless Motors", IEE Japan, The Papers of Joint Technical Meeting on Semiconductor Power Converter and Industry Electric and Electronic Application, SPC-01-102, November 8, 2001, pp. 49–54, at Ashikaga (*in Japanese*).

21

Applications and test machines

Akira Chiba

In this chapter some examples of bearingless drives that have been developed by various manufacturers are presented. The characteristics, requirements and development purposes are described. The applications for these drives are:

1. Canned pumps and drives;
2. Compact pumps;
3. Bubble bed reactor;
4. Blood pumps;
5. Spindle drives; and
6. Semiconductor processing.

21.1 Canned pumps and drives [1–4]

Canned pumps are used in pumping applications where a fluid leak should be avoided and contact with steel components will cause corrosion. For example, hydrochloric acid is pumped around paper mill plants using canned pumps because the processing of wood chips and recycled paper uses this corrosive liquid. Ball bearings cannot be used here because of steel corrosion. Usually carbon slide bearings are used instead. These are cylindrical in shape. During operation there is no mechanical contact because there is fluid between the carbon cylinder and the stationary bearing sheath when pumping at speed. However, there is contact during starting, causing wear. In addition, the bearings may fail if small particles contaminate the fluid. Hence the life time of a carbon slide bearing is short and it is very difficult to see if the pump has a small but developing bearing problem. Therefore it is known for them to fail and seize suddenly in which case they should be replaced in very short time to prevent the loss of production. (Conventional bearings usually wear in a more gradual manner which can be diagnosed – allowing replacement in scheduled down-time.) The suspension problem in conventional canned pumps may be overcome by the use of a bearingless drive and/or magnetic bearings.

Figure 21.1 shows a picture of a test canned pump and drive developed by Ebara Research Co. Ltd, Japan. Figure 21.2 shows a cross section of the canned pump. The pump impeller is at the right-hand end of the unit. Fluid enters the impeller from the right and then is pumped in an upward direction. The other half of the unit consists of the bearingless motor units and a magnetic thrust bearing. There are two bearingless motor units and these are responsible for the torque and radial suspension of the rotor. The four radial axes are actively suspended by the two bearingless motor units, and the magnetic thrust bearing, which is located at the left-hand end, generates an axial magnetic force to balance the significant thrust force caused by the pump impeller. The rotating component is surrounded by a metal can made of Hastelloy, which is a corrosion-resistant metal. The inner surface of the stationary component is also protected by a thin cylindrical Hastelloy can and the space between the Hastelloy cans is filled by the fluid. While the pump is small, the motor and suspension components are quite large. The manufacturer of this pumping unit specializes in the integration of mechanical and electrical components in this way.

Table 21.1 shows the specification of the canned pump. The rotor diameter is 68 mm and the axial length is 50 mm for one bearingless unit. The input power is

Figure 21.1 Canned pump with integrated bearingless units. (Courtesy: Ebara Research Co. Ltd)

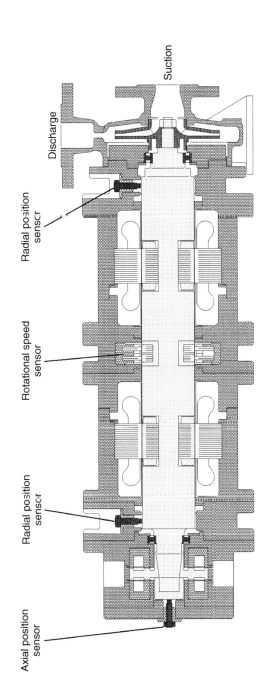

Figure 21.2 Cross section of canned pump. (Courtesy: Ebara Research Co. Ltd)

Suction

Discharge

Radial position sensor

Rotational speed sensor

Radial position sensor

Axial position sensor

Volute pump

Bearingless motor #1

Bearingless motor #2

Axial magnetic bearing

Table 21.1 Specification of canned pump.
(Courtesy: Ebara Research Co. Ltd)

Rotor diameter	68 mm
Rotor stack length	50 mm
Airgap length	0.6 mm
Rotating part length	600 mm
Rotating part weight	14 kg
Can thickness	0.2 mm
Head	24 m
Flow rate	180 l/min

about 2.2 kW. The head is 24 m and the flow of the pump is 180 l/min. Another
15-kW machine is also described in [2].

Figure 21.3 shows a picture of a 30-kW canned pump developed by Sulzer
Pumps Ltd and Sulzer Electronics Ltd, Switzerland [3–4]. Figure 21.4 shows
a cross section of this canned pump. The fluid enters from the left-hand side
and pumped in the upward direction by the pump impeller. On the right-hand
side of the impeller is a radial magnetic bearing. This takes the form of a
homopolar permanent magnet machine. The windings are arranged to operate in
differential mode so that no current is required when there is no radial force.
A 3-phase winding set is used in the magnetic bearing which is driven by a

Figure 21.3 30-kW canned pump. (Courtesy: Sulzer Pumps Ltd)

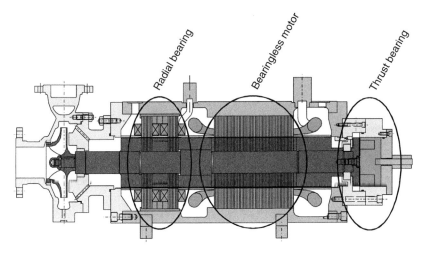

Figure 21.4 Cross-section of 30-kW canned pump. (Courtesy: Sulzer Pumps Ltd)

3-phase inverter. A controller from a general-purpose inverter is modified to act as the current regulators for the magnetic bearing, and only three power wires are required for two-axis active magnetic suspension. This is a nice development of the radial magnetic bearing and it should be noted that this is an application using a bearingless motor with a static magnetic field.

The middle part of the pump drive is a bearingless unit with an input volt-ampere rating of 45 kVA. An induction type of bearingless motor is used. The bearingless unit generates torque and 2-axis radial forces. On the right-hand side is a magnetic thrust bearing. In total, five axes are actively controlled by magnetic forces. The head and flow for the pump are 62 m and 30 l/s respectively, and the radial force is rated at 3500 N.

21.2 Compact pumps, bubble bed reactor

21.2.1 Compact pump [5]

Figure 21.5 shows two pictures of compact pumps from Levitronix GmbH. These have ratings of 50 and 300 W. The plastic upper components are the pump casings while the lower metal components are the frames for the bearingless drives. Figure 21.6 shows a cross section of one of these pumps. The fluid enters the inlet at the top of the pump and is pumped out of the radially orientated outlet. A plastic pump impeller is located on the rotor which also has integrated permanent magnets. The permanent magnets are magnetized in the radial direction allowing the generation of torque and two-axis radial forces. The stator iron teeth are located around the rotor on the outside of the plastic casing. The stator

Figure 21.5 Compact pumps integrated bearingless units. (Courtesy: Levitronix GmbH)

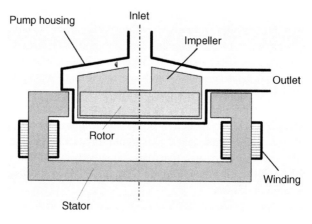

Figure 21.6 Cross section of a compact pump. (Courtesy: Levitronix GmbH)

teeth are bent down into a C-shape so that the diameter of the pump is minimized. The motor and suspension coils are wound around the stator teeth and the suspension coils are operated in differential mode to minimize the current requirements. Only two radial magnetic forces are actively regulated; however, the rotor is suspended passively to restrict the thrust and tilt movements, and also the rotor diameter and thickness are carefully designed to realize stable magnetic suspension with just two axes of active suspension. There is no mechanical seal in this pump so that no fluid leakage is expected. The applications for this sort of pump are external blood pumps, semiconductor process pumps, chemical process pumps, food process pumps and other pump applications where a mechanical seal should be avoided.

Table 21.2   Specification of compact pumps.
(Courtesy: Levitronix GmbH)

	DCP1.2	CP3.2
Flow (l/min)	18.0	50
Pressure (bar)	1.2	3
Motor power (W)	50.0	300
Diameter (mm)	87.0	131
Height (mm)	70.0	104
Voltage (V)	72.0	24

Table 21.2 gives the specifications of the two pumps. The head is 8 m and the flow is 18 l/min for the small pump and 25 m and 50 l/min for the large pump.

21.2.2 Bubble bed reactor [6]

Figure 21.7 is a picture of a bubble bed reactor [6]. In this, animal cells are grown in the reactor fluid. Oxygen is supplied from a nozzle at the bottom and there is a downstream fluid flow so that the oxygen bubbles stay in the fluid. The downstream is generated by a pump impeller. Usually, the pump impeller is driven by an external drive. The torque transfer mechanism requires a mechanical suspension structure to support the impeller, and also a shaft and a mechanical seal are required. If the cell is broken at the seal then the fluid can be contaminated. Alternatively the bearingless motor provides a magnetic suspension force and rotational torque to the pump impeller without a mechanical seal so that the possibility of contamination is reduced. The impeller is actively suspended in two perpendicular axes in the fluid. Other degrees of freedom for the motion is restricted by the careful design of the rotor diameter and thickness.

In the electro-mechanical structure, the permanent magnets are magnetized in the axial direction and placed on a ferromagnetic ring. The permanent magnets provide homopolar radial flux flow across the airgap between magnetic bearing iron poles and the rotor ring. Attractive forces are generated between the bearing iron poles and the rotor ring and these are regulated by the current in the magnetic bearing winding.

21.2.3 Blood pump [7,8]

Figure 21.8 shows a picture of an implantable blood pump under development in Ibaraki University, Japan. Figure 21.9 shows a diagram of the electromagnetic structure. The stator core, not shown in the figure, is located inside the ring-shaped rotor. The rotor ring is supported by two-axis active magnetic suspension. Several eddy current sensors are illustrated and used for test purposes. On the rotor ring, a plastic impeller is attached; the blood comes in through the inlet

Figure 21.7 Bubble bed reactor. (Courtesy: Levitronix GmbH)

on the top of the blood pump and exits through the outlet. The plastic impeller design is quite important to avoid haemolysis.

In blood pump applications, magnetic suspension is highly desirable. In conventional blood pumps used in surgery, the operation time is limited to several hours to avoid clogging problems. These are due to blood particles being destroyed in the blood pump. These broken particles adhere to the seal which may eventually clog because of this build-up. Also the broken particles can result in a thrombus. Therefore, supporting a pump impeller with magnetic suspension is an effective way to counteract these problems. A blood pump with integrated magnetic bearings has been shown to have a long operating life when tested in trials.

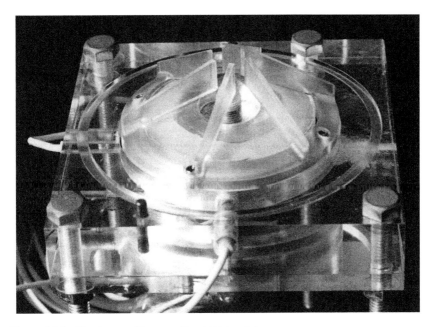

Figure 21.8 Blood pump. (Courtesy: Ibaraki University)

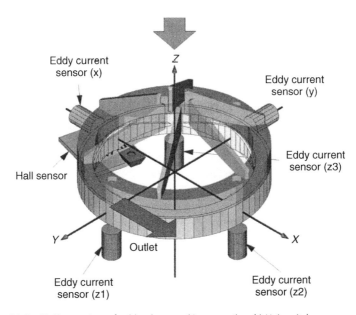

Figure 21.9 Bird's eye view of a blood pump. (Courtesy: Ibaraki University)

Another application is the artificial heart. Compact blood pumps implanted into the human body may help many people. There are many requirements for a blood pump; from an electromechanical point of view, these are

1. Compactness.
2. Low power magnetic suspension.
3. High efficiency operation.
4. Smooth blood flow and low wear rate.

Electric power is supplied through human skin via a transformer. The primary side of the transformer is fixed on the outside of the body while the secondary side is buried under patient's skin. If the required power is low then the power supply and pump can be compact. In normal human activity several watts are required to pump the blood, and a high efficiency motor and pump are required. Power losses in the magnetic suspension should be minimized to avoid a temperature rise – the pump should operate as close to body temperature as possible.

Another experimental blood pump has been manufactured in Switzerland and the USA [8]. The structure is basically similar to the compact pump described in this section. The diameter is 69 mm and thickness is 30 mm for the motor component. These compact structures provide a promising future for the development of the blood pump.

21.3 Spindle motor and semiconductor processing

21.3.1 Spindle motor [9–10]

Some computer-related machines are required to run at high speeds. In hard disk storage, disks are driven by an electrical motor at a constant speed, for example 15 000 r/min. To enhance the data transfer speed some hard disks have an even higher speed. High-speed hard disks can suffer from excessive heat generation. As the rotational speed increases, mechanical bearing loss also increases so that the motor power increases likewise. Hence the heat generation due to electrical and mechanical losses increases, resulting in raised temperature which shortens life time and increases the possibility of premature failure. If magnetic suspension is used then the drive loss can be reduced and hence lower motor power is required. There is now a need for even higher speed drives in DVD and CD applications.

Polygon mirror scanners are used in information-processing machines such as laser printers. These also require a high-speed drive. In order to enhance the processing speed, the rotational speed of the drive motor needs to be increased.

Figure 21.10 Computer storage spindle. (Courtesy: Sankyo Seiki Mfg Co.)

Figure 21.10 shows a drive developed by Sankyo Seiki Mfg Co., Japan. An aluminium sleeve is seen on the printed board. Figure 21.11(a,b) shows a cross section and side view of the spindle without the aluminium sleeve. The upper electromagnet is a hybrid type magnetic bearing while the lower one is a bearingless motor with radial magnetic suspension. The rotor is external to the stator. In addition, permanent magnet bias rings are seen; these produce the axial flux required for homopolar excitation of the stator cores. Radial displacement sensors are placed above the unit. Other radial displacement sensors are located on the printed board for the detection of radial movement of the lower end.

Another notable bearingless motor hard disk drive developed in Switzerland is described in [10].

21.3.2 Semiconductor processing [11]

In semiconductor wafer processing, the surface of the silicon wafer is chemically processed. Usually the silicon wafer is placed on a turntable during this chemical process, which includes the use of a corrosive chemical. The life expectancy of the turntable is limited because the supporting bearing is damaged by the corrosive chemical leaking in through the seal. If the turntable is suspended by magnetic suspension, the chemical process can be separated from the drive and bearing since the seal is eliminated. Moreover, both sides of the wafer can be processed simultaneously.

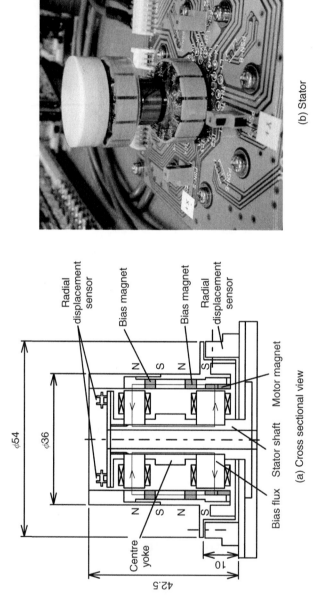

Figure 21.11 Details of computer storage spindle. (Courtesy: Sankyo Seiki Mfg Co.)

(a) Cross sectional view

(b) Stator

Radial displacement sensor

Bias magnet

Bias magnet

Radial displacement sensor

Motor magnet

Stator shaft

Bias flux

Centre yoke

φ54

φ36

42.5

10

N S N S

N S N S

(a)

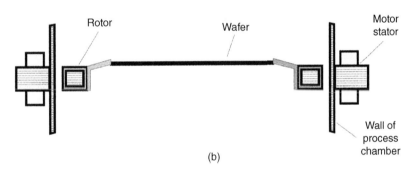

(b)

Figure 21.12 Semiconductor wafer suspension: (a) wafer suspension with a bearingless motor; (b) side view. (Courtesy: Levitronix GmbH)

Figure 21.12(a,b) show a bearingless system for semiconductor processing. In Figure 21.12(a), the rotor, stator and controller are shown. In Figure 21.12(b), a cross section of the silicon wafer placed on the rotor ring is illustrated. This structure is similar to the bubble bed reactor bearingless motor.

In this chapter, we have seen several examples of bearingless motors that are either commercial prototypes or early production versions. As described in Chapter 1, the possible applications for the bearingless drive are varied and wide and some notable developments have been included in this chapter. Space constraint prevents the inclusion of further examples. Most of the applications take advantage of non-contact suspension and low suspension losses. More applications will be seen in the future as the drive technology is further developed.

References

[1] T. Satoh, S. Mori and M. Ohsawa, "Study of Induction-Type Bearingless Canned Motor Pump", Institute of Electrical Engineering of Japan (IEEJ), International Power Electronics Conference (IPEC), Tokyo, Japan, April 3–7, 2000, pp. 389–394.

[2] M. Ohsawa, S. Mori and T. Satoh, "Study of the Induction type Bearingless Motor", Seventh International Symposium on Magnetic Bearings (ISMB), August 2000, Zurich, pp. 389–394.

[3] C. Redemann, P. Meuter, A. Ramella and T. Gempp, "Development and Prototype of a 30 kW Bearingless Canned Motor Pump", IPEC, pp. 377–382, Japan, April 2000.

[4] C. Redemann, P. Meuter, A. Ramella and T. Gempp, "30 kW Bearingless Canned Motor Pump on the Test Bed", Seventh International Symposium on Magnetic Bearings (ISMB), August 2000, Zurich, pp. 189–194.

[5] M. Neff, N. Barletta and R. Schoeb, "Bearingless Centrifugal Pump for Highly Pure Chemicals", ISMB – 8, August 2002, Mito, pp. 283–287.

[6] R. Schoeb, N. Barletta, M. Weber and R. von Rohr, "Design of a Bearingless Bubble Bed Reactor", ISMB, 1998, pp. 507–516.

[7] T. Masuzawa, T. Kita and Y. Okada, "An Ultradurable and Compact Rotary Blood Pump with a Magnetically Suspended Impeller in the Radial Direction", *International Society for Artificial Organ*, Vol. 25, No. 5, 2001, pp. 395–399.

[8] R. Schoeb, N. Barletta, A. Fleischli, G. Foiera, T. Gempp, H. G. Reiter, V. L. Poirier, D. B. Gernes, K. Bourque, H. M. Loree and J. S. Richardson, "A Bearingless Motor for a Left Ventricular Assist Device (LVAD)", Seventh International Symposium on Magnetic Bearings (ISMB), August 2000, Zurich, pp. 383–388.

[9] H. Kanebako and Y. Okada, "New Design of Hybrid Type Self-Bearing Motor for High-Speed Miniature Spindle", ISMB – 8, August 2002, Mito, pp. 65–70.

[10] R. Vuillemin, B. Aeschlimann, M. Kuemmerle, J. Zoethout, T. Belfroid, H. Bleuler, A. Cassat, P. Passeraub, S. Hediger, P. A. Besse, A. Argondizza, A. Tonoli, S. Carabelli, G. Genta and G. Heine, "Low Cost Active Magnetic Bearings for Hard Disk Drive Spindle Motor", ISMB, Boston, US, 1998, pp. 3–9.

[11] www.levitronix.com.

Index

Printed and bound by CPI Group (UK) Ltd, Croydon, CR0 4YY

08/05/2025

01864810-0001